The Corrosion of Copper and Its Alloys

A Practical Guide for Engineers

Roger Francis

NACE INTERNATIONAL

THE CORROSION SOCIETY

1440 South Creek Drive
Houston, Texas 77084

NACE INTERNATIONAL

THE CORROSION SOCIETY

Printed in the United States of America

Library of Congress Cataloging in Publication Data

Francis, Roger, Engineer.
The corrosion of copper and its alloys : a practical guide for engineers / Roger Francis.
p. cm.
Includes bibliographical references and index.
ISBN 978-1-57590-225-8 (hardback)
1. Copper—Corrosion. 2. Copper alloys Corrosion. I. Title.
TA480.C7F73 2010
620.1′8223—dc22 2010005806

ISBN: 978-1-57590-225-8

NACE PRESS

NACE International
1440 South Creek Drive
Houston, Texas 77084
http://www.nace.org

To all my colleagues at BNF

For your friendship, support, and the patience to pass on so much of your expertise

Contents

Preface

When I left university in 1973, I went to work for BNF Metals Research Association in Oxfordshire, as a junior research scientist in the corrosion department. (It had just moved to Oxfordshire from central London.) A year later, the organization changed its name to BNF Metals Technology Centre. The organization was founded in the 1920s and was set up to support the nonferrous metals industry. It carried out research and development work on production, processing, and corrosion of all nonferrous metals. By far the bulk of the work was carried out on copper and its alloys. The organization was a membership-based one and all members received copies of research reports relevant to their industry. Some, but not all, of the results of its research were published.

I left the organization in 1991 and less than two years later, the organization went bankrupt. Unfortunately, most of the library, including its collection of research reports going back to the first one in 1924, was consigned to a skip. When I left BNF, I had copies of quite a few reports to which I had contributed. I thought it a shame that all this knowledge should be lost, especially as there was (and is) no obvious successor organization with the practical expertise in copper and its alloys. Hence, I conceived the idea of an engineering review of the corrosion of copper and its alloys to present an up-to-the-minute snapshot, including as much of this unpublished data as possible.

During my quest for information, photographs, and case histories, I contacted friends and ex-BNF colleagues, many of whom had data that I thought was lost and who also gave me other unpublished data. I would like to take this opportunity to thank them all. In particular, I would like to thank Hector Campbell, David Davies, Bob Kain, Carol Powell, and Phillip Munn. I would also like to thank Angela Vesey of CDA, who gave me such good access to their records, including old BNF reports. A lot of data also came from old reports of the International Copper Research Association, now the International Copper Association. Finally, the contribution of the staff at TWI to the chapter on joining is gratefully acknowledged. Any differences between the information as you gave it to me and as it appears in print are my fault. I hope you feel the end result meets your expectations. Finally, I must thank my brother Ian, who did the original drawings.

CHAPTER 1

Introduction

Mankind first began using copper domestically some 6,000 to 10,000 years ago. Copper trinkets that date to the 9th millennium BC have been found in Iran. Copper is soft; however, when smelted with tin, it forms bronze, a harder, stronger metal. The period of widespread use of bronze is termed the Bronze Age, which began in northern Europe about 1500 BC. The fact that we are discovering copper and copper alloy implements that are thousands of years old, and in good condition, testifies to the metal's excellent corrosion resistance.

Bronze enabled the manufacture of weapons, tools, vessels, vases, and so on. The Romans were the first to use bronze for building ties in construction. Brass is an alloy of copper and zinc, which is less strong than bronze but widely used by the ancient Greeks and Romans for personal ornamentation and decorative metalwork. Many brass artifacts from this era are in good condition, again showing the metal's excellent resistance to corrosion.

During the Middle Ages, bronze was used for bells and cannon. The decorative use of copper on buildings also came to be appreciated. Figure 1.1 shows the copper clad roofs on the towers of the medieval cathedral in Bamberg, Germany.

The industrial revolution brought about big changes, with increased production, lower levels of impurities, and a wider range of alloys. At this time, the high electrical conductivity of copper and its use for both wires and lightning conductors were demonstrated. This period also saw the invention of the brass stamping press, which made brass forgings cheap and plentiful. Copper sheathing was also being used on wooden-hulled ships, particularly by the Royal Navy, to prevent attack of the timbers by marine borers. The cladding also reduced marine fouling, enabling higher water speeds.

FIGURE 1.1 Copper cladding on the towers of Bamberg Cathedral, Germany.

The latter half of the 19th century heralded the commercial production of electricity using copper wire for generators and transmission lines. A century later, the electronics era made extensive use of copper for printed circuit boards. Figure 1.2 shows copper wire being used in a modern, high-efficiency electric motor.

Copper has also been shown to be essential for human health, and the World Health Organization recommends a daily intake of 1–3 mg of copper to maintain health.

Copper is ever present in the modern world, for instance, as an electrical conductor in domestic appliances and electronics and as a vehicle for carrying our drinking water. Copper and its alloys are widely used for piping, pumps, and valves handling fresh water, seawater, and brines, and they are also used in heating systems. Copper alloys have also found numerous applications in industry, where corrosion resistance is important for handling more aggressive fluids.

Their high thermal conductivity and corrosion resistance means that copper alloys are widely used for heat exchanger tubes in power stations,

FIGURE 1.2 Copper wires being used in the construction of a modern, high-efficiency electric motor. (Photograph courtesy of CDA.)

desalination plants, oil and gas production, and so on. Corrosion resistance and the pleasing appearance of copper and its alloys mean that they are popular for fittings inside and outside buildings as well as for decoration.

The foregoing demonstrates the combination of properties that makes copper and its alloys so widely used today. The aim of this book is to draw together the large body of information on the corrosion of copper and its alloys into a single volume in a form that is useful to engineers who are trying to select the most suitable alloy for a specific application.

The next chapter classifies the commonly used copper alloys by group and discusses the merits and properties of each group. The corrosion resistance of copper and its alloys is dependent on the film that forms when they are first exposed to the service environment. Chapter 3 describes the current thinking about the composition and nature of the films on the commonly used alloys, what is needed to make them protective, and what can cause the formation of less protective films.

Chapters 4 to 14 describe different types of corrosion and how these can affect copper and its alloys. In particular, each chapter ends with a section on how to avoid that form of corrosion, both in new plants and in service. Chapters 15 to 17 deal with the specific environments of underground corrosion, atmospheric corrosion, and corrosion in waters. These chapters discuss the factors that affect corrosion and the types of corrosion that may occur, referring readers back to Chapters 4 to 14. The final chapter discusses corrosion issues associated with the three most common methods of joining copper alloys: soldering, brazing, and welding.

In addition, there are two appendixes. Appendix A summarizes the compositions of the commonly used alloys by group. Alloys are mostly referred to by their UNS numbers,[1] but national standards have been used for alloys for which no UNS number exists. Appendix B reviews the different test methods used to determine erosion corrosion resistance and assists the reader in comparing results from the wide variety of methods reported in the literature.

If this document is read from beginning to end, a certain amount of repetition is noticeable. This is because most readers will want to dip into sections relevant to their current problem and will not wish to keep following references to other chapters. Where the point at issue requires more lengthy discussion, references to other chapters are unavoidable.

Reference

1. *Metals and Alloys in the Unified Numbering System* (Warrendale, PA: SAE International and West Conshohocken, PA: ASTM International, 2004).

CHAPTER 2

Copper and Its Alloys

In this chapter, the composition and important properties of the common copper alloys are discussed. This chapter does not pretend to be an in-depth discussion of the metallurgy and physical properties of copper alloys; rather, it is intended to acquaint engineers with the important alloys and their properties. More detail can be found in the references throughout this chapter. The composition of all the alloys referred to in the text can be found in the tables in Appendix A. These tables can then be used as a ready reference when reading any of the subsequent chapters.

2.1 Coppers

The coppers are essentially 99.85% copper or higher, with other elements being mostly residuals from manufacturing. A major use of copper is as an electrical conductor, and high-conductivity copper contains a maximum of 0.04% impurities, although it may contain about 0.05% oxygen. When heated in a reducing environment, these coppers can become brittle, and for this reason, oxygen-free high-conductivity copper is recommended.

Copper is widely used in sheet, strip, and foil form, and may be either tough pitch copper or phosphorus-deoxidized copper (~0.01% P). Although tubing may be made from either form of copper, it is more commonly made of phosphorus-deoxidized material.

Copper is not very strong in the annealed form, as shown in Table 2.1. Where higher strength is required, copper may be cold worked, which greatly increases the 0.2% proof stress at the expense of ductility. For instance, the typical 0.2% proof stress of copper is around 70 MPa, with 55% elongation. When heavily cold worked, copper's proof stress increases

TABLE 2.1 Typical mechanical properties of copper and some of its alloys at room temperature in the annealed condition

Class	Common name	UNS no.	0.2% Proof stress (MPa)	Tensile strength (MPa)	Elongation (%)
Copper	Phosphorus-deoxidized (pdo) copper	C12200	70	325	60
	Arsenical copper	C14200	100	300	50
Brass	Commercial bronze	C22000	80	265	55
	Cartridge brass	C26000	120	325	70
	Aluminum brass	C68700	150	360	60
	Hot stamping brass	C38000	180	400	35
	Manganese bronze	CZ114(A)	195(B)	460(B)	15(B)
Cu-Ni/ Ni-Cu	90/10 copper-nickel	C70600	140	320	40
	70/30 copper-nickel	C71500	170	400	40
	Alloy 729	C72900	760(C)	900(C)	11(C)
	Marinel	C72420	700(C)	930(C)	17(C)
	Alloy 400	N04400	200	500	60
	Alloy K-500	N05500	800(C)	1,000(C)	25(C)
Tin Bronze	Phosphor bronze	C51000	120	340	60
	LG4 gunmetal	C93410	140	270	20
Silicon Bronze	Silicon bronze	C65500	100	340	40
Aluminium Bronze	Alloy D	C61400	230	540	40
	Alloy E	C63300	290	710	20
	Cast nickel Aluminum bronze	C95800	270	660	15
Nickel Silver	63/10/27	C74500	110	350	65
	63/20/17	NS108(A)	140	380	50

(A) British Standard quoted as no UNS number.
(B) Minimum mechanical properties.
(C) Typical properties in heat-treated condition.

to more than 320 MPa, with an elongation of about 4%. An alternative way to gain a more modest increase in strength is to use arsenical copper (~0.4% As), which also increases the creep strength at moderately elevated temperatures.

2.2 Brasses

Copper alloys containing zinc, commonly in concentrations from 5% to 40%, are known as brasses. Other elements may be added to modify properties such as strength, machinability, or corrosion resistance. The strength of brasses increases as the zinc content increases, as shown in Table 2.1, although greater increases can be obtained by adding additional elements.

Brasses containing up to about 37% zinc are single-phase alpha alloys, whereas alloys containing more zinc are two-phase alloys, with both alpha and beta phases. The quantity of the beta phase increases as the zinc content (or alloying additions with a high zinc equivalence) increases. Some elements, such as silicon and aluminum, have a high zinc equivalence, and so the beta phase will form at substantially lower zinc contents than in simple binary alloys. Aluminum brass, with 2% aluminum, has a zinc content of 22%, rather than 30%, to retain its all-alpha structure, which can then be protected from dezincification by arsenic additions.

Before the discovery of arsenic's ability to inhibit dezincification, it was discovered that tin reduced the rate of dezincification of the alpha phase. This is why alloys such as admiralty brass and naval brass contain 1% tin. The tin addition also improves the machinability of naval brass. The tin addition to admiralty brass has been observed to improve its corrosion resistance in moderately polluted seawater, compared with 70/30 arsenical brass or aluminum brass.[1] Aluminum was found to increase the resistance of alpha brass to erosion corrosion, but at concentrations significantly above 2%, an increased susceptibility to localized pitting occurs, particularly in polluted seawater.[2]

As discussed in detail in Chapter 9, arsenic is a required addition to many single-phase brasses to inhibit dezincification. Most specifications require 0.02–0.06 wt% arsenic. However, in the 1980s, the German DIN standard 17660 restricted arsenic to 0.02–0.035 wt%. The effect of arsenic on both intergranular corrosion and stress corrosion cracking (SCC) was reviewed by Campbell, who concluded that arsenic contents in the range 0.035–0.06 wt% caused no significant increase in susceptibility to either form of corrosion.[3] In fact, the presence of manganese or iron impurities can tie up part of the arsenic so that it no longer acts as an inhibitor for dezincification.[4] Hence alloys with arsenic at the lower end of the range may contain insufficient free arsenic to maintain dezincification resistance, and an arsenic content of 0.035 wt% should be regarded as a minimum, rather than a maximum. Because dezincification-resistant (DZR) brass often uses large quantities of scrap in the melting process, it is necessary to increase the arsenic range for this alloy to 0.02–0.15 wt%, with a typical concentration being 0.08 wt%.

The brasses traditionally used for hot stamping of fittings, valve bodies, etc., are duplex because all alpha-phase alloys tend to crack at forging temperatures. However, these duplex fittings are susceptible to dezincification in some waters (see Chapter 9). DZR brass was invented to provide a brass that is two-phased at hot stamping temperatures but can be converted to an all-alpha structure by a simple, poststamping heat treatment, typically 500°C for 2 hours.[4] The presence of arsenic then makes the alloy resistant to dezincification.

Lead is a common addition to brasses, usually 1% to 2%, that gives improved machinability. In addition, lead helps to make sand castings pressure-tight. *Bronze* is probably one of the most misused terms in metallurgy. Traditionally, bronzes are alloys of copper and tin. However, some brasses have been termed *bronzes*. Thus, 90/10 brass is sometimes called *commercial bronze*. Another widely used term is *manganese bronze,* for the family of high-tensile brasses. These alloys contain about 38% zinc and 0.5% to 5% manganese, with optional additions of iron, aluminum, and lead. These are duplex brasses that are heat treated to give moderate to high strength, with strength increasing as the beta content of the alloy increases. Because these alloys are duplex, they are susceptible to dezincification; and so service in aggressive environments, such as seawater, should be avoided unless the alloy is galvanically protected.

2.3 Bronzes

Traditionally, the bronzes are alloys of copper and tin, with smaller additions of other elements for specific properties. The alloys known as silicon bronze and aluminum bronze have been so named because they have many similarities with the tin bronzes with regard to metallurgical properties. Other alloys called bronzes, such as manganese bronze, commercial bronze, etc., are, in fact, brasses and are discussed in Section 2.2.

2.3.1 Tin Bronzes and Gunmetals

The wrought tin bronzes usually contain small additions of phosphorus used as a deoxidant in the melting process, and are sometimes called *phosphor bronzes*. These alloys have very good corrosion resistance, particularly in seawater, which increases as the tin content increases. The higher-tin bronzes have very good resistance to polluted seawater (see Chapter 7). The wrought phosphor bronzes are often used for corrosion-resistant springs.

When the tin bronzes are produced as castings, it is common to add some zinc, and then the alloys are traditionally known as gunmetals because of their early use for cannon barrels. To ensure good pressure tightness of castings, such as in valve bodies and pump casings, it is usual to add lead. This is why 10% phosphor bronze (C90700) may be used for pump impellers, but it is rarely used for pump cases because the alloy contains no lead. Admiralty gunmetal, also with 10% tin, offers the same corrosion resistance and can be made pressure-tight if lead is at the upper level of its specification range (~2%).

LG2 gunmetal is a very old alloy that has been cast for over 200 years. However, it is difficult to cast in thick, complex sections, and for such castings, LG4 gunmetal or M bronze is recommended.

If the lead content is increased further, to approximately 10%, the leaded tin bronzes make very good bearing alloys. For such applications, there is a long history of successful use in pumps and other rotating machinery.

Additions of tin in excess of ~10% result in the precipitation of the delta phase. This makes the alloy even stronger, but also significantly more brittle; thus, there are few commercial uses of such alloys.

2.3.2 Silicon Bronze

There are a number of silicon bronzes, but by far the most common is a silicon bronze that contains approximately 3% silicon and 1% manganese and produces an alloy with moderate strength. When manganese bronze is used for fasteners, it is susceptible to dezincification in seawater, and many other waters, as well as to stress corrosion cracking. Silicon bronze offers a more corrosion- and SCC-resistant alloy, which has a similar strength to manganese bronze. The mechanical properties in Table 2.1 are for annealed material, but in the as-manufactured condition, the 0.2% proof stress can be 400 MPa, while still retaining 20% elongation. The stress corrosion cracking resistance of silicon bronze is discussed in more detail in Chapter 10.

2.3.3 Aluminum Bronze

The copper-aluminum alloys combine high strength with good corrosion resistance and are widely used in both wrought and cast forms, particularly nickel aluminum bronze (NAB). For a detailed discussion of their metallurgy, properties, and corrosion resistance, the reader is referred to Harry Meigh's book on this class of copper alloy.[5] An extensive review of the corrosion resistance of aluminum bronzes was prepared by Campbell.[6]

Up to about 8% aluminum, the alloy is single-phase. With higher aluminum content, the beta phase is formed, which has good corrosion resistance and superior erosion corrosion resistance to the alpha phase. However, if these two-phase alloys are cooled too slowly, the gamma-2 phase can form, and this phase is highly susceptible to corrosion. However, iron contents of 2% or more suppress the formation of the gamma-2 phase and refine the grain structure.

When the aluminum content is 9% to 10%, additions of nickel and iron of 4% to 5% produce an alloy with high strength and corrosion resistance. The structure of the alloy is complex and was explained by Lorimer et al.[7] A simplified diagram of the structure is shown in Figure 2.1. As the alloy cools, the martensitic beta phase transforms to alpha and kappa phases. Figure 2.1 shows four kappa phases, all of which are enriched in iron and/or nickel. If the alloy cools too quickly, such as occurs during welding or with thin sections, not all the beta phase may transform. The residual beta phase (labeled "martensite" in Figure 2.1) is anodic to the alpha phase and it will rapidly corrode in seawater.[6-7] However, a short heat treatment will ensure that all the beta phase transforms to alpha and kappa. For castings, heating to 675° ± 25°C for 4–6 hours followed by air cooling is recommended. For wrought products, a higher temperature of 725° ± 25°C is preferred. For this reason, NAB should not be used in the as-welded condition in seawater, and Royal Navy specifications forbid it (see Chapter 18).

The kappa III phase can cause corrosion problems if it becomes anodic to the alpha phase because it is present in long, thin stringers. The origin of this problem and the best ways of avoiding it are discussed in Chapter 9.

Alloys with very high manganese content (>10%), such as CMA1 (see Table A.5), have an alpha-beta structure, but the beta phase has a very different composition to the beta phase of binary copper-aluminum

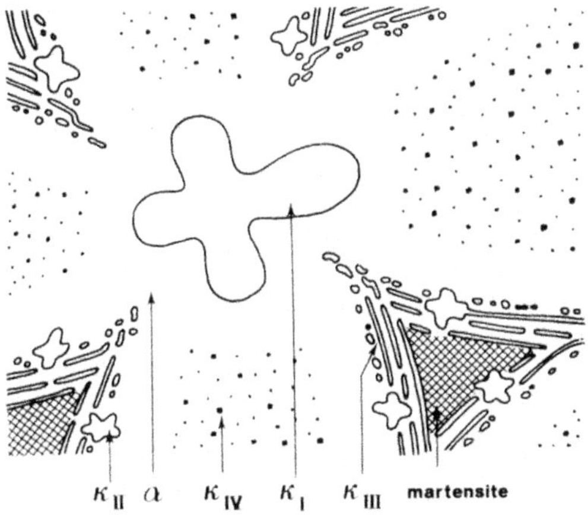

FIGURE 2.1 Schematic of the structure of nickel aluminum bronze.[7]

alloys, and it is more susceptible to corrosion. However, these alloys have good cavitation resistance and a high fatigue strength.

Generally, the corrosion resistance of the aluminum bronzes increases as the aluminum and other alloying additions increase. Hence, NAB, in both wrought and cast forms, is highly corrosion-resistant, but it is very susceptible to erosion corrosion in polluted seawater (see Chapter 7): in such applications, high-tin bronzes are preferred. In addition to clean seawater, NAB has excellent resistance to corrosion in a wide range of aggressive chemicals, including hydrofluoric acid (HF) and fluorine.[6]

2.4 Copper-Nickel and Nickel-Copper Alloys

Some common copper-nickel alloys are shown in Table A.2. Nickel and copper form a continuous solid solution, so all copper-nickel compositions are single phase, unless other elements are added. Manganese is usually present because it is used as a deoxidant in the melting process. Manganese also acts as a solid solution strengthener and increases the resistance to erosion corrosion.[8] However, iron is five times more efficient at increasing the resistance to erosion corrosion, and this is why alloys such as 90/10 and 70/30 copper-nickel contain iron.[8] With 30% nickel, iron is soluble at up to several percent; but with 10% nickel, the solubility limit of iron at room temperature is about 1%. Hence, commercial 90/10 copper-nickel, with 1.5% or more iron, must be quenched from the annealing temperature (~900°C) to keep the iron in solution. If this is not done, the corrosion resistance decreases significantly. Hence, it is important to check the quality of 90/10 copper-nickel with a simple magnetic permeability gauge to ensure that the iron is in solution. This is discussed in more detail in Chapters 7 and 17.

Copper-nickel alloys, particularly 90/10 and 70/30 copper-nickel, are widely used in marine applications because of their excellent resistance to both fouling and corrosion. They are used for heat exchanger tubes, piping for seawater and firewater systems, and cladding for larger steel items, such as water boxes.

There have been numerous evaluations of other alloying additions to copper-nickel alloys to improve their corrosion resistance. Kirk reviewed a large body of mostly unpublished data and there is a summary of this and other works in Chapter 7.[9] Although some of the alloying additions conferred benefits, there was usually a downside. For example, an increase in the resistance to erosion corrosion was often accompanied by increased sensitivity to crevice corrosion or stress corrosion cracking. Only one alloying element conferred increased strength and resistance to erosion corrosion without a significant negative impact, and that was chromium.

Table A.2 shows the composition of two wrought alloys and one cast alloy containing chromium. The cast alloy is an INCO experimental alloy, but a similar composition (with ~2% Cr) is used by the Royal Navy to replace NAB castings that have suffered corrosion problems. Although the author knows of no other users of the cast alloy, Alloy 722 (with 0.5% Cr) has been used for heat exchanger tubes, as an alternative to 90/10 or 70/30 copper-nickel, where there are suspended solids in the water (see Chapter 7).

The strength of the common copper-nickel alloys is only moderate, and there are applications that require much higher strength, combined with corrosion resistance. One way to boost the strength of copper-nickel alloys is to add tin (as in Alloy 729). This alloy displays spinodal strengthening through the development of submicroscopic chemical composition fluctuations. Table 2.1 shows the much higher strength obtainable with Alloy 729 compared with single-phase copper-nickel alloys. An alternative method of increasing the strength is to add aluminum and niobium (as in Alloy C72420) and heat treat to precipitate the gamma prime phase. After heat treatment, such alloys have similar strength to Alloy 729, as shown in Table 2.1. The principal use of these high-strength alloys is for marine fasteners, where high strength and corrosion resistance are required. Although both high-strength alloys have good resistance to corrosion and erosion corrosion in seawater, they can suffer SCC under some circumstances, in some fluids. This is discussed in more detail in Chapter 10.

Alloy 400 is a nickel-copper alloy with good resistance to corrosion and excellent resistance to erosion corrosion in seawater. However, the alloy is susceptible to crevice corrosion in clean seawater (see Chapter 6) and deep pitting in polluted seawater, despite its high nickel content (see Chapter 7). It is not galvanically compatible with high-alloy stainless steels and nickel-chrome-molybdenum alloys in seawater, but it can be safely coupled to copper alloys (see Chapter 14).

For higher-strength applications, Alloy K-500 is available, which has additions of aluminum and titanium so that gamma prime is formed during heat treatment. Table 2.1 shows the large increase in strength that is possible with this alloy compared to Alloy 400. Alloy K-500 behaves more or less like Alloy 400 in terms of corrosion.

2.5 Nickel Silvers

The name of this group of alloys is a little confusing because, although they contain nickel, they contain no silver. The name comes from their appearance. This class of copper alloy has excellent tarnish resistance, and they are essentially single-phase brasses with a high nickel content, as shown in Table A.6. Because they are not inhibited, these alloys will suffer dezincification, particularly in high-chloride waters. However, their resistance to atmospheric corrosion is excellent, although they may suffer SCC in the presence of moist ammonia (NH_3). Higher nickel content equates to greater resistance to SCC. Normally, alloys with 18% nickel have good resistance to tarnishing, and there is no real need for the higher-nickel alloys, unless the atmosphere is particularly aggressive.

Nickel silvers have excellent spring properties, and have been widely used in electrical equipment for both springs and electrical connectors. The higher-nickel alloys are somewhat stronger than the lower-nickel ones, as shown in Table 2.1, although greater increases in strength can be obtained by moderate cold working.

References

1. R.S. Bem, P.T. Gilbert, BNF Metals Research Association Report, A1095, September 1955.
2. R.S. Bem, BNF Metals Research Association Report, A1132, December 1956.
3. H.S. Campbell, BNF Metals Technology Centre, Miscellaneous Publication, MP600, December 1982.

4. J.E. Bowers, P.W. Oseland, G.C. Davies, "Development of a Hot-Stamping Brass Resistant to Dezincification," *Br. Corros. J.* 13, 4 (1978): p. 177.

5. H.J. Meigh, *Cast and Wrought Aluminium Bronzes, Properties, Processes and Structures* (London, U.K.: Institute of Materials, 2000).

6. H.S. Campbell, "Aluminium Bronze Alloys Corrosion Resistance Guide," Copper Development Association U.K. Publication, 80, July 1981.

7. G.W. Lorimer, F. Hasan, J. Iqbal, N. Ridley, "Observation of Microstructure and Corrosion Behavior of Some Aluminium Bronzes," *Br. Corros. J.* 21, 4 (1986): p. 244.

8. G.L. Bailey, "Copper-Nickel-Iron Alloys Resistant to Seawater Corrosion," *J. Inst. Met.* 79, 5 (1951): p. 243.

9. W.W. Kirk, "Evaluation of Alloying Effects on Corrosion Behaviour of Copper-Nickel Alloys in Seawater," INCRA Report, PRJ382, April 1986.

CHAPTER 3

Film Formation and Properties

In many environments, copper and its alloys corrode at low rates, which is due to the protective nature of the films that form. These films are not simple, and the mechanisms by which protection is conferred are often complex. In some cases, they are still not well understood. The films and their properties for some common copper alloys are reviewed in this chapter, along with the effects of some common additions to waters and some common pollutants.

3.1 Copper and the Brasses

At ambient temperature, the principal corrosion product on copper in aqueous solution is cuprous oxide (Cu_2O). In its pure form, it is bright red, but small quantities of impurities, such as iron, manganese, and nickel, can change its color to orange, dark red, brown, or nearly black. In hot waters, cupric oxide can also form, usually as a black layer on top of the cuprous oxide. The ratio of the two oxides cannot be used to infer the operating temperature because small quantities of impurities can affect the temperature at which cupric oxide first forms and also the ratio of the two oxides.

Diffusion through cuprous oxide is relatively easy because of the high number of cation vacancies in the lattice.[1] North and Pryor showed that the rate of corrosion is controlled by the combined electronic and ionic resistance of the Cu_2O.[1] They also showed that on exposure to water, the defect structure of cuprous oxide changes. This explains why air-formed cuprous oxide films on copper do not confer protection in water. The air-formed film is broken down, and then a new film forms with a modified structure.[1]

When copper is first placed in water, the initial corrosion rate is high but decreases as the cuprous oxide film forms and thickens.[2] The effect of chloride ions in the water is to increase the corrosion rate, both initially and after a Cu_2O film has formed. Once a stable cuprous oxide film forms, it slowly but continually dissolves at the film-water interface, and new oxide forms at the metal-film interface. The copper that is slowly released from the oxide can accumulate there, if the water flow is not too turbulent, until the solubility of copper in the water is exceeded and copper hydroxychloride ($Cu_2Cl[OH]_3 \cdot 4H_2O$) deposits on the surface of the oxide, if there is a significant chloride content in the water. This is usually as one of the atacamites, for example, paratacamite ($Cu_2[OH]_3Cl \cdot 4H_2O$). In hard waters of low chloride content, it is more common to get a layer of basic copper carbonate, for example, malachite ($CuCO_3 \cdot Cu[OH]_2$). In soft waters, it is common to get basic copper sulfate deposited over the cuprous oxide. These green outer films are usually not very protective and are easily removed if the water turbulence increases.

The other anions that are commonly found in waters are sulfate, bicarbonate, and nitrate. Sulfate is strongly adsorbed onto cuprous oxide, which can affect the nature of the film and the way that it breaks down.[2] This is discussed in more detail in Chapter 5. Nitrate is not adsorbed onto copper oxide, and although it can affect the pitting resistance of copper, its effect is much less than that of sulfate. Shalaby et al. found that bicarbonate had little effect on the corrosion of copper.[3] They added bicarbonate as the sodium salt, while Ives and Rawson investigated both sodium and calcium salts.[2] They found that sodium bicarbonate ($NaHCO_3$) did not adsorb onto cuprous oxide, whereas calcium bicarbonate ($Ca[HCO_3]_2$) adsorbed strongly. Bicarbonate is usually present in waters as calcium salt and this explains why Type I pitting occurs in hard and moderately hard waters (see Chapter 5). This is because the presence of adsorbed bicarbonate makes film breakdown more difficult so that it occurs at isolated spots, rather than at numerous sites.

Mor and Beccaria studied the effect of temperature on corrosion and found that the corrosion rate of copper decreased at higher temperatures in synthetic seawater.[4] This suggests that an oxide with fewer defects forms at higher water temperatures, following the model of North and Pryor.[1]

One property of cuprous oxide that is often not recognized is its photoelectric sensitivity. While this is of little concern inside pipes and closed tanks, it can influence corrosion where the copper is exposed to sunlight.[5] Depending on its electrochemical potential and the wavelength of light to which it is exposed, cuprous oxide can be either photopositive or photonegative.[5] Wilhelm et al. suggested that illumination during initial film growth could affect how protective the oxide was in service.[5]

When brasses corrode, they also form cuprous oxide. Zinc tends to be detected in the film as well—more so as the zinc content of the brass increases. It has not been determined whether zinc substitutes for copper in the oxide matrix or whether zinc compounds are formed for low zinc brasses. If the chloride content of the film is high, such as in seawater, many of the zinc compounds that might form are soluble.

Beccaria and Poggi investigated the corrosion of admiralty brass in seawater at 40°C.[6] After 15 days, they identified both cuprous and cupric oxides, zinc oxide (ZnO), and zinc sulfate ($ZnSo_4 \cdot 7H_2O$) as well as copper hydroxychloride. They also found a small quantity of tin oxide (SnO). Admiralty brass has been observed to perform better than both 70/30 brass (C26000) and aluminum brass in slightly polluted water. This was always believed to be due to the presence of the 1% tin in the alloy. The detection of tin oxide in the film on admiralty brass shows that tin is active in the corrosion process.

Albaya et al. also studied the films on admiralty brass, but they did not find compounds other than cuprous oxide and copper hydroxychloride.[7] However, their tests were in sodium chloride (NaCl), rather than natural seawater, and they only used x-ray diffraction to identify the corrosion products. Beccaria and Poggi used more sophisticated techniques that do not require the corrosion product to be crystalline.[6]

3.2 Aluminum Brass

Aluminum brass behaves somewhat differently to other brasses, and the films that form on aluminum brass can vary substantially. In some fresh waters, sodium chloride solution, and synthetic seawater, the film on aluminum brass is cuprous oxide, as seen on 70/30 brass. However, in certain fresh waters, aluminum brass has been observed to pit.[8] When this happened, the film on the nonpitted surfaces was identified as a mixed spinel of zinc and aluminum ($ZnO.Al_2O_3$).[8] The nature of this film and the absence of copper suggest that there was organic material or microbes in the water that are chelating the copper and preventing it from becoming incorporated into the film.

In natural seawater, the films on aluminum brass are different again. Usually, there is an outer orange/brown deposit of hydrated iron oxide (FeOOH), which is formed either from corroding iron or steel, or by the regular addition of ferrous sulfate ($FeSO_4 \cdot 7H_2O$) to the seawater. Beneath this is a thin, soft film, usually white in color, which is a mixed basic hydroxy carbonate of magnesium and aluminum of the hydrotalcite ($Mg_6Al_2[OH]_{16}CO_3 \cdot 4H_2O$) family of compounds

($Mg_6Al_2CO_3[OH]_{16}.4H_2O$). This compound was identified by several laboratories, all investigating a spate of heat exchanger tube failures in very large crude carriers in the early 1970s. The nature of the film and the mechanism by which it protects the metal surface was the subject of extensive work by Castle and his coworkers at the University of Surrey, in the United Kingdom.[9-13]

Hydrotalcite contains two metal cations: one a minor element in the alloy and the other a minor constituent of seawater. For this compound to form, both copper and zinc must be chelated by microbial action to prevent their incorporation into the film. Further work by Castle et al. showed how aluminum brass reacts with the various biological layers that form on aluminum brass in natural seawater.[14-15]

The mechanism by which hydrotalcite protects the surface involves its ability to vary its composition widely from the nominal composition. As the pH decreases, hydrotalcite sheds magnesium and takes up aluminum. As the pH increases, the reverse happens.[9-13] Thus, the hydrotalcite is able to buffer the pH at the film surface and prevent accelerated attack.

Hydrotalcite is a soft compound with little mechanical strength, and so it cannot resist turbulent water flow. Castle and Parvisi showed that the hydrotalcite is not only present as a film on the surface, but it is also present in the pores in the FeOOH layer above it.[11] Thus, the hydrated iron oxide is able to provide an erosion-resistant network within which the hydrotalcite is fixed.

3.3 Copper-Nickel Alloys

The two most studied copper-nickel alloys are 90/10 (C70600) and 70/30 copper-nickel (C71500). These studies have used sodium chloride solution, synthetic seawater, and natural seawater. The results in all three media are similar, but the films formed in natural seawater have some properties that those formed in synthetic solutions do not.

North and Pryor showed that cuprous oxide forms on copper-nickel alloys, and the resistance of this film is improved by the incorporation of nickel into vacant lattice sites and as a substitute for copper ions.[1] The nickel content of the film increases as the nickel content of the alloy increases. However, later work has shown that the films on copper-nickel alloys are more complex than this in solutions with significant chloride concentrations. Ijsseling et al. and Drolenga et al. were among the first to detect a thin layer next to the metal surface that was rich in copper and chloride ions.[16-17] Detailed surface analysis by Kato et al. and Castle and Parvisi showed that it is this thin layer of copper chloride at the metal

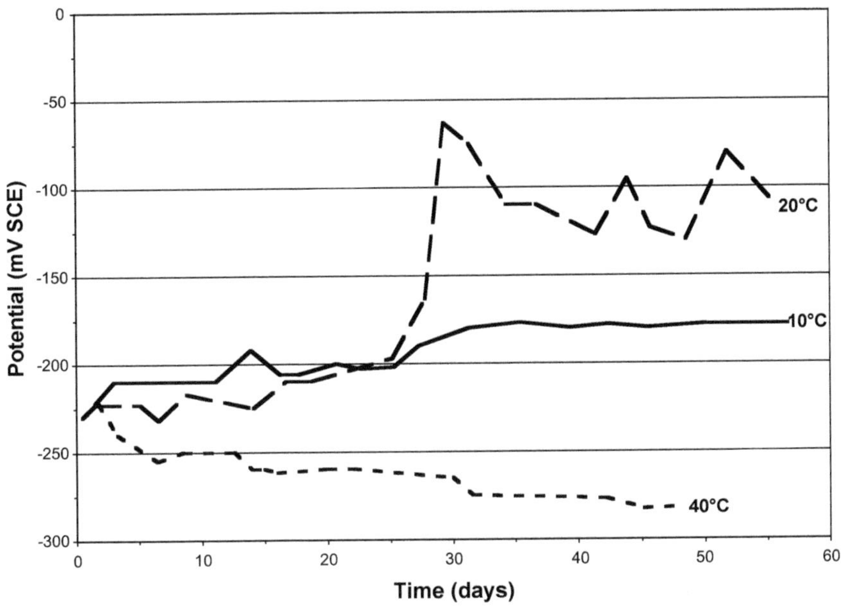

FIGURE 3.1 Potential vs. time for 90/10 copper-nickel at various seawater temperatures.[20]

surface that provides the high level of cathodic polarization seen with 90/10 copper-nickel.[13,18-19] The high resistance to corrosion of 90/10 copper-nickel is achieved primarily by inhibition of the cathodic process: the reduction of dissolved oxygen. In contrast, the outer layer of cuprous oxide enriched in nickel and iron contributes only a small amount of anodic polarization. The iron in 90/10 copper-nickel is essential for resistance to erosion corrosion, and Castle and Parvisi suggested that under turbulent water conditions, the iron forms an outer layer of hydrated iron oxide (FeOOH), which provides a mechanical defense in the same way that ferrous sulfate additions do for aluminum brass (Section 3.2).[13]

The enrichment of the oxide layer with nickel and iron appears to be temperature-dependent.[13,16,20] At ambient temperature, the oxide is mostly enriched with nickel, while at elevated seawater temperatures (around 40°C), an iron-rich film develops. Ijsseling et al. showed that there is a potential change at elevated seawater temperatures, compared with that observed at 10°C.[20] At 20°C, the potential moved more positive after ~25 days, finishing at around –100 mV SCE (saturated calomel electrode) (Figure 3.1). At 40°C, the potential moved steadily more negative, finishing at about –280 mV SCE. The film at 40°C gave the lowest corrosion rate, and it appears that the changes in the film at 20°C were different from those at 40°C.

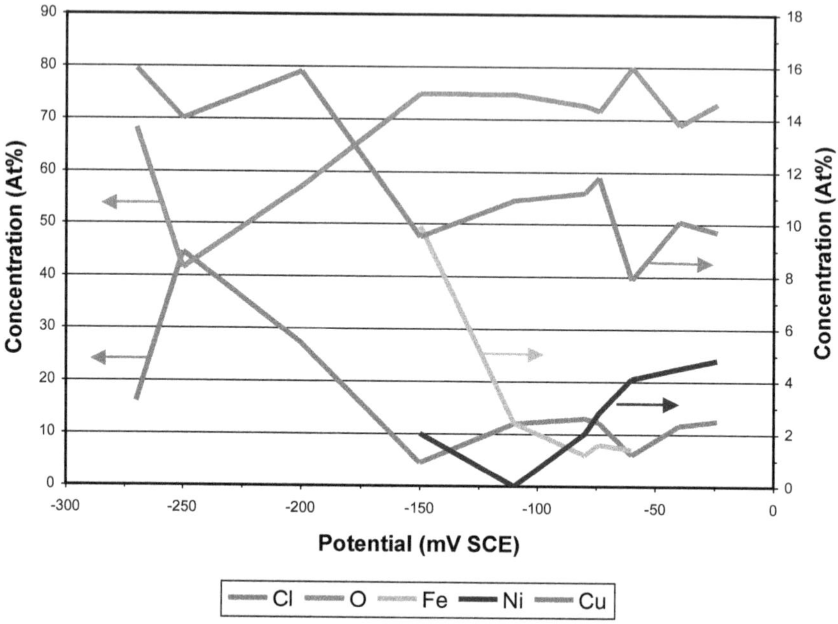

FIGURE 3.2 Concentration of elements in the film on 90/10 copper-nickel vs. potential.[13]

The potential at which 90/10 copper-nickel is first exposed has a strong effect on the nature of the protective film. Blundy and Pryor demonstrated that nickel enrichment is strongly a function of the potential.[21] More detailed work by Castle and Parvisi (Figure 3.2) showed that the chloride-rich layer formed only over a narrow potential range, while nickel and iron enrichment occurs at somewhat more positive potentials.[13]

Less work has been carried out on the films that form on 70/30 copper-nickel, but Castle and Parvisi suggest that the films are similar in nature to those on 90/10 copper-nickel.[13] It is certainly well established that the cuprous oxide film is enriched in nickel.[1,22] Castle and Parvisi failed to find any iron enrichment in the film, although their exposure time was only 3 hours.[13] However, 90/10 copper-nickel showed iron enrichment after the same length of time in seawater. Both Efird and Francis showed that over a period of days, the potential of 70/30 copper-nickel moves considerably in the positive direction.[23-24] However, this potential change is a function of temperature, and at temperatures of 10°C and below, no change of potential occurred (Figure 3.3).[24] Simple film analysis showed that the films on tubes with positive potentials were enriched in iron, and sometimes in nickel as

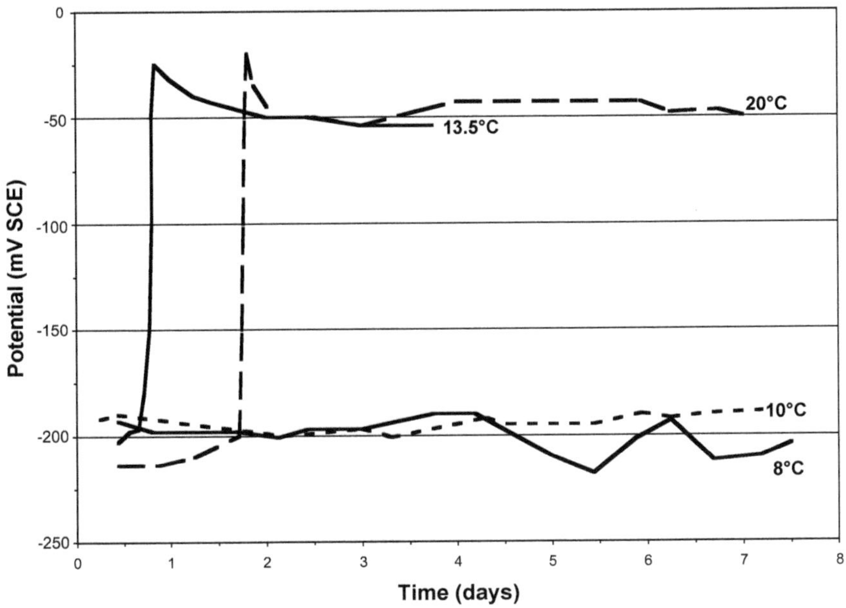

FIGURE 3.3 Potential vs. time for 70/30 copper-nickel in seawater at various temperatures.[24]

well. Francis postulated that iron was continuously dissolving from the alloy through the film, and an iron-rich film formed when the iron concentration at the film-liquid interface exceeded the solubility limit.[24]

Efird and Francis also examined the 66/30/2/2 Cu/Ni/Fe/Mn alloy (UNS C71640) and found a similar effect of temperature on potential ennoblement.[23-24] The films were very thin, and the simple analysis that was carried out by both workers showed no iron enrichment in the films with positive potentials. However, they argued that more detailed analysis by techniques such as x-ray photoelectron spectroscopy (XPS), might reveal iron enrichment. Castle and Parvisi suggest that the protective layer on 70/30 copper-nickel is similar to 90/10 copper-nickel and contains an inner layer rich in copper and chloride ions, but they present no detailed analysis.[13]

Ross reported tests on 90/10 copper-nickel in treated seawater at elevated temperatures, as is found in parts of multistage flash (MSF) desalination plants.[25] Copper oxide was the primary corrosion product, but there was also a considerable concentration of magnesium found in the film at both 60°C and 107°C. Some of this may have been magnesium hydroxide, but pyroaurite ($Mg_6Fe_2CO_3[OH]_{16}.4H_2O$), which is related to hydrotalcite, as seen on aluminum brass (Section 3.2), was also detected on some samples. It was not determined whether this compound acts as a pH buffer, as hydrotalcite does. Ross also found deposits of calcium carbonate ($CaCO_3$) on the oxide films, which seemed to be a factor further reducing corrosion by acting as a barrier.[25] Carbonate scaling does not usually occur in MSF plants because the water is treated to prevent this, as the scaling causes a reduction in plant efficiency.

3.4 Aluminum Bronze

The air-formed film on aluminum bronze is mainly alumina (Al_2O_3), and this gives the alloys very good corrosion resistance in air up to 600°C or higher.[26] In seawater, alumina has been found in the films on aluminum bronze, but this was after a short exposure to seawater, when alumina from the air-formed film could still be expected on the metal surface.[27-28] Francis and Maselkowski carried out tests on cast nickel aluminum bronze (C95800) for periods of 1–12 weeks in natural seawater.[29] The films were analyzed after 4 and 12 weeks' exposure at 10°C. Samples were exposed at both 0.1 and 2.5 m/s flow velocity. After 4 weeks, there was a thin, greenish outer layer, which was presumed to be copper hydroxychloride. Beneath this was a black layer, which was tentatively identified as a derivative of Fe_3Al by x-ray diffraction. The black layer was slightly enriched in aluminum compared to the base metal but was greatly enriched in nickel and iron. The results after 12 weeks were a little more confused because of the much thicker layer of copper hydroxychloride. However, not only was the film enriched in aluminum, nickel, and iron, but there was also a small amount of magnesium present. It is possible that this was present as hydrotalcite, as was found on aluminum brass (Section 3.2). However, it could not be identified by x-ray diffraction. The exact nature of the films and the method by which they confer protection were not determined, but unpublished work by Castle on NAB also found iron and nickel enrichment in the film and a lower level of aluminum than in the base metal.[30]

3.5 Tin Bronzes

These alloys have received even less attention than the aluminum bronzes in terms of the nature of the protective films, but their excellent resistance to corrosion in seawater is well recognized. Atsumi et al. examined the films on a commercial 8% tin bronze alloy (C52100) exposed in seawater.[31] From an energy-dispersive x-ray analysis of the film, they concluded that the very high tin content of the film suggested the presence of a tin oxide as a distinct phase. Oxides of tin are both robust and stable, and the presence of a layer of essentially tin oxide would explain the good corrosion resistance of the tin bronzes.

3.6 Ferrous Sulfate

Throughout the first half of the 20th century, admiralty brass (C44300) and aluminum brass (C68700) were the most commonly used alloys for tubes in seawater-cooled heat exchangers. In the 1950s, there was a move to replace bare carbon steel and cast iron components with more corrosion-resistant alloys or to apply coatings to increase reliability. Following this, there was an increase in the number of heat exchanger tube failures due to erosion corrosion. In 1961, Bostwick described how daily additions of ferrous sulfate to the cooling water replaced the iron-rich films that had previously been formed by corroding steel, with a large reduction in the number of tube failures.[32] Shortly after, Lockhart described a similar success story at a U.K. power station.[33] Since that time, additions of ferrous sulfate solution (typically 1 mg/l Fe^{2+} for 1 h/d) have been made at many plants to solve heat exchanger corrosion problems.

As described in Section 3.2, the FeOOH layer that is formed by ferrous sulfate additions appears to offer mechanical protection in turbulent water to the more fragile films adjacent to the metal. The use of iron additions to combat the corrosion of copper alloy heat exchanger tubes was surveyed for U.K. power stations, and the published literature was examined to determine when ferrous sulfate dosing would be effective, and for which alloys.[34]

Although intended originally for brass tubes, ferrous sulfate has occasionally been used successfully to solve corrosion problems with 90/10 and 70/30 copper-nickel. The original corrosion problem that iron additions were used to overcome was erosion corrosion (impingement attack), but they have also been used to try and solve other corrosion problems with varying degrees of success. In Japan, ferrous sulfate dosing solved problems of localized attack of aluminum brass tubes that had a cathodic film of manganese dioxide (MnO_2) on them.[35] Manganese usually is not present at high concentration in seawater. However, this film was produced by the oxidation of manganese in solution in seawater by chlorine, added to control fouling in the local seawater.

Ferrous sulfate dosing has also solved corrosion problems caused by ammonia (NH_3) pollution. The author used ferrous sulfate dosing for 2 months at the start-up of a MSF desalination plant to form more protective films on 90/10 copper-nickel tubes. The water was intermittently polluted with ammonia, which had caused low-temperature hot spot attack in the heat exchangers when the first unit was started. These were retubed, and ferrous sulfate dosing was started from new at a level of 1 ppm Fe^{2+} for 1 hour every day. The plant has now been in operation since 1994, with no further failures. At another MSF plant in the Middle East, the ammonia pollution was almost continuous because the water intake was near the outfall of a fertilizer plant. Daily ferrous sulfate dosing throughout its life was recommended for this plant as it was tubed with aluminum brass. There had been failures of 90/10 copper-nickel tubes in an earlier plant on the same site. The plant has now been in operation since 1998, with no problems.

Ferrous sulfate does not work in water polluted with hydrogen sulfide (H_2S), unless protective films can first be formed in clean water.[36] A U.K. power station, operating in a tidal estuary, used ferrous sulfate dosing to prevent corrosion of 70/30 copper-nickel tubes, where H_2S was an intermittent pollutant at certain states of the tide.[34]

All these experiences suggest that the initial formation of a hydrated iron oxide layer, under relatively clean seawater conditions, enables a more protective film to form beneath, at least in some cases. The mechanism by which this occurs is not known but could simply involve the FeOOH layer acting as a barrier to harmful species in the bulk water.

It is well documented that chlorine/hypochlorite dosing to control fouling interferes with the efficacy of ferrous sulfate dosing.[37-38] The recommendation is that if the chlorination is continuous, it should be turned off while ferrous sulfate dosing is taking place. If the chlorination is intermittent, then it should be staggered so that it does not occur during ferrous sulfate dosing.

Service experience with Baltic and Arctic waters suggests that aluminum brass does not perform quite as well as might be expected, and some companies prefer copper-nickel alloys for heat exchangers. The reason for this may be connected to the fact that ferrous sulfate dosing is common with aluminum brass to improve protective film formation and corrosion resistance. Effertz and Fichte examined the formation of iron-rich films on copper alloys due to ferrous sulfate dosing and found that the thickness of the film was proportional to the quantity of FeOOH colloid produced by adding ferrous sulfate to the cooling water.[39] They found that the colloid yield doubled for every 10°C increase in temperature and that the yield at low temperatures (around 5°C) was very low. Hence, the iron-rich films on aluminum brass will not be very thick and protective in cold water, increasing the susceptibility to erosion corrosion. This explains the superior performance of copper-nickel alloys, which do not normally require ferrous sulfate dosing to ensure good corrosion resistance. Much of Effertz and Fichte's work was carried out with river water, but they found no difference in FeOOH colloid formation in fresh water and seawater.

3.7 Chlorine

Chlorine is added to seawater cooling systems to control macrofouling such as by weeds and shellfish. It also has the effect of keeping microfouling (slime) at acceptable levels. Chlorine is usually generated electrolytically in a bypass line and is then injected into the seawater intake, although in some land-based plants, tanks of sodium hypochlorite (NaOCl) may be used instead. Chlorine is a powerful oxidizing agent, and it reacts rapidly with organic material in seawater. Hence, the concentration decreases rapidly from the point of dosing. To control fouling, it is only necessary to have a very small residual (for example, 0.1 mg/L) of free chlorine at the outfall/discharge.[40] The chlorine content at the inlet will depend on the organic activity of the water (biological chlorine demand) and the length of piping from intake to outfall. For large, land-based plants, chlorine concentrations from 5 to 25 mg/L may be necessary at the seawater intake at the height of the fouling season, while a ship may manage with 1 mg/L or less.

Chlorine interferes with the formation of FeOOH films from ferrous sulfate dosing. It oxidizes the FeOOH colloid, making it more difficult for it to be attracted to the tube wall. Hence, ferrous sulfate dosing is usually not very effective with continuous chlorination of the water, unless the chlorination is stopped during the ferrous sulfate dosing.[38]

Chlorine also affects film formation on copper-nickel alloys, if the concentration is high enough. As the loss of corrosion resistance is seen most under turbulent water conditions, it is presumed that the chlorine prevents the formation of a mechanically protective, iron-rich layer.

Francis reported the effects of continuous chlorine additions to seawater on the films on aluminum brass that was also dosed with ferrous sulfate.[41] In clean seawater and seawater containing 1 mg/L chlorine, a white film of hydrotalcite formed, as usual. However, with the addition of 4 mg/L chlorine, the film was thin and golden, and XPS analysis showed it to be predominantly monovalent copper and chloride. The film also showed

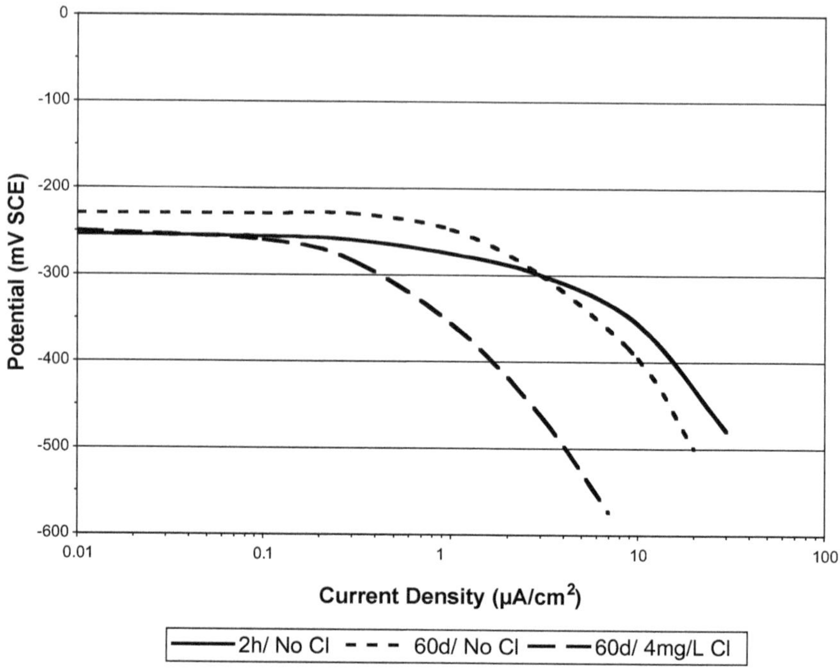

FIGURE 3.4 Cathodic polarization curves for aluminum brass in seawater.[41]

a high degree of cathodic polarization (Figure 3.4) compared with the film produced after exposure to clean seawater. The inhibition of the cathodic reaction by a film rich in chloride and copper was seen for 90/10 copper-nickel (Section 3.3), and it appears that high concentrations of chlorine can cause its formation on other copper alloys. Unfortunately, the copper chloride layer is not mechanically strong, and it is easily disrupted in turbulent water. The presence of high concentrations of chlorine would prevent the formation of a FeOOH layer to give mechanical protection.

Francis et al. also investigated the films formed on cast nickel aluminum bronze in seawater at 10°C containing 0.5 mg/L chlorine.[29] Their results in clean seawater were presented in Section 3.4. In chlorinated seawater, the films were generally of the same appearance as in clean seawater. While the films formed in chlorinated seawater were also enriched in nickel and iron compared with the base metal, enrichment was not found to the same extent as in clean seawater. In addition, magnesium and chloride were present in small but significant quantities in the films formed in chlorinated seawater, although their compounds were not established.

3.8 Sulfide

When hydrogen sulfide is present in seawater, it is almost always produced by the action of sulfate-reducing bacteria (SRB). These bacteria need anaerobic conditions to thrive and are normally inactive in aerated seawater. When the water is stagnant or access to oxygen is restricted, for example, in bottom mud, the aerobic bacteria gradually consume the dissolved oxygen. Once the oxygen is gone, the anaerobic bacteria, including SRB, become active. They reduce the sulfate in seawater to H_2S as part of their metabolic processes. In deaerated seawater containing H_2S, the corrosion rate of copper alloys is low, similar to the corrosion rate in clean, aerated seawater.[42] However, a mixture of deaerated seawater containing H_2S and aerated seawater is very aggressive to most copper alloys.[42-43] This can happen when stagnant bottom water is mixed with aerated top water by tidal action or incorrect siting of the suction intake.

This mixture of aerated seawater containing H_2S is not stable, as the dissolved oxygen and the H_2S react together:

$$2H_2S + O_2 \rightarrow 2H_2O + 2S \tag{3.1}$$

However, this reaction is slow, with a half-life of 15 to 20 minutes, which is more than enough time for the reacting mixture to pass through most cooling systems. If the H_2S and O_2 have time to fully react, say, in a holding tank or basin, then the elemental sulfur that is produced is still aggressive to copper alloys, but not as much as H_2S.[43]

Eiselstein et al. explained the corrosion of copper-nickel alloys using the Evans diagram.[43] The effect of sulfide is to move the anodic curve electronegatively, but this does not cause a large corrosion rate in deaerated water because the cathodic reaction is hydrogen ion reduction. However, when dissolved oxygen (DO) is present, the cathodic reduction of DO is the cathodic reaction, and this occurs at much more electropositive potentials. The result is a very high corrosion rate. As corrosion by sulfides usually involves localized pitting under aerated conditions, corrosion rates exceeding 10 mm/y are possible.[43]

The protection of copper-nickel alloys by a thin layer rich in copper and chloride adjacent to the metal surface was described previously.[13] Castle and Parvisi also showed that there is a difference in the properties of this film, depending on whether it is formed in synthetic solutions or natural seawater. Films formed in synthetic solutions showed no significant loss of cathodic polarization when the solution was deaerated (Figure 3.5A); however, films formed in natural seawater showed a substantial loss of

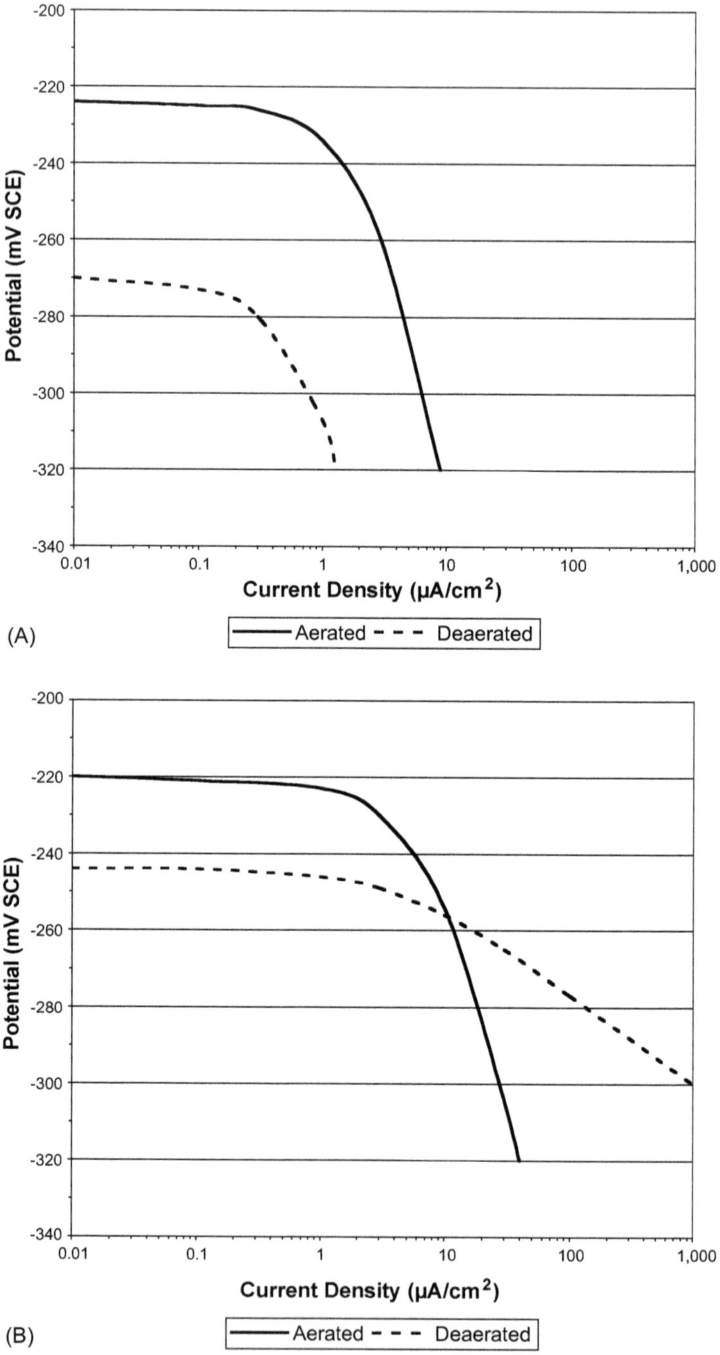

FIGURE 3.5 Cathodic polarization curves for 90/10 copper-nickel after exposure to (A) sodium chloride solution and (B) natural seawater.[13]

polarization when the solution was deaerated (Figure 3.5B). This was observed for both 90/10 and 70/30 copper-nickel. Clearly, there is an organic compound present in natural seawater that aids the cathodic polarization reaction and that is deactivated in deaerated seawater. Castle and Parvisi presented some preliminary results of identifying the type of organic compound that causes this effect.[13]

The presence of sulfides in seawater is well known to cause accelerated corrosion of copper alloys, and the outer layer is usually a thick, porous layer of copper sulfide (CuS). Beneath this is a thin layer of cuprous oxide, containing both nickel and sulfur. These layers are clearly not protective, and the copper chloride layer is absent. Corrosion rates are observed to be highest where the water is most turbulent, although the attack is corrosion, rather than erosion corrosion. This presumably means that the transport of oxygen and sulfide to the metal surface is controlled by its availability in solution, rather than by diffusion through the sulfide-oxide film. Kato et al. observed what appeared to be enhancement of the cathodic reaction; that is, depolarization, by films on 90/10 copper-nickel formed in sodium chloride solution containing H_2S.[44]

When the conditions are restored to clean seawater, the copper sulfide gradually spalls off, and the normal chloride and oxide films are slowly restored on copper-nickel alloys.[42] It can take days or weeks before corrosion rates decrease to the levels normally found in clean seawater.

Nickel aluminum bronze is particularly susceptible to attack by sulfides. Schussler and Exner observed that unlike the copper-nickel alloys, NAB did not reform protective films in clean seawater following exposure to sulfides, even after 50 days.[45]

The high-tin bronzes (Sn $>7\%$) are well known to have good resistance to corrosion in sulfide-polluted seawater.[31] Although the films that form on these alloys in clean and polluted seawater have not been studied in detail, it is known that tin oxides are thermodynamically more stable than tin sulfides. Hence, the more protective oxide should form, provided that there is some DO present to aid its formation.

References

1. R.F. North, M.J. Pryor, *Corros. Sci.* 10 (1970): p. 297.
2. D.J.G. Ives, A.E. Rawson, *J. Electrochem. Soc.* 109 (1962): p. 462.
3. H.M. Shalaby, F.M. Al-Kharafi, A.J. Said, *Brit. Corros. J.* 25 (1990): p. 292.
4. E.D. Mor, A.M. Beccaria, *Corrosion* 31 (1975): p. 275.

5. S.M. Wilhelm, Y. Tanizawa, C.-Y. Liu, N. Hackerman, *Corros. Sci.* 22 (1982): p. 791.

6. A.M. Beccaria, G. Poggi, *Analyst* 111 (1975): p. 275.

7. H.C. Albaya, J.B. Bessone, P.M. Taberner, O.A. Cobo, *Corrosion* 30 (1974): p. 437.

8. H.S. Campbell, BNF Metals Technology Centre Miscellaneous Publication, MP574, August 1972.

9. J.E. Castle, D.C. Epler, *Surf. Sci.* 53 (1975): p. 286.

10. D.C. Epler, J.E. Castle, *Corrosion* 35 (1979): p. 451.

11. J.E. Castle, M.S. Parvisi, "Minor Alloying Elements and the Marine Corrosion of Copper Alloys," International Colloquium on Choice of Material for Condenser Tubes and Plates and Tube and Tightness Testing (Avignon, France: SFEN, 1982).

12. Z.B. Luklinska, J.E. Castle, *Corros. Sci.* 23 (1983): p. 1163.

13. J.E. Castle, M.S. Parvisi, *Corros. Prev. Control* (February 1986): p. 5.

14. J.E. Castle, D.C. Epler, "The Development of Inorganic and Biological Fouling Layers on Copper Based Alloys," Progress in the Fouling of Industrial Plant Conference (Nottingham, U.K.: Institute of Corrosion, 1981).

15. J.E. Castle, M.S. Parvisi, A.H.L. Chamberlain, *Microbial Corrosion*, Book 303 (MET SOC, 1983), p. 36.

16. F.J. Ijsseling, J.M. Kroughman, L.J.P. Drolenga, "The Corrosion Behaviour of the System CuNi10Fe/Seawater: The Protective Layer of Corrosion Products," Fifth International Congress on Marine Corrosion and Fouling (Barcelona, Spain: 1980), p. 146.

17. L.J.P. Drolenga, F.J. Ijsseling, B.H. Kolster, *Werkst. Korros.* 34 (1983): p. 167.

18. C. Kato, B.G. Ateya, J.E. Castle, H.W. Pickering, *J. Electrochem. Soc.* 127 (1980): pp. 1890, 1897.

19. C. Kato, H.W. Pickering, *J. Electrochem. Soc.* 131 (1984): p. 1219.

20. F.J. Ijsseling, L.J.P. Drolenga, B.H. Kolster, *Brit. Corros. J.* 17 (1982): p. 162.

21. R.G. Blundy, M.J. Pryor, *Corros. Sci.* 12 (1972): p. 65.

22. G.E. McGuire, A.L. Bacarella, J.C. Griess, R.E. Clausing, L.D. Hulett, *J. Electrochem. Soc.* 125 (1978): p. 1801.

23. K.D. Efird, *Corrosion* 33 (1977): p. 347.

24. R. Francis, *Brit. Corros. J.* 18 (1983): p. 35.

25. R.W. Ross, CORROSION/77, paper no. 94 (Houston, TX: NACE International, 1977).

26. H.S. Campbell, *Aluminium Bronze Alloys Corrosion Resistance Guide*, Publication 80 (Potters Bar, U.K.: Copper Development Association, 1981).

27. A. Schussler, H.E. Exner, *Corros. Sci.* 34 (1993): p. 1793.

28. B.G. Ateya, E.A. Ashour, S.M. Sayed, *Corrosion* 50 (1994): p. 20.

29. R. Francis, C.R. Maselkowski, BNF Metals Technology Centre Report, R506, May 1985.

30. J.E. Castle, University of Surrey, correspondence to author, 1985.

31. T. Atsumi, K. Sudo, K. Nagata, *Sumitomo Light Met. Tech. Rep.* 26 (1985): p. 13.

32. T.W. Bostwick, *Corrosion* 17 (1961): p. 12.

33. A.M. Lockhart, *Proc. Inst. Mech. Eng.* 179 (1964–1965): p. 459.

34. R. Francis, INCRA Project, 289, January 1979.

35. S. Sato, M. Okawa, *Sumitomo Light Met. Tech. Rep.* 17 (1976): p. 17.

36. S. Sato, T. Nosetani, Y. Yamaguchi, K. Onda, *Sumitomo Light Met. Tech. Rep.* 16 (1975): p. 23.

37. R. Francis, *MP* 21 (1982): p. 44.

38. E.B. Shone, G.C. Grimm, *Trans. Inst. Marine Eng.* 98 (1985), Paper II.

39. P.H. Effertz, W. Fichte, *VGB Kraftswerktech.* 57 (1977): p. 116.

40. F. Inui, *Bull. MESJ* 1 (1973): p. 21.

41. R. Francis, *Corros. Sci.* 26 (1986): p. 205.

42. B.C. Syrett, *Corros. Sci.* 21 (1981): p. 187.

43. L.E. Eiselstein, B.C. Syrett, S.S. Wing, R.D. Caligiuri, *Corros. Sci.* 23 (1983): p. 223.

44. C. Kato, H.W. Pickering, J.E. Castle, *J. Electrochem. Soc.* 131 (1984): p. 1225.

45. A. Schussler, H.E. Exner, *Corros. Sci.* 34 (1993): p. 1803.

CHAPTER 4

General Corrosion

General corrosion involves more or less uniform metal dissolution from a surface. The rate at which this occurs varies with pH, temperature, dissolved oxygen (DO), and the presence of aggressive ions such as halides and sulfides etc. In the following section, data have been collected from a variety of common environments found in service, including the effect of environmental variables.

4.1 Natural Waters

4.1.1 Seawater and Fresh Water

Seawater typically contains about 19,000 mg/L chloride, but this may vary from 5,000 to 10,000 mg/L when dilution with fresh water occurs and may rise up to 26,000 mg/L when concentrated by evaporation (for example, the Arabian Gulf). The corrosion rates of some common copper alloys in seawater from several different sources are shown in Table 4.1,[1-7] in which most of the data was collected at local ambient temperature (10° to 25°C), but that in reference 5 was obtained at 40°C, hence the slightly higher corrosion rates compared with the other references.

All the data in Table 4.1 show the low general corrosion rate of copper alloys in seawater. The corrosion rate of pure copper is somewhat higher but still does not exceed 0.1 mm/y under the most aggressive conditions.

Efird and Anderson carried out exposure tests on 90/10 and 70/30 copper-nickel for up to 14 years, under quiet, flowing (0.6 m/s), and tidal

TABLE 4.1 General corrosion rate of some common copper alloys in seawater at ambient temperature[(A)]

	Corrosion rate (μm/y)						
Alloy	Ref. 1	Ref. 2	Ref. 3	Ref. 4	Ref. 5	Ref. 6	Ref. 7
Copper	–	15–100	10–40	50–100	–	–	40
Brass	–	10–30	8–13	2–7	50–300	20	50
90/10 Cu-Ni	<25	3–15	12	–	70	18	40
70/30 Cu-Ni	<25	3–15	6	–	40	25	25
NAB	55	25–50	5–10	1.3–2	–	–	60
Gunmetal	18	25–50	–	–	–	–	25–40
Alloy 400	<25	<25	–	–	8	13	25

[(A)] Most of these data were collected at local ambient temperatures (10–25°C), but the data from reference 5 were obtained at 40°C (hence the slightly higher corrosion rates compared with the other references). References 1–6 are data as collected, whereas reference 7 is design data taken from the U.K. Ministry of Defence *Guidelines for Naval Service*. The higher numbers in reference 7 are probably because of the high tidal water velocity at the U.K. test site that was used to generate data for the guidelines.

conditions.[8] The metal loss with time is shown for 90/10 and 70/30 copper-nickel in Figures 4.1A and 4.1B, respectively. Although the initial rate of corrosion was highest, it was still only about 12 μm/y for both alloys—not a high value. After 14 years, the rate for 90/10 copper-nickel in all three environments was the same, at 1.3 μm/y. For 70/30 copper-nickel, the rate was 0.8 to 1.9 μm/y. These are all very low figures, and the data show that the film becomes more protective with time, with steady state not being reached for several years. The results also show that the films form most slowly under flowing conditions. Over the 14-year exposure, none of the samples of either alloy showed any significant localized corrosion.

As the temperature increases, so does the corrosion rate of copper alloys, but not dramatically. Aluminum brass and copper-nickel alloys have been used successfully for many years in seawater-cooled heat exchangers, where discharge temperatures of aerated seawater may be up to 50 or 60°C. Data have been published for higher temperatures, but these are mostly at reduced DO levels, as are found in multistage flash (MSF) and multiple effect distillation (MED) desalination plants. Table 4.2 shows data from May et al. at 41 and 86°C. The corrosion rates at 41°C were generally higher than those at ambient temperature (Table 4.1), except for Alloy 400. The value for admiralty brass at 41°C seems anomalously high, but other authors also report increased corrosion rates for this alloy in warm seawater. Increasing the temperature from 41 to 86°C has little effect on the corrosion

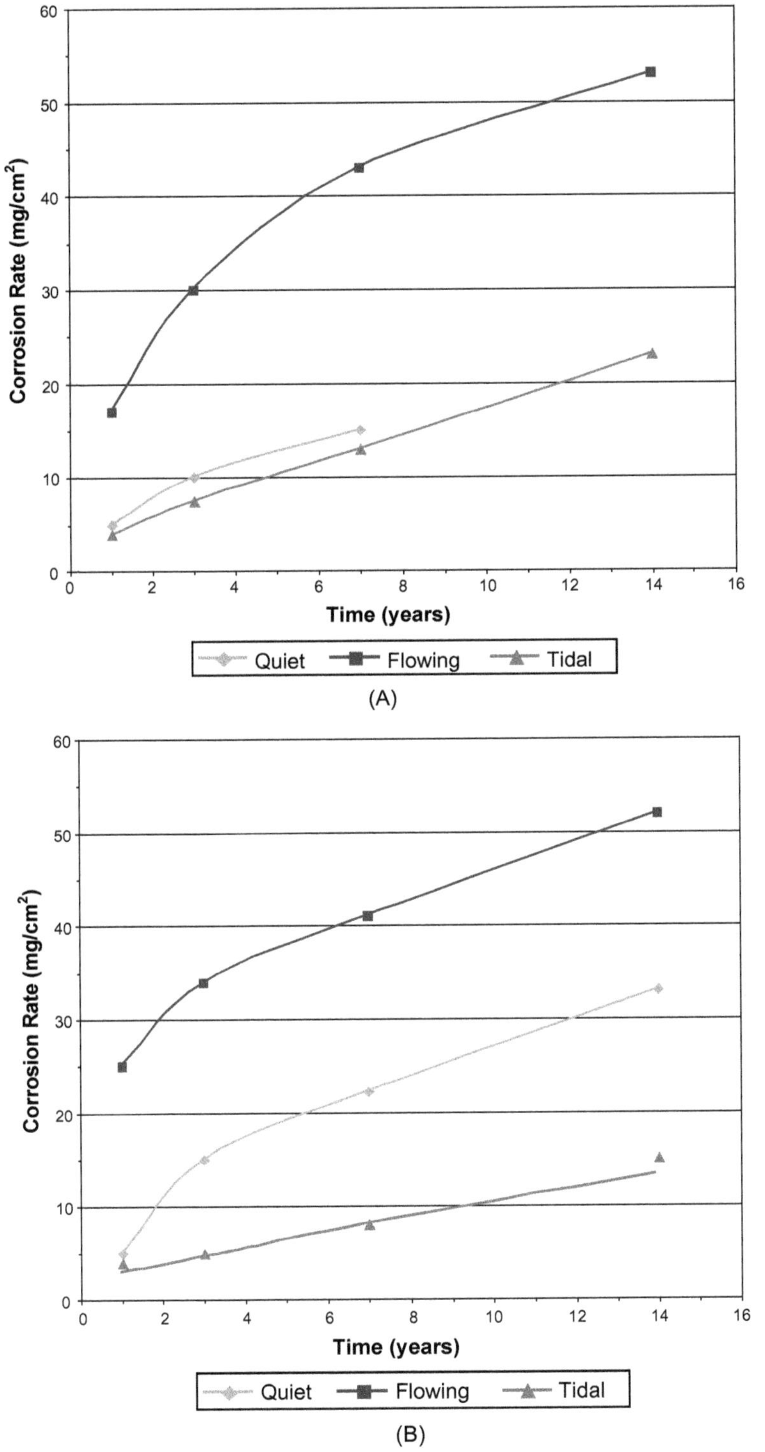

FIGURE 4.1 Effect of time on the corrosion of copper-nickel alloys in ambient temperature seawater: (A) 90/10 copper-nickel; (B) 70/30 copper-nickel.[8]

TABLE 4.2 Effect of temperature on the corrosion of various copper alloys in fully aerated seawater[5]

	Corrosion rate (μm/y)	
Alloy	41°C	86°C
Admiralty brass	305	48
Aluminum brass	48	64
90/10 Cu-Ni	71	30
70/30 Cu-Ni	38	36
Alloy 400	8	10

rate, except for admiralty brass, whose rate reduces to levels similar to the other copper alloys. The reason for this is probably that although the kinetics of corrosion will be more rapid at higher temperatures, the reduced oxygen concentration at 86°C will slow down the cathodic reaction, which is the reduction of DO.

Copper alloys are widely used in MSF and MED desalination plants, where temperatures can be high and oxygen concentrations are low. Schreiber et al. showed the effect of DO concentration at 107°C on copper and copper-nickel alloys (Figure 4.2).[9] The corrosion rates

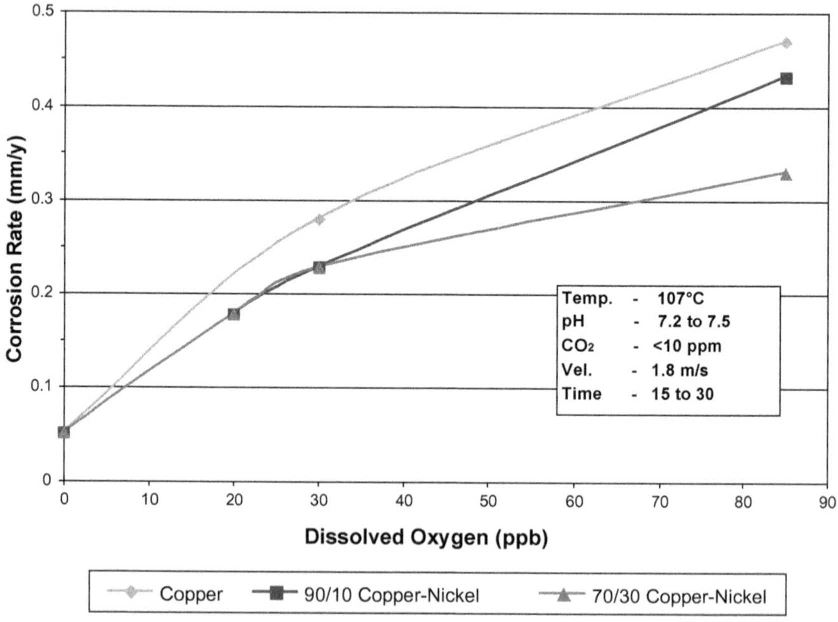

FIGURE 4.2 Effect of DO on the corrosion of copper and some alloys in seawater.[9]

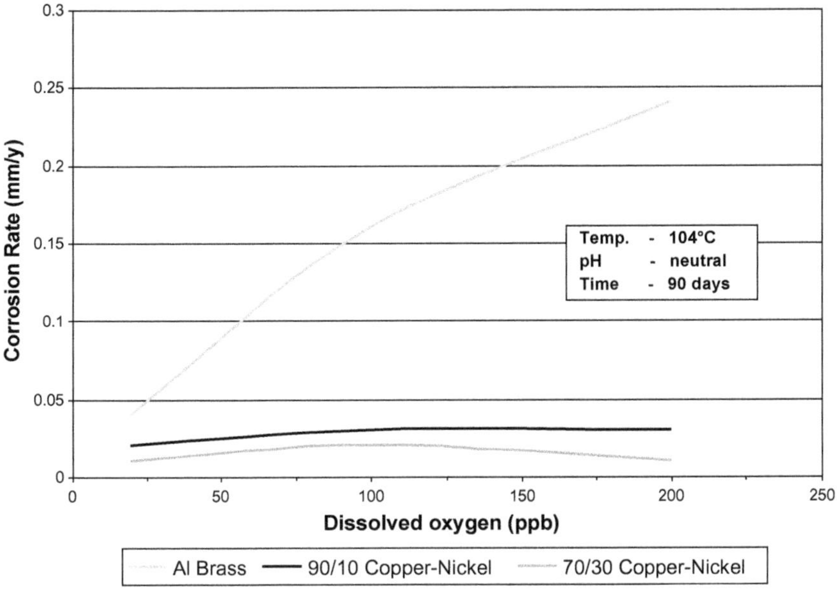

FIGURE 4.3 Effect of DO on the corrosion of some copper alloys.[10]

are high compared with those of Oldfield et al., at 104°C (Figure 4.3).[10] The difference may be due to the longer exposure time used by Oldfield et al. as corrosion rates decrease with time, or it may be that the velocity in Schreiber et al.'s tests was higher, which also increases corrosion rates.

Figure 4.4 shows the effect of temperature on the corrosion rates of copper alloys with 200 ppb DO; the corrosion rates for the copper-nickel alloys are very low at all temperatures. Figure 4.5 shows the effect of temperature measured in a plant in Texas, USA.[6] The results show the opposite effect to that shown in Figure 4.4, with corrosion rates decreasing with increasing temperatures. The reason for this is the variable oxygen concentration in a real plant, with the highest oxygen content at the lowest temperatures, and vice versa, whereas the data in Figure 4.4 are at a constant oxygen concentration.

Ross investigated the effect of water composition, pH, and temperature on the corrosion of copper-nickel alloys over a range of DO concentrations.[11] Ross found that there was no indication of protective film formation on copper-nickel alloys at 32°C and $pH < 6$ with $O_2 > 500$ ppb. Reducing the DO to 100 ppb or increasing the $pH > 6$ improved the formation of protective films. This is explained by the Pourbaix

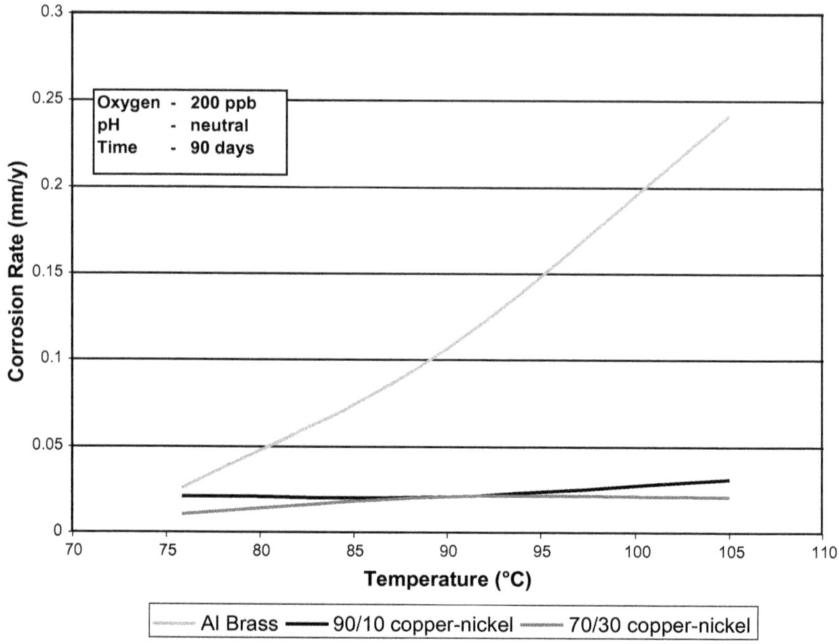

FIGURE 4.4 Effect of temperature on the corrosion of some copper alloys.[10]

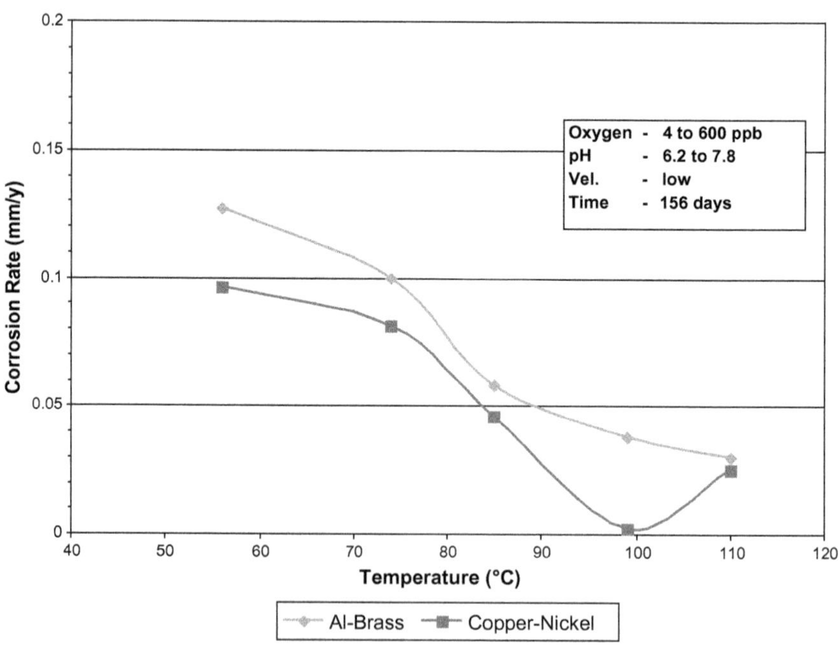

FIGURE 4.5 Effect of temperature on the corrosion of copper alloys.[11]

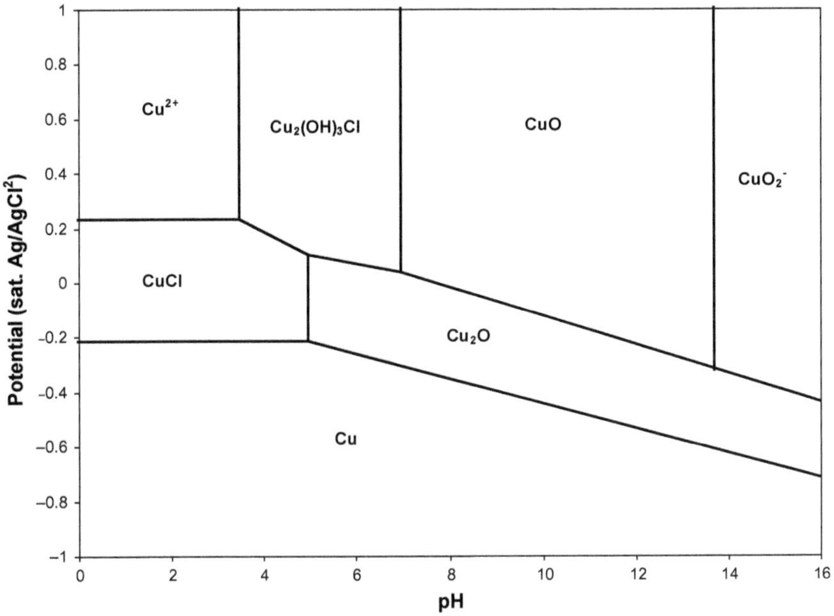

FIGURE 4.6 Pourbaix diagram for seawater (copper-chloride-water system).[12]

diagram for copper in seawater (Figure 4.6).[12] This shows that cuprous oxide (Cu_2O) is only stable down to pH 5 at normal potentials, but this is under equilibrium conditions. Under flowing conditions, Ross's observation that protective films did not form below pH ~ 6 is not surprising, and copper and its alloys are not normally recommended for continuous service in fresh water or seawater at pH < 6.

Ross also found a synergistic effect between pH, DO, and total alkalinity (bicarbonate) on the corrosion of copper-nickel alloys at elevated temperatures in the range 54 to 77°C.[11] At pH ≥7.5, increasing oxygen concentration (>100 ppb) accelerated corrosion when the bicarbonate was low (25–50 mg/L). When the bicarbonate was 65–90 mg/L and the pH ≥ 7.5, corrosion was inhibited regardless of the oxygen concentration. At higher seawater temperatures (107°C), protective film formation appeared complete, irrespective of bicarbonate and DO concentration, when the pH ≥ 7.5.

In fresh waters, the lower chloride content, typically <200 mg/L, means that the corrosion rate of copper and its alloys is even lower than for seawater. Corrosion rates of 1 to 10 μm/y are typical of copper in fresh water at temperatures from 10 to 60°C after a year or more.

Vik et al. reported the results of a monitoring exercise on copper pipes in Oslo, Norway.[13] The supply water is a soft, surface-derived type with the following characteristics: pH 5.0 to 7.0; total alkalinity 3 to 30 mg/L (as $CaCO_3$); calcium 1 to 5 mg/L. This is clearly an aggressive water, in which the formation of a protective film on copper tubes will be difficult. Over a period of 46 weeks, the corrosion rate at both 0.17 and 0.6 m/s was 76 ± 26 μm/y. While higher than normally seen in fresh waters, the corrosion rate was still not excessive.

4.1.2 Cuprosolvency

The slight general dissolution of copper that takes place in some fresh waters has little effect upon the copper pipes or tanks but may affect water quality. Water containing 1.5 mg/L or more copper can cause green stains on sanitary fittings. Smaller concentrations of copper can cause accelerated corrosion or pitting of other metals downstream of copper pipes. As little as 0.1 mg/L copper is sufficient to increase the corrosion rate of galvanized steel, and higher concentrations can cause pitting, particularly in hot waters. Francis describes such a case from a block of flats in Denmark.[14] For this reason, it is always recommended that galvanized steel be used upstream of copper pipes, but never downstream. Copper concentrations as low as 0.02 mg/L can cause pitting of aluminum in some waters.[15]

Normally, the copper concentration drops rapidly in the first few days as a protective film forms on the copper surfaces. However, in some waters, high concentrations of copper can persist. These waters are generally soft and have a low pH (associated with a high free carbon dioxide [CO_2] content). Cuprosolvency is more of a problem in hot waters as copper dissolves more readily at higher temperatures; however, some hard waters that are cuprosolvent when cold may be less of a problem when hot because the precipitation of calcium carbonate scale reduces copper dissolution. It has been suggested that under some conditions, bicarbonate in fresh waters can increase cuprosolvency, but this has never been verified by other workers.[16] Figure 4.7 shows the appearance of a typical copper pipe in a cuprosolvent water. The tube has a thick green deposit that is loosely adherent, with little copper oxide beneath it.

A number of authors have reported the conditions that cause cuprosolvency. Gilbert found that all the cuprosolvent waters that he investigated had a pH from 4.9 to 6.9.[17] Nuttall stated that cuprosolvent waters have a $pH < 6$, while the EEC mandatory limits for potable water are $6.5 < pH < 9.5$.[18] Hence, waters within these limits would not be expected to cause

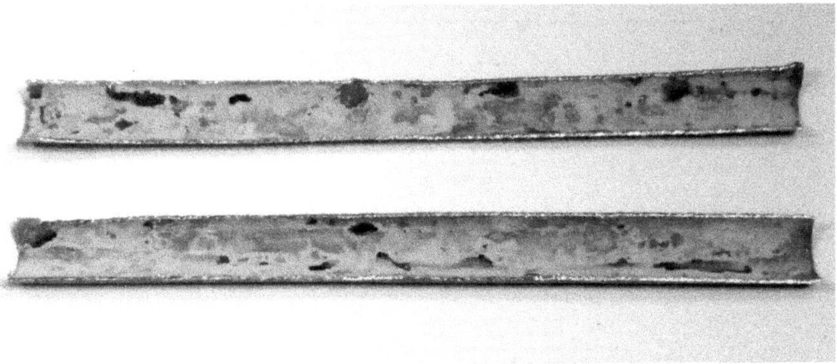

FIGURE 4.7 Appearance of films on copper tube exposed in a cuprosolvent water.

cuprosolvency. Acidic soft waters, whose pH has been increased, say, by caustic soda additions, are not buffered, and some slight corrosion can cause the local pH to drop to low values, allowing the water to become cuprosolvent in those regions.

However, there are other ways that high copper levels can be produced. Long periods of stagnation, particularly when piping is new, have been observed to result in poor film formation and high copper levels.[19] Maselkowski reviewed reported cases of cuprosolvency, including several for which there had been long periods of stagnation after a new piping system had been filled.[20] These problems are usually associated with large buildings that have long pipe spurs that are used infrequently and involve soft waters of low pH.

The levels of copper that are permitted in waters vary; for instance, in the United Kingdom, the permitted level is 2 mg/L. It is important to realize that copper is essential for human health; the World Health Organization recommends a daily intake of 2.0 to 3.0 mg of copper. Gilbert found levels of 2.9 to 5.3 mg/L of copper in water that had been standing overnight.[17] In Norway, good water has ≤ 0.1 mg/L copper when flowing, while 0.2 to 0.3 mg/L is acceptable, but copper >0.3 mg/L is unacceptable. For standing water, copper concentrations >1.0 mg/L are unacceptable. Figure 4.8 shows the copper concentration monitored by Vik et al. in Oslo tap water over a period of 42 weeks from 1987 to 1988.[13] The pH varied from 5.0 to 7.0, and the total alkalinity varied from 3 to 30 mg/L (as calcium carbonate [$CaCO_3$]). It can be seen that the flowing water was acceptable most of the time, but the standing water was unacceptable most of the time. This demonstrates the increasing risk of cuprosolvency in acidic, stagnant water.

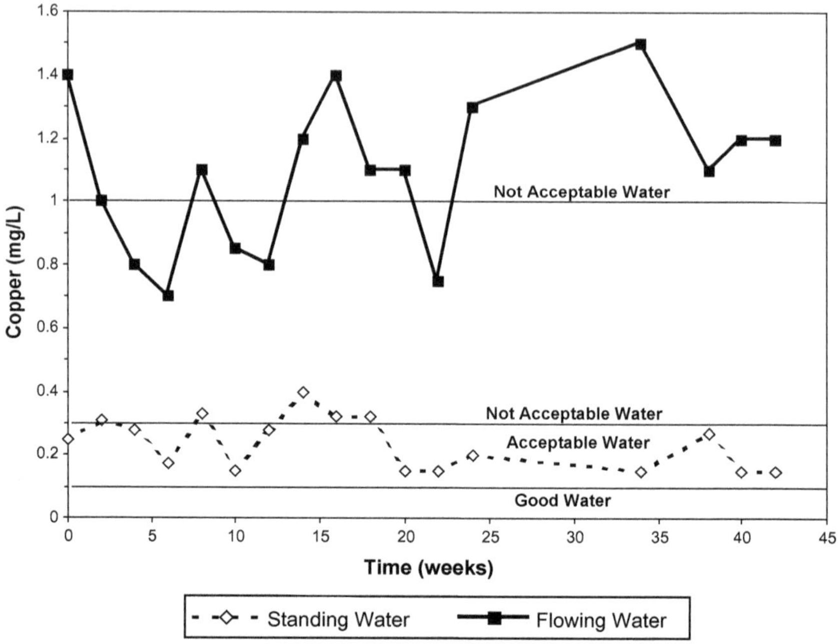

FIGURE 4.8 Copper concentration in Oslo tap water during 1987–1988.[13]

A similar problem is often referred to as blue water. This is caused when an insoluble copper corrosion product, believed to be a form of copper hydroxide ($Cu[OH]_2$), is formed but does not deposit on the tube walls. It is carried along in water, giving it a blue or bluish green color. Oliphant lists some of the causes of blue water[19]:

1. Waters with bicarbonate contents <20 mg/L (as $CaCO_3$), provided that other factors are present
2. Waters with $pH > 9$, where copper hydroxide is more stable and is more likely to be formed
3. Microbiological films that inhibit access of DO to the metal surface and prevent the formation of copper oxide
4. Water stagnation during the early life of piping

Critchley et al. compared the dissolution of copper from pipes exposed to sterile water with pipes exposed to sterile water to which bacteria from a problem water had been added.[21] Much higher rates of copper dissolution occurred on tubes exposed to the water with bacteria. This suggests that bacteria can contribute to the problem of blue water and that its causes need not be solely chemical, as already mentioned by Oliphant.

FIGURE 4.9 Appearance of a copper tube carrying so-called blue water. (Photograph courtesy of P. Munn.)

Munn noted that blue water can sometimes occur with waters that would not normally be thought to cause this problem.[22] He detected large quantities of phosphorous in the blue-green deposit on the tubes—presumably copper phosphate. A typical example is shown in Figure 4.9. These waters are being dosed with phosphates to reduce plumbosolvency (see next section), although the waters are not obviously at high risk to cause this. It appears that the presence of phosphates interferes with the formation of protective films. Sometimes the phosphate corrosion products can cause significant blockage of the pipe.[22]

4.1.3 Rosette Corrosion

Although this is really a type of general corrosion,[19] it has some of the appearances of pitting and is associated with the use of aluminum protector rods to prevent Type I pitting of copper water cylinders and tanks. Hence, the topic is discussed in more detail in Chapter 5.

4.1.4 Plumbosolvency

Copper alloys are frequently used for fittings in potable water systems; the most commonly used ones are forging brasses (e.g., C38000 and CZ132) and LG2 gunmetal (C83600). All these alloys contain lead (2 to 5%) to improve machinability and to produce pressure-tight fittings. The concern is that in some waters, unacceptable quantities of lead may leach from fittings.

Campbell discussed the water compositions that are likely to lead to high dissolution rates of lead.[15] Hard waters, and those that produce scale (positive Langelier Index), are the least likely to cause leaching of lead, whereas waters with high nitrate and/or chloride contents lead to increased dissolution. The worst waters for lead dissolution are soft waters of low pH or nonscaling waters with a high dissolved CO_2 content. The quality of these waters can be improved by additions of lime ($Ca[OH]_2$).

Paige and Covino investigated lead dissolution in distilled water under static conditions as a function of both time and temperature for a number of commonly used copper alloys.[23] They found that most of the lead dissolved in the first 24 hours, and it exceeded the U.S. limit of 15 μg/L under all conditions. This is hardly surprising given the aggressive nature of distilled water toward lead dissolution, as defined by Campbell.[15] They also found that a pretreatment in acetic acid (CH_3COOH) to remove lead smeared by machining did not significantly affect lead dissolution. Increasing the water temperature either had no effect on lead dissolution or reduced it only slightly.

Tests by Oliphant showed that there was no correlation between the lead content of the alloy and lead release into natural, fresh waters.[24] Oliphant also reported test results to the draft standard DD 256:2002 for a range of leaded alloys.[19] A round-robin evaluation of this test method found wide variation in the results obtained from laboratory to laboratory, and there is currently no European standard test for lead release from copper alloy fittings.

The limits on lead release in the United Kingdom were 50 μg/L up to 2003, when the limit was reduced to 25 μg/L. From 2013, the limit is to be reduced to 10 μg/L. Oliphant found that all the alloys met the 25 μg/L limit, but compliance with the 10 μg/L limit might be difficult for some alloys.[19]

4.1.5 Disinfection

When potable water systems are installed in large public buildings, it is common to disinfect the tubes with a strong solution of sodium hypochlorite (NaOCl), at anywhere from 20 to 100 mg/L free chlorine. Francis and Saxton investigated the effect of strong disinfection regimes on the subsequent performance of copper tubes.[25] Tests for up to 30 days under both stagnant and flowing conditions (up to 3 m/s) showed no effect of a pretreatment of 100 mg/L chlorine for 2 hours or 20 mg/L chlorine for 16 hours.

The tests included both a hard well water and a soft, surface-derived water at room temperature and at 60°C. Under continuous chlorine dosing at 1 mg/L, there was no effect on copper corrosion at room temperature under any flow condition. At 60°C, there was a marked increase in general corrosion rates in both waters, an increase in the susceptibility to Type II pitting under slow flows in the soft water, and an increased susceptibility to erosion corrosion at high flow velocities. The effect of chlorine on Type II pitting is discussed in more detail in Chapter 5.

Bond et al. reported the results from tests lasting up to 12 months using the same disinfection regimes, flow rates of 0.1 and 2 m/s, and ambient and 60°C temperatures.[26] After 12 months, the corrosion rates of all the tubes were in the range 1 to 3 μm/y—a very low rate. The corrosion rates after 4 months were somewhat higher, but all were <0.01 mm/y—still a low corrosion rate. Type II pitting was not seen in the hot soft water with continuous chlorine dosing because the soft water used was different from the water used in the 30-day tests and did not support Type II pitting. The conclusion was that disinfection of copper tubes under these regimes had no detrimental effect on the performance of copper tubes in subsequent service.

4.2 Acids

The data in this section are taken from Rabald,[27] and those seeking more detailed data or corrosion rates in acids other than those mentioned here are urged to consult Rabald's book.

4.2.1 Sulfuric Acid

Sulfuric acid (H_2SO_4) is the most commonly used of the mineral acids, and copper has moderate resistance to dilute acid, but the corrosion rate increases when the solution is aerated or oxidizers are present, and also when the temperature is increased. The following example in 6% sulfuric acid shows the effect of aeration:

Temperature (°C)	Corrosion rate (mm/y)	
	Static	Aerated
60	0.45	1.18
80	0.61	1.51
90	0.61	1.67

Brass is not recommended for use in sulfuric acid under any conditions, but tin bronzes (without zinc) have been used at room temperature from

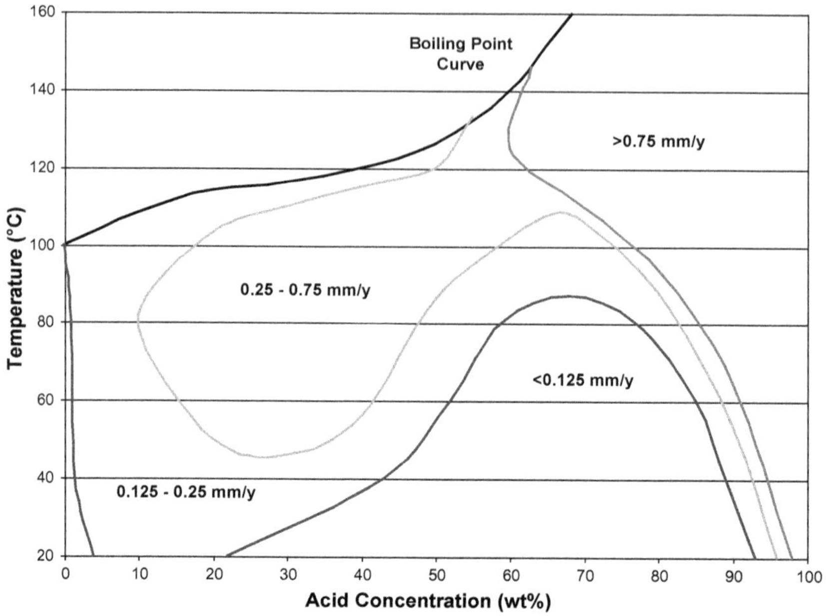

FIGURE 4.10 Corrosion rate of nickel aluminum bronze in sulfuric acid.[27]

1% to 80% acid. The corrosion rate varies from 0.04 mm/y (deaerated) to 0.4 mm/y (aerated).

Aluminum bronze has good resistance from 20 to 90°C up to 80% acid concentration. Corrosion rates are in the range 0.08 to 0.14 mm/y, but this increases under strong aeration. Under nonoxidizing conditions, nickel aluminum bronze (C63200) has good resistance to sulfuric acid over a wide range of temperatures and acid concentrations,[7] as shown in Figure 4.10. The copper-nickel alloys are inferior in performance to both NAB and tin bronzes in sulfuric acid.

Alloy 400 has some uses in sulfuric acid, particularly at room temperature, when air is not present, as shown in Table 4.3. The corrosion rate increases in the presence of air, particularly at elevated temperatures.

4.2.2 Hydrochloric Acid

In acid concentrations less than 30%, copper corrodes in the range 0.08 to 0.85 mm/y, provided that the solution is free of air or other oxidizers. With oxidizers present or with stronger acid, the corrosion rate increases substantially.

Brass should not be used in hydrochloric acid (HCl) under any conditions, but tin bronzes corrode at 0.33 to 2.5 mm/y in dilute acid in the

TABLE 4.3 Corrosion of Alloy 400 in sulfuric acid[27]

Acid concentration (%)	Temperature (°C)	Corrosion rate (mm/y)	
		Deaerated	Aerated
2	20	0.12	0.80
2	20	0.19	1.0
10	20	0.15	0.90
30	20	0.10	0.41
50	20	0.09	0.30
70	20	0.05	0.28
80	20	0.02	0.27
90	20	2.8	3.9
5	50	0.38	1.7
5	80	0.37	2.9
5	90	0.20	2.5
5	100	0.05	0.4
25	260	Unsuitable	Unsuitable

temperature range 15 to 100°C. Aluminum bronze is similar to tin bronze, and oxidizers greatly affect corrosion rates. The following data are from Rabald[27]:

Temperature (°C)	Acid concentration (wt%)	Corrosion rate (mm/y)	Comment
20	3.6	0.18	–
100	15	2.29	–
100	30	0.57	–
20	30	0.57	–
20	30	1.71	0.8% chlorine
20	30	4.70	1% $FeCl_3$

This indicates that aluminum bronze would not normally be suitable for service in HCl, but it has been used for hooks in pickling/descaling baths for steel, where the galvanic couple with steel prevents corrosion.

The copper-nickel alloys corrode at about 0.3 mm/y, provided that there is no aeration or oxidizers. Alloy 400 has some uses in hydrochloric acid, but corrosion rates increase dramatically in the presence of oxidizers (e.g., oxygen, iron, copper). Table 4.4 shows some typical corrosion rates.

4.2.3 Nitric Acid

Copper and its alloys have unacceptably high corrosion rates in nitric acid (HNO_3) under all conditions, as does Alloy 400, and should not be used.

TABLE 4.4 Corrosion of Alloy 400 in hydrochloric acid[27]

Acid concentration (%)	Temperature (°C)	Corrosion rate (mm/y)	Comments
0.5	100	0.73	Unaerated
1.0	30	0.08	Deaerated
1.0	70	0.13	–
2.0	27–50	1.80	–
3.6	60	0.90	Unaerated
5.0	30	0.18	Deaerated
5.0	30	2.10	Aerated
5.0	70	1.20	Deaerated
5.0	70	1.30	–
30.0	30	6.7	Aerated

4.2.4 Hydrofluoric Acid

In liquid hydrofluoric acid (HF) at 15 to 20°C, copper corrodes at ~0.5 mm/y, but this increases under agitation. Brasses perform similarly to copper.

Tin bronzes corrode at 0.5 mm/y in liquid HF at 15 to 40°C, rising to 1.5 mm/y at 80 to 90°C. The aluminum bronzes have low corrosion rates in HF, for example, 0.11 mm/y in 40% HF at 20°C. This corrosion resistance has resulted in the use of nickel aluminum bronze for the nozzles that spray HF to produce the frosting of light bulbs and also for the acid collection trays beneath the nozzles.[7]

The 70/30 copper-nickel alloy also has good resistance to HF. In liquid HF, the corrosion rate varies from 0.18 to 0.25 mm/y in the temperature range 15 to 90°C. In 98% HF at 38°C, the corrosion rate is 0.05 mm/y.

Alloy 400 is, in general, very resistant to attack by HF, but its corrosion rate increases when oxidizers or reducing sulfur compounds are present. Following are some typical corrosion rates in pure HF:

Acid concentration (%)	Temperature (°C)	Corrosion rate (mm/y)
30	21	0.203
48	21	0.003
70	21	0.003
93	21	0.152
100	150	0.076

TABLE 4.5 Corrosion of nickel aluminium bronze and 15% copper-nickel in phosphoric acid [27]

Acid concentration (%)	Corrosion rate (mm/y)					
	NAB			15% Cu-Ni		
	15°C	50°C	75°C	15°C	50°C	75°C
20	0.06	0.10	0.25	0.61	0.17	0.58
40	–	–	–	0.03	–	0.11
60	0.01	0.01	0.00	0.01	–	0.03

4.2.5 Phosphoric Acid

As with other mineral acids, the corrosion rate of copper is moderate at low acid concentrations and temperatures. The presence of oxidizers greatly accelerates attack. The brasses perform similarly to copper, while the tin bronzes are more corrosion-resistant. For example, bronze with 8.5% tin corrodes in 80% acid at 0.1 mm/y at 60°C, at 0.21 mm/y at 110°C, and at 0.6 mm/y at 120°C.

Nickel aluminum bronze is very resistant to corrosion in phosphoric acid, whereas copper-nickel alloys are somewhat less resistant, as shown in Table 4.5.[27]

4.2.6 Organic Acids

Copper can have low corrosion rates in acetic acid, provided that air and oxidizers are excluded. Following are some typical corrosion rates at 20°C:

Acid concentration (%)	Corrosion rate (mm/y)	
	Deaerated	Aerated
6	0.03	0.58
50	0.08	1.84
99	0.05	0.05

The corrosion rate also varies with temperature, as shown in the following, for solutions in contact with air:

Acid concentration (%)	Corrosion rate (mm/y)	
	25°C	Boiling
20	0.40	1.67
60	0.14	0.31
100	0.20	6.74

Brass can also be acceptable in acetic acid if air and oxidizers are excluded. The bronzes have low corrosion rates in acetic acid, with tin bronze corroding at 0.08 mm/y in 33% acid at the boiling point and aluminum bronze corroding at 0.06 mm/y under the same conditions.

Alloy 400 has good resistance to acetic acid at 20°C at all concentrations. The corrosion rate is 0.08 mm/y when little air is present, rising to 0.6 mm/y with strong aeration and agitation. The greatest corrosion rate occurs in 80% acetic acid. At elevated temperatures, corrosion rates can rise up to 0.61 mm/y.

In formic acid (HCOOH), the performance of copper and its alloys is similar to its performance in acetic acid, with high corrosion rates when air or oxidizers are present. Alloy 400 is less sensitive to the presence of air than other copper alloys, but the corrosion resistance is also similar to the corrosion resistance in acetic acid.

In 50% citric acid ($C_6H_8O_7{\cdot}H_2O$), the corrosion rate of copper increases from 0.57 mm/y at 20°C to 0.69 mm/y at 70°C in aerated solution. When deaerated, the corrosion rates are 0.10 and 0.12 mm/y, respectively. Brass performs similarly to copper in citric acid, as do the tin bronzes, although the latter are less sensitive to aeration.

Aluminum bronze is very resistant to citric acid, with a corrosion rate of ~0.03 mm/y in 33% to 36% acid from 20°C to the boiling point; 70/30 copper-nickel also has a low corrosion rate in all strengths of citric acid, typically, 0.04 mm/y at room temperature and 0.18 mm/y at 60°C.

4.3 Alkalis

The most common alkali is sodium hydroxide (NaOH) (caustic soda), and copper corrodes at modest rates from room temperature to 200°C up to 75% concentration, provided that air is excluded and there are no oxidizers. Brasses have been used at room temperature in up to 33% caustic soda with corrosion rates $\leq$0.13 mm/y. Tin bronzes are somewhat better than the brasses under the same conditions, with a corrosion rate of $\leq$0.10 mm/y.

Aluminum bronzes are satisfactory in dilute solutions at around room temperature, but they corrode rapidly at elevated temperatures or in strong solutions; 70/30 copper-nickel has been used at up to 75% concentration from 2 to 175°C. In 65% to 75% caustic, the corrosion rate is 0.10 mm/y at 150 to 175°C.

Alloy 400 has excellent resistance to caustic soda and is frequently used for numerous components in its manufacture. Figure 4.11 shows the isocorrosion curve (0.1 mm/y) for Type 316L stainless steel (UNS 531603),

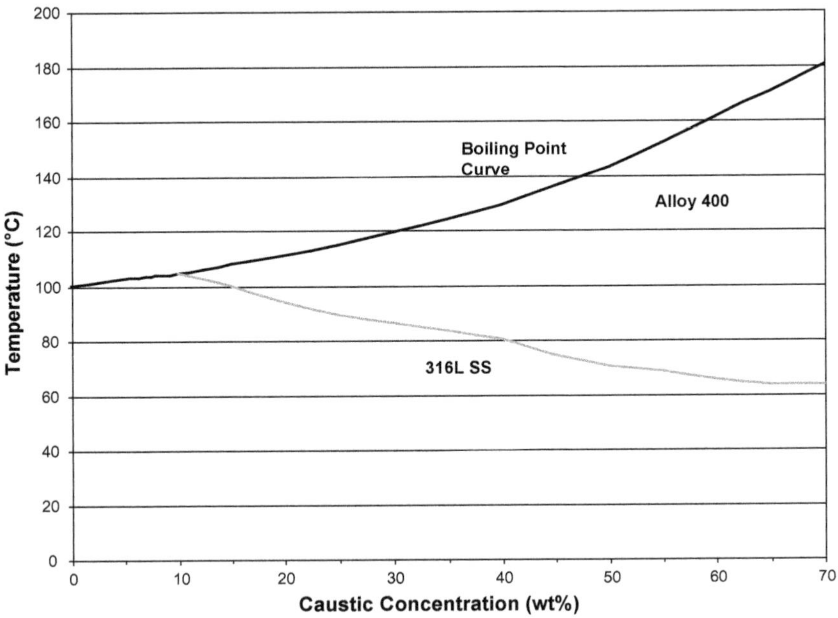

FIGURE 4.11 Isocorrosion curve (0.1 mm/y) for 316LSS and alloy 400 in caustic soda.

and it can be seen that Alloy 400 can be used up to the boiling point, even in strong solutions.

The performance of copper and its alloys in calcium hydroxide ($Ca[OH]_2$) is very similar to its performance in sodium hydroxide.

Ammonia (NH_3) is very alkaline and forms cuproammonium complexes, which tend to accelerate corrosion rates. Copper is not suitable for use in ammonia, and brasses are liable to suffer stress corrosion cracking (SCC; see Chapter 10). The tin bronzes are much more resistant to SCC; they corrode at ~0.53 mm/y in 5% aerated ammonia at room temperature. Aluminum bronzes are not so resistant as the tin bronzes, with a corrosion rate of 0.82 mm/y under the same conditions.

Ammonia was once used as a method of pH control in boiler feed water in power stations, and it would carry over with the steam into the condensers. This caused SCC of brass condenser tubes in some power stations, but copper-nickel tubes performed much better. Caruso and Michels carried out tests to study the effect of ammonia on some common condenser tube alloys.[28] In the fog test, the alloys were exposed to an ammoniacal mist that condensed onto tubes through which cooling water at 12°C was passed. The mist was produced from distilled water containing 1,000 mg/L ammonia at

TABLE 4.6 Corrosion in ammonia fog and spray tests for some copper alloy condenser tubes [28]

	Average corrosion rate (mm/y)	
Alloy	Fog test	Spray test
Phosphorized copper	0.17	>10
Admiralty brass	0.28	13.2
Aluminum brass	0.10	4.3
Al-bronze (C60800)	0.18	NT[(A)]
90/10 Cu-Ni	0.10	0.05
70/30 Cu-Ni	0.01	<0.01
Alloy 400	0.01	<0.01

[(A)] NT, not tested

30°C, and the test lasted 100 days. In the spray test, a solution containing 2,000 mg/L ammonia was sprayed onto tubes at 21°C so that it cascaded over several tubes mounted one above the other. The test lasted 170 days, and the results of both tests are shown in Table 4.6. It can be seen that while the corrosion rates of copper, the brasses, and aluminum bronze were high, the copper-nickel alloys performed better, with 70/30 copper-nickel and Alloy 400 performing the best.

Rabald reported the corrosion rate of Alloy 400 in 60% ammonia at 20°C as 0.2 mm/y without air, rising to 0.6 mm/y with aeration and agitation.[27]

4.4 Other Chemicals

The comments given in this section are only general; the reader requiring data for a specific compound is urged to consult Rabald's book.[27]

Corrosion rates in neutral salts (6< pH<8) are generally low (the specific example of chlorides was discussed previously). Less aggressive salts, such as sulfates and nitrates, cause lower corrosion rates than chlorides.

Metal salts with a low pH, such as iron, copper, and manganese, produce higher corrosion rates, particularly if the anion is chloride; however, the dissolution rate will be higher when the pH is below the stability range for cuprous oxide—typically, pH 5.0–5.5. However, these corrosion rates need not be excessive if the temperature is close to ambient, concentrations are low, and there is little agitation. Remember that the commercial versions of these salts often contain some free acid, which makes them more aggressive than the pure salt.

Alkaline salts, such as sodium silicate (Na_2SiO_3), phosphate (Na_2PO_4), and carbonate (Na_2CO_3), are not as aggressive as the strong alkalis, such as sodium and calcium hydroxide. Copper and its alloys have acceptable corrosion resistance over a wide range of concentrations and temperatures. Note that the presence of oxidizers tends to increase corrosion rates, although not necessarily to an unacceptable level. Testing in specific environments is advised, especially when more than one compound is present.

The other important group of salts are those that can form complexes with copper such as the acetates, citrates, and tartrates. General corrosion in solutions of these salts is not usually very high, but over certain potential ranges, it is possible to get SCC, particularly with brasses. Tin bronzes and the copper-nickel alloys are much more resistant to SCC in these environments; susceptibility to SCC is discussed in more detail in Chapter 10.

4.5 Methods of Prevention

Low levels of corrosion in fresh water and seawater can be maintained, provided that flow velocities are low enough to avoid erosion corrosion (see Chapter 7) and pollutants are not present (see Chapter 3). These low corrosion rates are maintained at elevated temperatures, provided that the pH is >6.

Where cuprosolvency is a problem, water treatment to increase the pH and bicarbonate content will significantly improve matters. Where the problem has been exacerbated by a long period of stagnation soon after first filling, methods need to be adopted to encourage the formation of protective cuprous oxide. Maselkowski described a case of cuprosolvency examined by BNF Metals Technology Centre at a school in Torquay, in the United Kingdom.[20] The copper pipes had been left full of stagnant water for some time before the school was occupied. Copper levels were mainly in the range of 3 to 4 mg/L, with excursions to 6 to 8 mg/L. The problem was solved by allowing water to run slowly but continuously through the system, particularly long spurs, for about 3 weeks. Thereafter, the copper concentration in the water returned to more normal levels.

Problems of blue water also require water treatment to encourage the formation of protective oxide films. If microbial action is also contributing to this problem, the water treatment will be more complicated and may involve biocides. Shock treatments with sodium hypochlorite (NaOCl) did not hinder protective film formation on copper tubes, and the oxidizing power of hypochlorite should discourage bacterial action, while encouraging oxide formation.[25,26]

Where plumbosolvency is a problem, water treatment with lime to increase the pH has been found to reduce lead dissolution rates. Where water treatment is not possible, then a switch from brass compression fittings to copper fittings, soldered with lead-free solder (see Chapter 18), is an alternative.

When selecting an alloy to handle an aqueous chemical solution, it is important to pick one of the more resistant alloys. As was described previously, the resistance of copper alloys varies according to the chemical, the pH, the temperature, and the degree of aeration. This latter variable is most important as copper alloys will resist corrosion well in a wide variety of chemicals when the solution is deaerated or of low oxygen content, but will corrode rapidly when the solution is aerated or there are oxidizers present.

As with waters, it is important to keep flow velocities down to avoid erosion corrosion problems.

Avoid the selection of copper alloys for environments in which they are susceptible to SCC, unless stresses are very low or are compressive. SCC can lead to premature failure, despite a low general corrosion rate (see Chapter 10). Again, select the more resistant alloys for specific environments.

Where multiple chemicals are present, particularly when the pH or the redox potential is uncertain, it is advisable to carry out corrosion tests to ascertain an alloy's suitability. When a copper alloy is connected to a less noble material, such as carbon steel, it will often be galvanically protected and will not corrode, even if the environment is aggressive (see Chapter 14). The use of nickel aluminum bronze hooks for lowering carbon steel plates into hydrochloric acid pickling solution is a typical example (see Section 4.2.2).

References

1. B. Todd, "Factors in the Choice of Materials for Marine Engineers," *Trans. Inst. Marine Eng.* 80 (1968): p. 161.
2. INCO, "A Guide to the Selection of Marine Materials," 2nd ed. (INCO, 1973).
3. F.M. Reinhart, "Corrosion of Metals in Hydrospace, Part IV, Copper and Copper Alloys," Naval Civil Engineering Laboratory Technical Note, N-961, April 1968.
4. C.W. Hummer, C.R. Southwell, A.L. Alexander, *Mater. Prot.* 7 (1968): p. 41.

5. T.P. May, E.G. Holmberg, J. Hinde, "Seawater Corrosion at Elevated Temperatures," Dechema-Monographian Bank, 47, 1962, p. 253.
6. E.M. Mattson, R.M. Fuller, "A Study of Materials of Construction in Distillation Plants," Joint Progress Report by Office of Saline Water and INCO, No. 163, October 1965.
7. H.S. Campbell, *Aluminium Bronze Alloys Corrosion Resistance Guide*, Publication 80 (Potters Bar, U.K.: Copper Development Association, 1981).
8. K.D. Efird, D.B. Anderson, *MP* 14 (1975): p. 37.
9. C.F. Schreiber, O. Osborn, F.H. Coley, *MP* 7 (1968): p. 20.
10. J.W. Oldfield, G. Swales, B. Todd, "Corrosion of Metals in Deaerated Seawater," Second Middle East Corrosion Conference (Bahrain: BSE-NACE, 1981), p. 48.
11. R.W. Ross, CORROSION/77, paper no. 94 (Houston, TX: NACE International, 1977).
12. G. Bianchi, P. Longhi, *Corros. Sci.* 13 (1973): p. 853.
13. E.A. Vik, T.L. Nilsen, P. Hallberg, "Corrosion Monitoring Techniques Used in Water Supply Systems," 11th Scandinavian Corrosion Congress (Stavanger, Norway, 1989).
14. R. Francis, *Galvanic Corrosion—A Practical Guide for Engineers* (Houston, TX: NACE International, 2001).
15. H.S. Campbell, "Corrosion of Metals in the Water Supply Industry," BNF Miscellaneous Publication, MP544, October 1968.
16. L. Trostad, R. Veimo, *J. Inst. Met.* 66 (1940): p. 17.
17. P.T. Gilbert, *Proc. Soc. Water Treat. Exam.* 15 (1966): p. 165.
18. J. Nuttall, "Copper," in *Corrosion and Related Aspects of Materials for Potable Water Supplies*, P. McIntyre, A.D. Mercer, eds. (London U.K.: Institute of Materials, Minerals, and Mining, 1993), p. 65.
19. R. Oliphant, "Causes of Copper Corrosion in Plumbing Systems," Foundation for Water Research Report, FR/R0007, May 2003.
20. C.R. Maselkowski, BNF Research Report, S495/2, February 1984.
21. M. Critchley, R. Taylor, R. O'Halloran, *MP* 44 (2005): p. 56.
22. P. Munn, "Effect of Phosphates on Corrosion of Copper in Domestic Hot and Cold Water Systems," U.K. Corrosion (Sheffield, U.K.: Institute of Corrosion, 2007).
23. J.I. Paige, B.S. Covino, *Corrosion* 48 (1992): p. 1040.
24. R. Oliphant, "Contamination of Potable Water by Lead from Leaded Copper Alloys," Second International Conference on Water Pipeline Systems, Conference Series Publication 10 (BHR Group, 1994).

25. R. Francis, H. Saxton, BNF Metals Technology Centre Research Report, R658/6, January 1991.

26. S. Bond, R. Francis, H. Saxton, BNF Metals Technology Centre Research Report, R658/12, August 1992.

27. E. Rabald, *Corrosion Guide*, 2nd ed. (New York, NY: Elsevier, 1968).

28. L. Caruso, H.T. Michels, *MP* 20 (1981): p. 35.

CHAPTER 5

Pitting Corrosion

When copper and its alloys are immersed in water and form a film of cuprous oxide (Cu_2O), it is usually protective, and corrosion rates are low. However, if this film breaks down locally, or a nonprotective film forms, particularly in localized areas, then rapid local attack can occur. Once this develops, the pits can be very narrow and may propagate rapidly. Failures by pitting attack may occur in a few months or only after 10 or more years. There are several different types of pitting attack that can occur on copper and its alloys, and they are discussed in the following sections, along with the actions necessary to prevent them.

5.1 Type I Pitting

5.1.1 Copper

In the late 1940s and 1950s, there was a great increase in failures in the United Kingdom by pitting of copper piping carrying fresh water. This corresponded to a time of massive rebuilding following the Second World War, but also to a change by the major copper tube manufacturers from open hearth annealing to bright annealing. Under the reducing conditions during bright annealing, any residual drawing lubricant could crack, leaving thin carbon films on the surface. The film was not visible to the naked eye, but it was suggested as early as 1950 that these films could play a part in the pitting mechanism.[1] In the 1960s and 1970s, failures were seen across Europe, as all the copper tube manufacturers adopted bright annealing. Today, Type I pitting, as it came to be called, has been seen on all continents.

Figure 5.1 shows a typical Type I pit in a copper pipe. The green mound over the pit is basic copper carbonate and calcium carbonate

FIGURE 5.1 Typical appearance of Type I pitting. (Photograph courtesy of H.S. Campbell.)

($CaCO_3$), with a red crystalline layer of cuprous oxide beneath it. Beneath the oxide is usually a thin layer of cuprous chloride (CuCl), but it is not always visible to the naked eye. The metal surrounding the pit also has a cuprous oxide film, often with some basic copper carbonate above it, but less than is seen above the pit (Figure 5.1).

Type I pitting occurs in cold, hard well waters, and failures have occurred in as little as 3 months. Type I pitting does not occur in warm waters, probably because other copper compounds become stable. Pitting does occur at elevated temperatures, which is discussed in the following sections on Types II and III pitting. Type I pitting does not occur in surface-derived waters, and Campbell suggested that this is because surface-derived waters contain traces of organic compounds, some of which could act as "natural inhibitors."[1]

Early work showed that the cuprous oxide film that formed in surface-derived waters was a dull brown color and adherent.[1-3] The oxide formed in well waters that caused pitting was red and flaked off easily. Waters containing natural inhibitor fluoresced under ultraviolet light, while pitting waters showed little or no fluorescence. The fluorescence can be quantified in a fluorometer against a solution of sodium chloride (NaCl) in distilled water.[4] An alternative way of assessing a water is to pass a current of 10 $\mu a/cm^2$ between two copper wires, one of which is in the water under test and the other of which is in the same water plus sodium hydroxide (NaOH) and sodium chloride.[4] After 14 days, the nature of the oxide on the wire in the test water will be as described previously for pitting and nonpitting waters, that is, red and flaky or brown and adherent, respectively.

The early work showed that the inhibitor was a negatively charged, acidic colloidal substance that was stable up to 100°C and could be extracted with chloroform.[1-3] It was thought that the inhibitor was possibly a delta lactone. BNF Metals Technology Centre revisited this area in the 1980s and extracted organic material from activated charcoal used at a water treatment works in the London area. This was separated by chromatography into six major groups, the smallest of which contained the inhibitor. Repeated separations of this part of the extract always showed the inhibitor to be in the smallest fraction. Eventually, the smallest fraction was too small for further separation or analysis, but when added to a pitting water, the inhibitor concentration was $\leq$1 mg/L, and it completely prevented pitting. Similar work by Edwards et al. also showed the effectiveness of the natural inhibitor in small quantities.[5] It is clear that a very large sample of organic material will need to be extracted to identify the inhibitor, and this work still remains to be done.

There is increasing attention by water companies to water treatment, including the removal of organic material. It is becoming increasingly common to pass water through activated charcoal or similar substances that remove much of the organic material present in surface-derived waters, including the natural inhibitor. Campbell reports several instances in which such treatments have turned nonpitting waters into pitting waters.[4]

The copper tube that is usually used for fresh water transport is tough pitch or phosphorous deoxidized (PDO), and both types are susceptible to Type I pitting. Devroey and Depommier found in field tests that pdo copper was slightly more resistant to pitting than tough pitch.[6] Sometimes copper also has small additions of arsenic, and field trials found that arsenical coppers were more susceptible to pitting than nonarsenical coppers.[7]

Copper tubing is produced in three conditions: hard, annealed, and half-hard. Hard tube requires no annealing and so should contain no carbon films. Half-hard tube is produced by cold finishing annealed tube. Devroey and Depommier compared uncleaned PDO copper tubes in all three tempers, exposed for 1 year in a pitting water.[6] The results, in Table 5.1, show that soft tubes were more susceptible to Type I pitting compared with half-hard tubes with a similar carbon content. Hard tubes were virtually immune to pitting, with only very shallow attack on all tubes.

Cornwell et al. carried out extensive field trials in the United Kingdom using two pitting waters and one nonpitting water.[8] They found that bright

TABLE 5.1 Effect of copper tube temper on the susceptibility to Type I pitting of uncleaned tubes[6]

		Maximum pit depth (after 1 year)		
Temper	No. of tubes	<0.1mm	0.1–0.45 mm	>0.45 mm
Soft	13	3 (23.1%)	2 (15.4%)	8 (61.5%)
Half-Hard	62	41 (66%)	13 (21%)	8 (13%)
Hard	33	33 (100%)	0	0

annealed tubes with no postanneal internal cleaning pitted readily in both pitting waters, with high tube potentials, while tubes that had been internally cleaned suffered no pitting and had low potentials. Both types of tube had low potentials in a nonpitting water, and no pitting was visible after testing. Transferring pitting tubes to a nonpitting water resulted in much lower potentials and the cessation of pitting, while the transfer of uncleaned tubes from a nonpitting water to a pitting water resulted in pit initiation and high potentials. Cornwell et al. defined three potential zones:

1. No pitting, $E < 100$ mV SCE (saturated calomel electrode)
2. Pitting possible, $100 < E < 170$ mV SCE
3. Pitting propagating, $E > 170$ mV SCE

Work by Pourbaix et al. found similar thresholds for pitting when comparing thermodynamic and potential equilibria.[9-13]

The work of Cornwell et al. as well as the work of other laboratories around Europe clearly showed that tubes that were properly cleaned internally were much less likely to pit, even if uncleaned tubes did.[8] This demonstrated the importance of carbon films in Type I pitting of copper tubes. Other possible causes of Type I pitting are described later. Type I pitting has also occurred in copper tanks and low-pressure hot water cylinders (in the United Kingdom), and these do not contain carbon films. Pits form at the bottom of tanks and cylinders because in a hot water cylinder, the water is cold in the bottom region. Lucey attributed the pitting to the formation of streaming corrosion products during early exposure such that large quantities of cuprous chloride are produced on the vessel sides and stream down to the lower regions.[14] This stimulates corrosion there and enables a pit to form, when otherwise, it would not.

Edwards et al. have taken this argument further and have suggested that any environment where cuprous chloride is produced at a rate that exceeds the rate of its oxidation or hydrolysis to copper oxide or other copper salts can encourage the formation of Type I pits.[5] This need not be in

FIGURE 5.2 Type I pitting at a solder flux run. (Photograph courtesy of P. Munn.)

tanks or cylinders, but could be in tubes, such as when the water is stagnant for a long time after first filling. The pipe shown in Figure 4.7 was suffering from cuprosolvency due to excessive stagnation after installation. However, a few pits had also formed under the loose corrosion products, and one of them was sufficiently deep to cause penetration and water leakage. Hence, the carbon film is just one way in which the metal surface can be shielded to allow the formation of sufficient cuprous chloride for pitting to initiate.

Other means by which high levels of cuprous chloride may be produced locally have been examined. Riedl and Klimbacher reported Type I pitting associated with flux residues from soldering, even though the tube was free of carbon films.[15] An example is shown in Figure 5.2. Flux residues can be acidic and frequently contain high levels of chloride, added as an activator, and so the production of higher than normal quantities of cuprous chloride locally is perfectly understandable. This is sometimes regarded more as a form of crevice corrosion (see Chapter 6), but there have been cases in which pitting under flux runs has occurred that was associated with pitting waters, while none has occurred in a nonpitting water. It clearly depends on the level of residues left in the flux as well as on the water composition.

Munn reports an increased tendency for Type I pitting in waters that are on the borderline of supporting pitting when they are dosed with phosphates.[16] Phosphates are added to waters by water companies to reduce lead dissolution from old lead pipes. This is frequently done, even

FIGURE 5.3 Type I pitting in a water with high phosphate. (Photograph courtesy of P. Munn.)

with hard waters that do not have a great tendency to be plumbosolvent. Munn attributes the problem to the formation of loosely adherent copper phosphate as well as the more normal basic carbonate. Figure 5.3 shows a typical pipe with a layer giving a strong signal for phosphorus in the scanning electron microscope and having Type I pits underneath it. The porous phosphate layer is able to shield the metal surface and encourages an increased concentration of cuprous chloride, which increases the likelihood of Type I pitting.

Munn also suggested that pitting could result in pipes left stagnant after a high chlorine disinfection treatment if the chlorine was inadequately washed out. High levels of chlorine do cause substantial copper dissolution, initially, under stagnant conditions (see Chapter 4).

5.1.2 Water Composition

Not all hard well waters will support Type I pitting, and a number of workers have investigated the water compositions that will cause pitting. Devroey and Depommier looked at a large number of waters from Belgium and Luxembourg and concluded that the most important factors governing Type I pitting were the calcium carbonate saturation index, the pH, and the organic content, as determined by fluorescence.[17] Aggressive waters had a positive saturation index, a pH slightly above 7, and very slight or no fluorescence.

A survey of waters in the United States concluded that pitting waters contained over 5 mg/L dissolved carbon dioxide (CO_2) (89% had between 10 and 40 mg/L).[18] The pH was in the range 7.0–7.8, and the dissolved oxygen (DO) content was high (10–12 mg/L). The pitting waters also had a high sulfate to chloride ratio, usually 3 or greater.

Lucey constructed a nomogram, originally based on 17 U.K. water compositions, that considered six factors in determining a pitting propensity rating, or PPR.[19] This nomogram was later refined using data from 103 U.K. waters.[20] The factors that were included were sulfate, chloride, nitrate, sodium, pH, and DO. Increasing sulfate, sodium, and DO increased the pitting propensity, while increasing chloride, nitrate, and pH decreased it. While sufficient chloride to produce cuprous chloride is required, large chloride concentrations decreased the pitting propensity.

More recent work by Edwards et al. suggested that increasing sulfate promotes pitting, while increasing bicarbonate (which is linked to pH) decreases pitting propensity.[5] Additional work showed that increasing chloride decreased the pitting propensity, while nitrate increased it.[21] The finding with nitrate is the only one that disagrees with Lucey's findings.

Jacobs et al. considered the role of sulfide on pitting in fresh waters.[22] This is uncommon in fresh waters in Europe, but it is prevalent in some areas of the United States, such as Florida, where concentrations from 0.01 to 0.3 mg/L are possible. They found that sulfide could depolarize the cathodic reaction and increase corrosion rates. It was suspected that pitting problems in several U.S. states where sulfide was present were exacerbated by its presence.

Lucey's nomogram has no basis in theory and was developed empirically from the known behavior of a number of waters; its usefulness lies in being able to predict the performance of other waters. Campbell reported the successful use of the nomogram for waters not included in compiling the nomogram, including some Icelandic waters.[4]

The nomogram is shown in Figure 5.4A, and Figure 5.4B shows how it is used. A little explanation will be helpful. First, a horizontal line is drawn from the chloride concentration on line PK to the curve PQ. Where it intersects, a vertical line is drawn (K′L). The sulfate concentration is fixed on line DE, and a horizontal line is drawn to the curve RS. The intersect, G, is moved up or down, according to whether the nitrate is in excess of the sodium or vice versa (point H). Line FH is then extended to the "effective

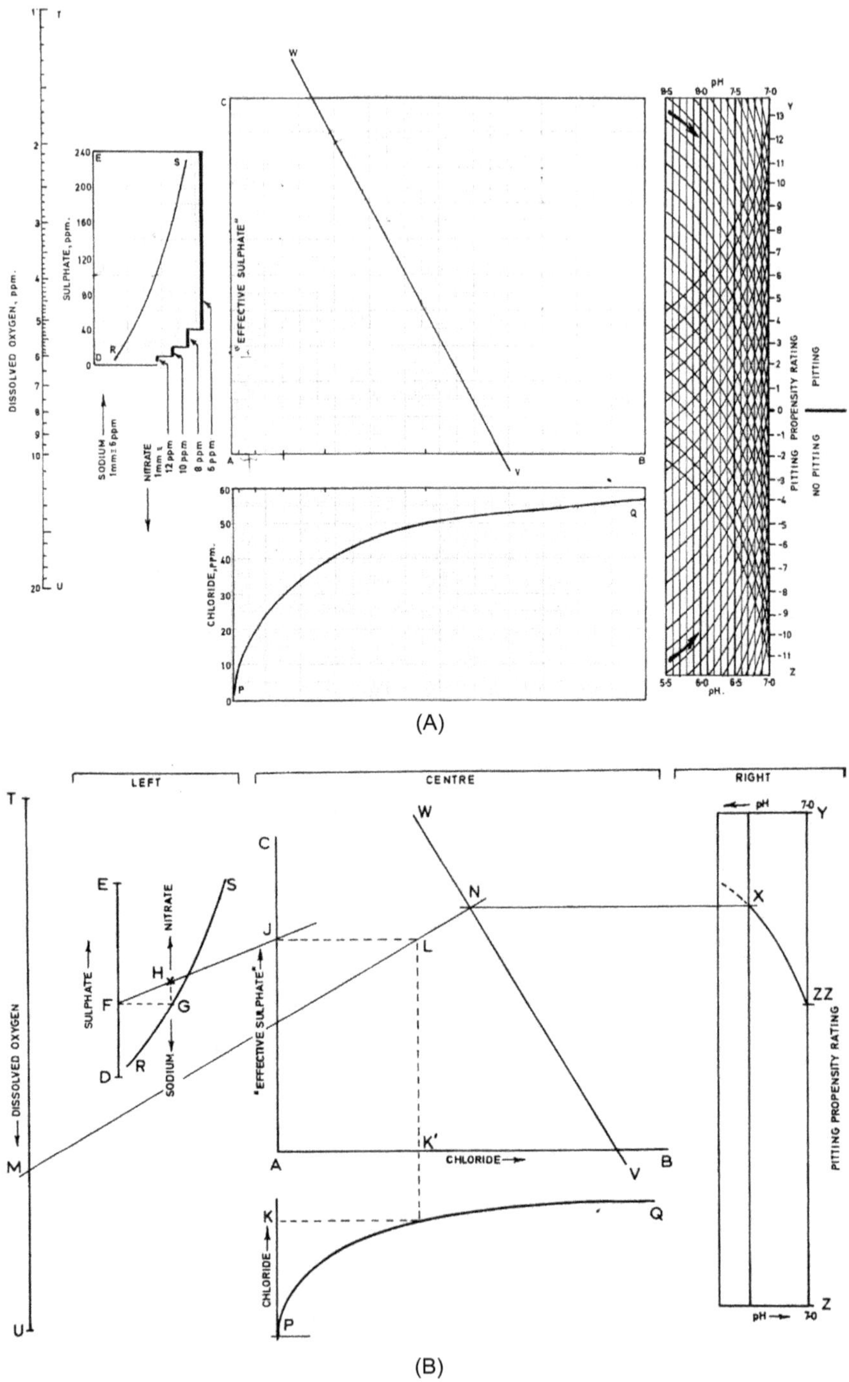

FIGURE 5.4 Lucey's nomogram for Type I pitting: (A) the nomogram; (B) how it is used.[14]

sulfate" axis (point J). The intersect of effective sulfate and chloride is point L. A line is then drawn from the DO concentration through point L to line VW (point N). A horizontal line is then drawn from point N until it intersects the curve of the pH of the water (curve XZZ). The curve is then followed to its intersection with YZ, where the PPR can be read off. This is a complicated procedure, and it was incorporated into a simple computer program to make its use easier. The nomogram produces a PPR that indicates the likelihood of failure, as follows:

1. No failures, PPR < 0
2. Borderline, failure unlikely (10 to 20 years), $0 <$ PPR < 1
3. Moderately aggressive, failure in 3 to 10 years, PPR $= 2$; and
4. Very aggressive, failure in less than 3 years, PPR > 2

Note that PPR values very much greater than 2 do not necessarily mean that the time to failure will be drastically reduced below 3 years.

The nomogram has a number of limitations and should only be used within a certain range of compositions. For example, it cannot handle chloride contents greater than ~50 mg/L, and it also suggests that a water with zero chloride would be very aggressive, while it is fairly certain that pitting would not occur.[4] Similarly, waters with a very low pH appear to be aggressive, when in fact, increased general corrosion would occur. One area in which the nomogram underestimates the PPR is for waters with low sulfate contents (<10 mg/L). There were insufficient waters of this type in the original sample to take proper account of this. The inclusion of sodium seems unusual, but following publication of the nomogram, several Type I failures were seen in base exchange softened waters, when untreated water was not aggressive. Hence, increasing the sodium content can increase susceptibility to Type I pitting.

One area in which the nomogram has been useful is in predicting the effect of mixing waters. This is not uncommon across Europe, and the nomogram can show the PPR at different mixing ratios. For many waters, the curve is smooth, almost linear, from 100% of the water with the highest PPR to 100% of the water with the lowest PPR. In some cases, however, mixtures can be more aggressive than either of the waters alone. One such case was reported by Lucey, and pitting failures occurred with a mixture close to the worst case scenario (Figure 5.5).[17] The author has also found the nomogram useful for this type of analysis on several occasions.

Lucey's nomogram does not appear to consider whether the water is surface-derived or from a well. However, it has been established that surface-derived waters frequently contain humic acid, which can hold in

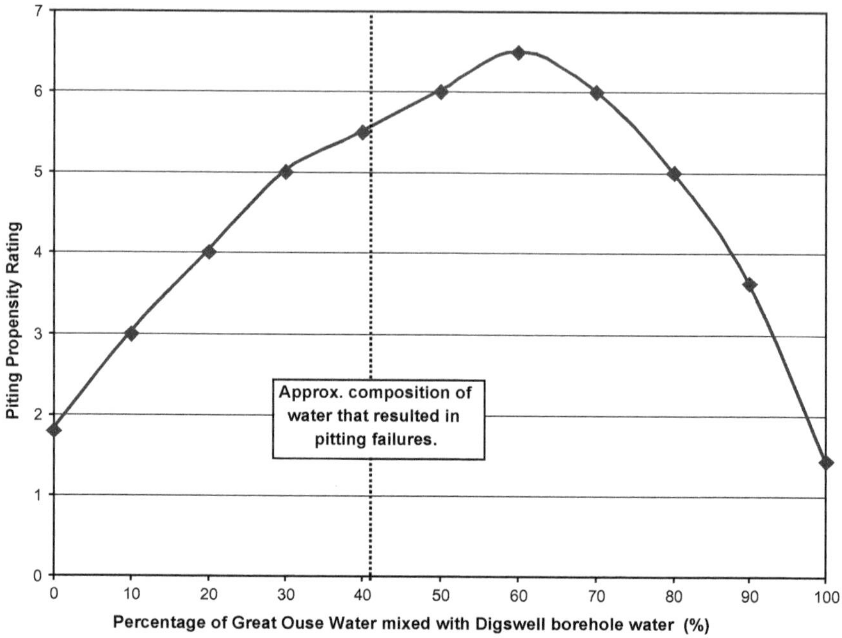

FIGURE 5.5 Effect of water mixing on propensity for Type I pitting.[19]

solution concentrations of calcium carbonate well above the equilibrium quantity. Consequently, these waters have an unusually high, stable pH. This is true for many surface waters, and a high pH gives a low PPR value on Lucey's nomogram.

A study in 1980 tried to rationalize the differences between waters causing Type I pitting in the United States and the United Kingdom.[23] One hundred eighteen water compositions from the United Kingdom and 38 from the United States were used in an automatic interaction detector computer program. Using only the U.K. waters, the critical parameters were CO_2 content, sulfate-chloride ratio, sulfate, and DO. Using only waters from the United States, the most important parameters were carbonate hardness followed by CO_2, nitrate (NO_3), and temperature. Taking all the waters together, total alkalinity, pH, DO, and calcium were the most important parameters. One possible cause of the discrepancies was changes in the water compositions of the U.S. waters between the original failures and the water sampling years later.

One factor that has turned out to be important is which cations are associated with which anions. This came to light when research at BNF Metals Technology Centre was initiated to manufacture a synthetic pitting water.[24] All the waters were tested using the BNF Metals Technology Centre accelerated pitting test cell, which is described in Section 5.8. Initial thought was that a water with a moderately high bicarbonate content (150 to 200 mg/L) and a sulfate-chloride ratio of approximately 2 would initiate pitting. Tests with a synthetic water made solely of sodium salts performed just like distilled water, with no pitting. As pitting waters all contain calcium salts, a water was made with calcium chloride ($CaCl_2$), sodium sulfate (Na_2SO_4), and sodium bicarbonate ($NaHCO_3$). This, too, performed just like distilled water. Finally, a synthetic water was made from the salts as they are present in natural pitting waters, that is, calcium bicarbonate, sodium sulfate, and sodium chloride. This water behaved just like an aggressive pitting water. All these waters had a similar pH (7.3 to 7.5), and no explanation of why the addition of bicarbonate as calcium bicarbonate is necessary has been offered. It is possible that it was the CO_2 content that made the water aggressive, but the calcium bicarbonate solution was purged with air to remove the excess CO_2 and increase the pH into the desired range.

5.1.3 Mechanism

It is not the intention to go into a full discussion of the mechanism of Type I pitting, but a brief review of the current position will be useful.

The first detailed study of the pitting of copper was published by May,[25] and this was followed by work by Pourbaix and his colleagues at CEBELCOR and Lucey at BNF Metals Research Association. In 1967, Lucey published a theory of pitting corrosion based on a detailed examination of service failures and extensive laboratory experiments.[14]

The theory requires an excess cuprous chloride concentration at the pit initiation site. All previous theories postulated a large cathodic area surrounding the pit, which is the anode. The production of hydroxyl ions at the cathode should, in hard waters, result in the precipitation of calcium carbonate scale on the cathodic areas. Lucey found, however, after examining a large number of service failures from Type I pitting, in both pipes and tanks, that there was no more carbonate scale around the pit than at areas remote from any pitting. There was, however, a large amount of calcium carbonate in the mound above the pit. This suggested that the cathodic reaction was taking place principally above the pit, and not on the surrounding surface. Sosa et al. have suggested that the white product was not carbonate and that the cathode was really in the area around the pit.[26] However, Lucey painstakingly removed the white crystals from many pit

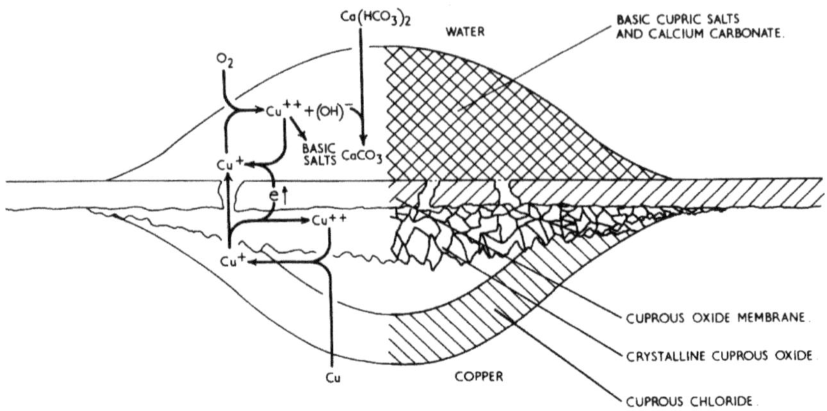

FIGURE 5.6 Schematic of Type I pitting according to Lucey.[14]

mounds, and they all effervesced when tested with cold, 10% sulfuric acid (H_2SO_4), showing the presence of carbonate.

Lucey also showed that there is a cuprous oxide membrane across the mouth of the pit, and he demonstrated that it acts as a bipolar membrane, with the lower surface being anodic and the upper surface being cathodic.[14] Cuprous chloride within the pit is anodically oxidized to cupric on the undersurface of the membrane, and the cupric ions thus produced can further attack the copper metal within the pit to reform cuprous ions. The principal cathodic reaction on the upper surface of the membrane is the reduction of cupric ions formed from cuprous ions by reaction with the DO in the water. The cuprous ions for this reaction come partly from diffusion through the pores in the oxide membrane from within the pit and partly by recycling cuprous ions produced by the cathodic reduction on the upper surface of the membrane. A mound forms above the pit, which is composed partly of basic copper salts and partly of calcium carbonate, by reaction between the cupric ions, oxygen, and calcium bicarbonate. Figure 5.6 shows a schematic of the structure of a corrosion pit and indicates the nature and site of each of the reactions involved.[14]

Lucey's theory was revolutionary in that it rejected the classical small anode surrounded by a large cathode theory. Sosa et al. have questioned Lucey's model and have repeated some of his experiments.[26] They claimed that they have demonstrated that pitting in copper follows the classical model, with a large cathode around a small anode. However, their model does not explain the crystals of calcium carbonate seen in the mounds above the pits. They suggested that the white product was not calcium carbonate, but it is hard to think of another possible compound that is white, crystalline, and effervesces vigorously in 10% sulfuric acid.

Sosa et al. suggested that it is a copper ion concentration cell that provides the driving force for pitting.[26] Lucey's model requires the formation of a cuprous oxide membrane across the mouth of the pit, and one way of producing this is a copper ion concentration cell, as described in the next chapter, on crevice corrosion. The author has cross-sectioned pitting at crevices on copper alloys and seen copper oxide membranes across the pit mouth; however, once established, the crevice (the source of the copper ion concentration gradient) can be removed, and the corrosion continues, with a mound of basic copper salts on top of the pit, just like a Type I pit. Hence, a copper ion concentration gradient is one way of establishing a pit. As Edwards et al. suggested, anything that creates an excess of cuprous chloride may initiate pitting;[5] a carbon film is just one possible route for this to happen. Once the pit is established, Lucey's model describes what is found in service very well. The mechanism of Type I pitting is described in more detail in the literature cited previously.

5.1.4 Copper Alloys

There are very few reports of pitting failures due to Type I pitting for copper alloys, and all of the ones that do exist are for heat exchanger tubing. Breckon and Baines reported some in the 1950s in marine heat exchangers.[27] Retief reported failures due to the presence of carbon films of both admiralty brass and 90/10 copper-nickel tubes in heat exchangers in South Africa handling brackish water.[28] Marsden reported similar failures with 90/10 copper-nickel in a recirculating water containing 100 to 1,000 mg/L chloride and 200 to 1,700 mg/L sulfate, also believed to be caused by carbon films in the tubes.[29]

Grassiani reported failures of admiralty brass tubes in a recirculating cooling water containing the following:[30]

100–250	mg/L	chloride
250–450	mg/L	sulfate
200–300	mg/L	bicarbonate
15–30	mg/L	silicon
130–160	mg/L	calcium
40–70	mg/L	magnesium
pH = 7		

In addition, the water was dosed with a variety of chemicals, including chromate (5 mg/L), phosphonates, polysacrylates, and azoles.

Heavy carbon films were found in the corrosion products and also on new tubes of both admiralty brass and 90/10 copper-nickel. Strong acid cleaning was required to remove the carbon films from new tubes.

Sato et al. carried out a study of the effect of surface finish on the corrosion of aluminum brass tubes carrying seawater.[31] They concluded that surface finish had no effect on long-term corrosion behaviour. However, they did not characterize the carbon contamination level by any of the recognized tests, and they did not include heat transfer in their tests.

Shone and Grimm reported experiences with corrosion over 25 years with the Shell tanker fleet.[32] This included failures of 66/30/2/2 Cu-Ni-Fe-Mn heat exchanger tubes due to deleterious films from the manufacturing process. Shone et al. did not specifically call the films carbon, but it seems likely that this is what they were.

The author conducted tests on a range of copper heat exchanger alloys with various surface finishes and also heat transfer.[33] All tests were conducted with once-through seawater for 3 months. The only tubes that suffered pitting were those with a heavy carbon film. Tubes with carbon films that would be classed as unsatisfactory for fresh water service showed no pitting and looked just like the tubes with clean bores. The pitting in the tubes with heavy carbon films was shallow without heat transfer (0.04 mm), but on the tubes with heat transfer, the maximum depth of attack was 0.2 mm on one tube and 0.73 mm on a duplicate. Both tubes had lots of pits in the zone subject to heat transfer. It is clear that carbon films can cause pitting in copper alloy tubes in seawater, but the level of carbon to cause pitting is much higher than for fresh water, and heat transfer is also required.

5.2 Type II Pitting

In the early 1950s, Campbell conducted a survey of pitting corrosion failures of copper tubes in the United Kingdom. About 65% of the failures were in cold, hard well waters—what later came to be called Type I pitting attack. A further 30% of failures were in soft, hot waters of low pH. This type of corrosion is now called *Type II pitting*, and it is mostly found in countries that have predominantly soft waters such as Scandinavia and Japan. However, failures have also been seen in the United Kingdom, Germany, and Canada. Type II pitting is slower than Type I, and failures usually occur after 8 to 12 years.

Type II pitting is characterized by deep pits of small cross section containing very hard, crystalline cuprous oxide and basic copper sulfate.

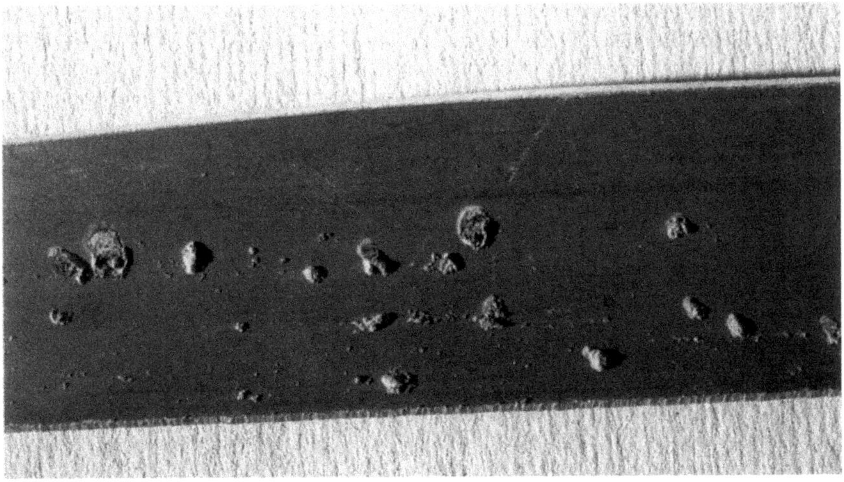

FIGURE 5.7 Typical appearance of Type II pitting. (Photograph courtesy of H.S. Campbell.)

The surface between the pits is a nearly black layer of water-formed oxide beneath a thin layer of silt deposited from the water.[4] Typical Type II pitting is shown in Figure 5.7. Sato et al. investigated a large number of failures of copper hot water pipes from the Tokyo area of Japan and found similar pits to those described by Campbell.[34] They also found a thin film of cuprous chloride in the base of the pit, although this was not visible to the naked eye. A schematic of a Type II pit, according to Sato et al., is shown in Figure 5.8.[34]

Mattson and Fredriksson examined failures of copper tubes in Sweden and concluded that Type II pitting would occur when the pH <7.4, the bicarbonate-sulfate ratio is less than 1, and the water temperature is 60°C or more.[35] Campbell reported data from Canada, Germany, and the United Kingdom that follow this trend.[4] A variation of Type II pitting can occur when the water contains manganese:[4] as little as 30 μg/L has caused problems, although the manganese concentration is usually higher. The black film between the pits then contains large quantities of manganese dioxide (MnO_2), which is a well-known cathodic depolarizer. The manganese dioxide appears to stimulate Type II pitting in waters that might otherwise be regarded as borderline.[4] These often have a pH >7.4 because they have been treated, but they have a low bicarbonate content, so any corrosion could rapidly lower the local pH below 7.4.

Kasahara et al. investigated failures of copper pipes from recirculating hot water systems in the Tokyo area of Japan, with a variety of supply

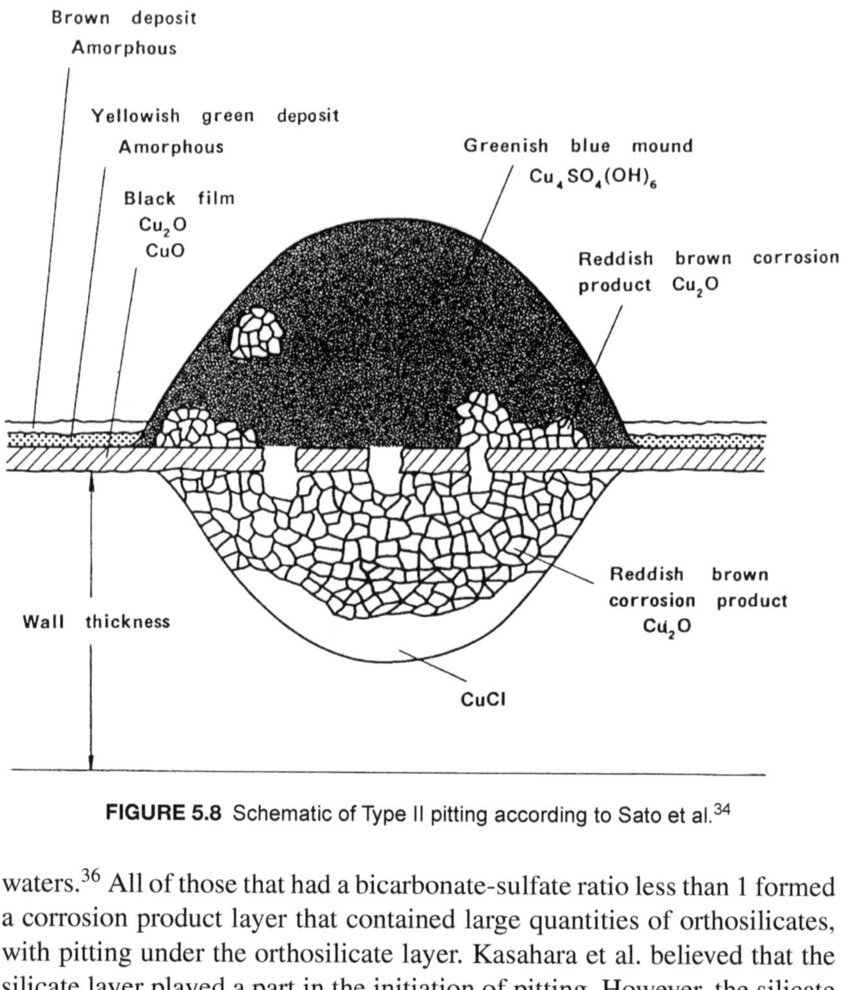

FIGURE 5.8 Schematic of Type II pitting according to Sato et al.[34]

waters.[36] All of those that had a bicarbonate-sulfate ratio less than 1 formed a corrosion product layer that contained large quantities of orthosilicates, with pitting under the orthosilicate layer. Kasahara et al. believed that the silicate layer played a part in the initiation of pitting. However, the silicate layer was also seen in waters with a bicarbonate-sulfate ratio greater than 1, and pitting did not occur in these tubes.

In the 1980s, concern about Legionnaires' disease led many large buildings in Japan to increase the chlorination of their recirculated hot water supply. Chlorine residuals of 1 to 3 mg/L were common, and there was a big increase in the failures by Type II pitting. Some of these were in waters in which the bicarbonate-sulfate ratio exceeded 1, and research was undertaken to investigate the effects of chlorine on Type II pitting.

Baba et al. reported that residual chlorine in hot water was a cause of pitting in copper pipes.[37] They also showed, in accelerated tests, that residual chlorine concentrations of a few milligrams per liter in potable water, along with a high water temperature and flowing conditions, affected the electrode potential of copper tubes.[38] The potential rose to levels in excess of 150 mV SCE, which is above the pitting potential at 60°C. This is not dissimilar to the potential of 170 mV for Type I pitting, allowing for differences in pH and temperature. After several days' exposure, at this potential, numerous small pits appeared on the tube surface, whereas under the same conditions, but with chlorine-free water, the tube potential remained below 80 mV SCE, and no pits were observed, even after prolonged exposure. Baba et al. concluded that chlorine, a strong oxidizing agent, plays an important role in the initiation of Type II pitting.

Suzuki et al. showed that the rise in the tube potential occurred in two stages: the first stage reflected the formation of a cuprous oxide layer at potentials up to 60 mV SCE, while the second stage, from 60 mV to 150 mV SCE, reflected an absorption reaction of the hypochlorite ions with the cuprous oxide layer.[39] They demonstrated that hypochlorous acid (HOCl) and hypochlorite ions, produced by the chlorination of the water, accelerated the formation of the cuprous oxide layer, and anions, such as ClO^-, interchanged with OH^-, increasing the electrode potential of the copper.

Baba et al. conducted further laboratory tests to reexamine the critical potential for the onset of pit growth on copper tube in flowing synthetic hot water by a potentiostatic method.[40] They also reexamined the effect of chlorine on the potential of copper tubes. The polarization tests were not successful in that no increase in current occurred even when pits formed, so no critical potential could be identified. However, examination of the tube surfaces confirmed the value of 150 mV SCE as the critical one for the initiation of pitting in flowing hot water. The change in corrosion potential, indicating the initiation of pitting, in water at pH 7 containing 3, 5, and 10 mg/L chlorine was observed to occur after 80, 60, and 37 hours, respectively. The structure and morphology of the pits that formed under these conditions were the same as that of Type II pitting described previously.

Francis and Saxton also investigated the effect of chlorine on copper tubes in Sheffield water, a soft, surface-derived water with a bicarbonate-sulfate ratio less than 1, at 60°C.[41] In nonchlorinated water, tubes that had been shock dosed with either 100 mg/L chlorine for 2 hours or 20 mg/L chlorine for 16 hours developed protective films, similar to those on tubes that had not been shock dosed. None of these tubes showed any pitting. However, tubes that were continuously dosed with 1 mg/L chlorine developed pits in the 30-day tests. The pits were broader than the normal Type II pits described by Mattson and Fredriksson and Campbell.[4,35] However, Hamamoto et al. also reported a similar pit morphology in an independent study with a Japanese soft water with a continuous chlorine residual of 1 mg/L.[42] Chlorine not only modified the shape of the pits, but also changed the rate of propagation. Type II pitting normally does not lead to leakage in less than 8 years, but both Francis and Saxton and Sato et al. found more rapid penetration rates in chlorinated water, with failures in less than 3 years in some cases.[34,41]

There has been less research to determine the mechanism of Type II pitting, but Mattson and Fredriksson carried out tests in an artificial pit cell and investigated the effects of pH and water composition on the oxide that formed on a copper anode.[35] At pH values from 7 to 10, three types of green corrosion product formed over a period of 4 days. Type A was adherent and formed in all waters containing bicarbonate but free of sulfate and chloride. In waters with 10 mg/L sulfate or more, a Type B crust formed, which was less adherent. This was identified as basic copper sulfate. In waters with chloride, and also in waters with sulfate and 280 mg/L bicarbonate, a Type C crust formed, which was adherent. The greatest corrosion occurred with a Type B crust, and if this is poorly adherent and has cracks or defects in it, the crust will act as a screen to enable the establishment of a copper ion concentration cell, in a similar manner to Lucey's model for Type I pitting (see Section 5.1.3). This would then create the conditions for pit initiation, with hydrolysis and pH reduction under the crust.

5.3 Type III Pitting

The occurrence of Type III pitting was not really recognized until the 1970s and 1980s, when there were a large number of failures, many in large public buildings such as hospitals, offices, and hotels. There was concern about the occurrence of a new type of pitting and why it had not been seen previously. However, when Campbell reviewed a large number of copper tube failures by pitting in the United Kingdom in the early 1950s, 5% of them were neither Type I nor Type II attack. Most of the 5% were in soft waters of low pH, either in warmer cold water pipes or in cooler hot water pipes. It seems likely that these were Type III pitting, but at that time, the preponderance of the other two types of pitting meant that it was lost in the statistics.

The author reviewed Type III pitting failures and found that they had been reported in Germany, Sweden, Australia, Saudi Arabia, the United

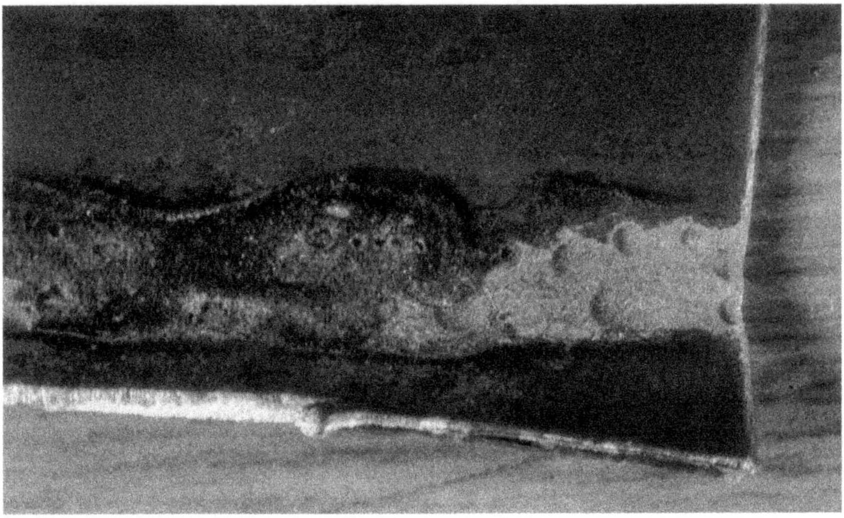

FIGURE 5.9 Typical appearance of Type III pitting. (Photograph courtesy of H.S. Campbell.)

Kingdom, Japan, and the Republic of Ireland.[43] Some of these were in cold water pipes, where they were in a warm environment and water temperatures were 25°C or more. Others were in hot water pipes, but far removed from the heater, with water temperatures typically ≤50°C. All the waters were soft, with a low pH, but other factors also appeared to be contributing.

Figure 5.9 shows typical Type III pitting, with broad deposits of basic copper sulfate covering multiple pits filled with cuprous oxide. Where the multiple pits perforate the tube, the typical pepper-pot appearance of Type III failures is created. The dark film between the pits is mainly cupric oxide, often with deposits of organic material above it.

The first instance of Type III pitting was reported by von Franque, and about the same time, an epidemic of failures occurred in one part of Sweden.[44-45] The failures in Sweden concerned cold water pipes in domestic properties, and the failures appeared to coincide with a change in water supply. Up to 1970, the water came from the Skallsjo works, in Floda, but thereafter, increasing amounts of water from the Oxsjo works, in Lerum, were added. The water compositions are shown in Table 5.2; the main difference is the low bicarbonate content of the Oxsjo water, particularly as a raw water. This is because the source was heavily affected by acid rain, and the water had to be treated with sodium hydroxide to increase the pH. Even so, the bicarbonate-sulfate ratio for the Oxsjo water was less than 1, while

TABLE 5.2 Water composition from two Swedish sources from 1970 to 1980[45]

	Concentration (mg/l)			
	Skallsjo (Floda)		Oxsjo (Lerum)	
Factor	Raw	Supply	Raw	Supply
pH	6	8.3	6.3	7–9
Total Hardness ($CaCO_3$)	28	28	8	10
Bicarbonate (HCO_3)	16	48	1	10
Chloride	16	13	12	12
Sulphate	10	13	17	18

that of the Skallsjo water was greater than 1. Tests in both waters showed that it was the Oxsjo water that caused the Type III pitting.[45] Linder and Lindman suggested that a natural inhibitor, as discussed in Section 5.1 for Type I pitting, might be present in Skallsjo water and was absent in Oxsjo water.[45] However, they did not conduct any tests to positively identify the inhibitor's presence.

In the 1980s, there was a series of failures in large buildings, mostly hospitals, in several countries around the world. Campbell et al. summarized the cases that they had investigated from several countries.[46] Because the pitting was somewhat of a cross between Type I and Type II pitting, Campbell et al. called it *Type 1½* pitting, but its appearance is clearly the same as that reported by other authors for Type III pitting.

Interestingly, Campbell et al. reported that none of the domestic properties adjacent to the hospitals had any pitting problems.[46] This suggests that stagnation and/or slow flow may play a part in pit initiation, as hospitals frequently have long runs of pipe, in which water flow is low or intermittent, with long periods of stagnation. All the waters had similar compositions:

Total hardness (as $CaCO_3$)	25–40 mg/L
Total alkalinity (as $CaCO_3$)	10–20 mg/L
Chloride	8–20 mg/L
Sulfate	10–30 mg/L

In most of the waters, the bicarbonate-sulfate ratio was less than 0.5, and Type II pitting was seen in the same waters at temperatures of 60°C or more. However, where the leaks occurred, the water temperature was $\leq$50°C, with a pH from 7 to 7.5.

Campbell et al. found that tubes with pits also contained polysaccharide films.[46] This is an exopolymer produced by many bacteria as part of their metabolic processes, and Campbell et al. identified three species of pseudomonad in the film on failed tubes. These bacteria were found in all the tubes that were suffering pitting and also caused pitting when injected into a sterile water of the same composition, when no pitting occurred in the sterile water.[46]

Fischer et al. described tests carried out on the biofilms (polysaccharides) removed from failed copper tubes from a German hospital.[47] These carried water at approximately 50°C and showed the typical pepper-pot appearance of perforations. The biofilm was present between the metal and the oxide-hydroxide layer. Pitting occurred where the biofilm was disrupted. This was believed to then cause a copper ion concentration cell to become established due to the high copper ion concentration in the exopolymeric layer. Peroxide and hydroperoxides were detected in the biofilm, and these may have provided a more rapid cathodic reaction than the reduction of DO, thus accelerating the corrosion process. The biofilm also contained localized concentrations of amino acids, particularly glycine ($C_2H_5O_2N$) and gluconic acid ($C_6H_{12}O_7$), which are both complexants for copper and which would increase local dissolution rates, thereby helping to establish the copper ion concentration gradient.

It is thought that biofilms are not essential for Type III pitting, but they can be a contributing factor. Nuttall reported that of the hospital failures in the United Kingdom, biofilms were only detected in approximately 50% of the failed tubes.[48] This does not mean that there was no biological activity in those tubes, only that the tests used detected no films. He also stressed that the presence of humic acid in the organic material in the water could be a contributory factor.[44] This was reiterated by Bond, who conducted tests on copper tubes in a Scottish hospital that was experiencing problems with Type III pitting in the hot water circuit.[49] Bond operated one rig on the normal hot water supply, and the other received water that was filtered to 0.5 μm. This reduced the incoming solids content considerably but did not produce a sterile water, which requires finer filtering. The water flow was 0.1 m/s for 6 hours, followed by 6 hours' stagnation, with a typical water temperature of approximately 50°C when the water was flowing, decreasing to about 25°C after 6 hours' stagnation. The tubes in the rig receiving filtered water suffered less pitting than the ones receiving the normal hot water supply, but the pitting was not totally suppressed. This suggests that the bacteria producing the biofilm were the critical component, and filtering the water just reduced their food supply. The role of biofilms in the localized corrosion of copper tubes carrying fresh water was recently reviewed by Keevil.[50]

Siederlarek et al. studied the formation of films on copper in waters simulating those that were present in a German hospital that had suffered numerous Type III pitting failures.[47,51] As with studies of Type I and Type II pitting, they found that the type of oxide that formed influenced the propensity for pitting. Waters with a significant sulfate content had a high propensity for pitting. They also found that water flow conditions significantly affected the nature of the film. The propensity for pitting increased when the water was stagnant or the flow was intermittent, with long periods of stagnation. These results mirror those reported for Type I and Type II pitting.

5.4 Aluminum Brass

Aluminum brass is not usually used to carry fresh water, but there are some instances of this. Campbell reports pitting of aluminum brass tubes carrying drinking water on board ship.[4] The pits were similar in structure to Type I pits in copper, apart from the presence of zinc corrosion products in the mounds. However, the pits were smaller and more numerous than Type I pits in a copper tube. The surface of the tube around the pits carried a film that was identified by x-ray diffraction as a mixed zinc-aluminum oxide of the spinel type $ZnO{\cdot}Al_2O_3$. The presence of a film free of copper suggests the presence of bacteria that are able to chelate copper.

This type of pitting has received little attention. The author was unable to reproduce it in the laboratory using alternate flow conditions at ambient temperature using both a hard well water and a soft, surface-derived water. The principal corrosion product in both cases was cuprous oxide.

5.5 Alloy 400

Alloy 400 has been observed to pit in very slow moving or stagnant seawater. Figure 6.1B shows crevice attack on Alloy 400 panels exposed to quiescent seawater, but pitting is also visible away from the crevice region.[52] At higher seawater velocities, no pitting occurs. This is thought to be due to a shortage of oxygen to keep the passive film intact at very low water flow rates. The pitting is exacerbated when sulfides are present in the seawater. This is thought to be due to sulfide inhibiting the formation of the passive oxide film and causing less protective sulfide films to form instead (see Chapter 3). Pitting can occur even at higher water flow rates if sufficient sulfide is present in the water.

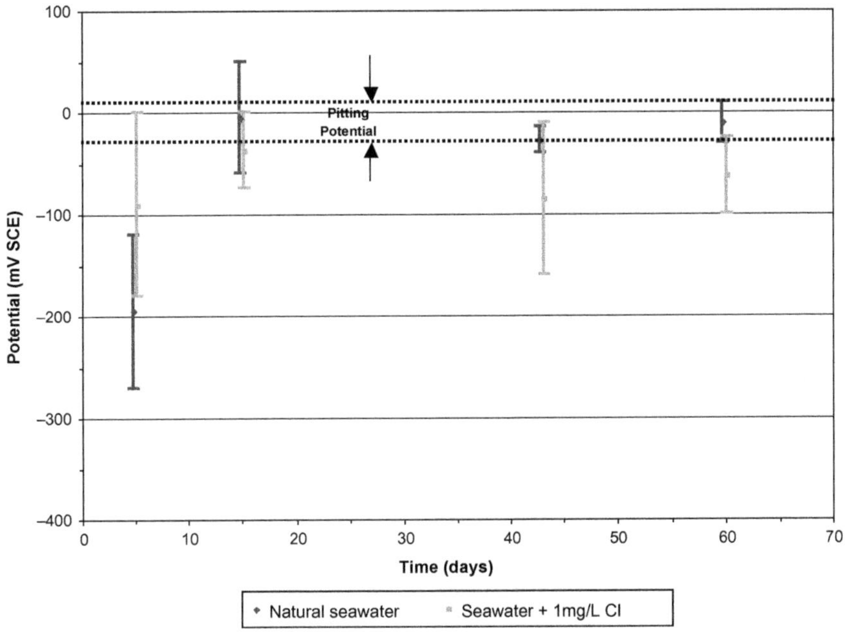

FIGURE 5.10 Corrosion potential vs. time data for Alloy 400 in seawater.[52]

Klein et al. investigated the crevice corrosion resistance of Alloy 400 in quiescent natural and chlorinated (1 mg/L) seawater at ambient temperature, but they also found that the alloy pitted outside the crevice region, as shown in Figure 6.1B.[52] They found that the pitting was most severe in natural seawater and was less severe in seawater containing 1 mg/L chlorine. They related this to the pitting potential of Alloy 400, as previously reported by Tipton and Kain, who found that the pitting potential was in the range −20 mV to +10 mV SCE at temperatures from 20 to 28°C.[53] Klein et al. plotted the potential of their Alloy 400 samples at various times of exposure up to 60 days, as shown in Figure 5.10. Pitting was seen on all the samples for which the potential exceeded the pitting potential.

Gallagher et al. found pitting in a seawater piping system in Alloy 400 that had seen mostly natural seawater due to the chlorination system working only intermittently. They investigated the potential in natural seawater and found that it exceeded the pitting potential soon after exposure.[54] Samples exposed in seawater with 0.5–0.8 mg/L chlorine had potentials that were just below the pitting potential and did not suffer attack.

This suggests that a biofilm of some kind forms on Alloy 400 in natural seawater, similar to that found on high-alloy stainless steels. This is sufficient to raise the potential into the pitting zone. A level of chlorine that is just sufficient to prevent the biofilm formation keeps the potential below the pitting potential and free of attack. However, chlorine is a powerful oxidizer, and excess chlorine will also elevate the potential above the pitting potential. The variable results of Klein et al. suggest that the chlorine demand of the water varied during their tests so that sometimes there was excess chlorine sufficient to cause pitting, and sometimes there was not. It appears that chlorine concentrations around 0.5 mg/L offer the safest fouling control with Alloy 400.

5.6 Formicary Corrosion

Formicary corrosion was named because the corrosion takes the form of tunnels of corrosion products. In cross section, the attack has the appearance of an ants' nest; hence the name. The surface around the leaks is grayish brown or bluish purple, and the pinhole leaks are extremely tiny so that they can only be seen under a microscope. The attack may occur either from the inside or from the outside of the tubes, depending on which side the aggressive conditions necessary for formicary corrosion are present.

Formicary corrosion was first described by Keyes in the 1960s and by Edwards et al. in the 1970s, but it was not properly understood until extensive research was undertaken in Japan in the 1980s.[55-56]

Yamauchi et al. and Notoya et al. examined over 20 service failures by formicary corrosion of copper refrigeration and air-conditioning tubes.[57-58] They had all failed on final test, prior to dispatch, or soon after entering service. Most of the failures were from the inside, but a few were from the outside. They appeared to be associated with the presence of carboxylic acids, either formic or acetic ($HCOOH$ or CH_3COOH). The internal failures were mostly associated with formic acid, while the external failures were from sushi bars, where they use a lot of vinegar (acetic acid). The carboxylic acids inside the tubes were thought to be produced during the final cleaning operation, and the corrosion initiated during storage. The service failures suggested that formicary corrosion requires oxygen, moisture, and a carboxylic acid.

Notoya et al. carried out tests on copper exposed in desiccators containing 100 mL of a 1% v/v solution of various carboxylic acids for 125 days.[58] The attack on the samples exposed over formic acid is shown in Figure 5.11 and was similar to that seen in service failures. Figure 5.11A shows the presence of voids due to attack within the metal, whereas Figure 5.11B shows microcracks in the copper oxide within the pits. Somewhat

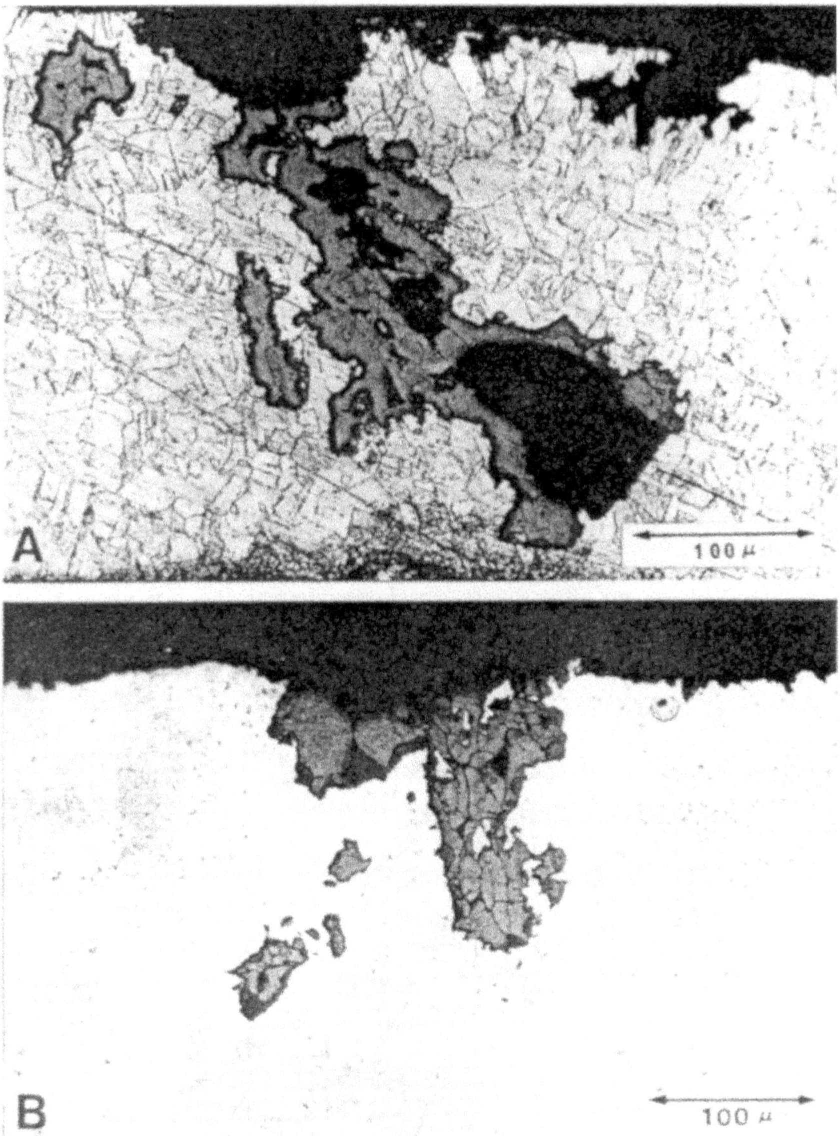

FIGURE 5.11 Microsections showing formicary corrosion: (A) presence of voids; (B) microcracks in oxide.

broader attack occurred with acetic acid, while the attack with propionic and n-butyric acids was much finer. No attack at all was seen with n-valeric and n-pentanoic acids. Clearly, the simpler the acid structure, the greater was the attack, and this is reflected in the dissociation constants for the acids, with the relative strength being greatest for formic acid.

Notoya et al. proposed a mechanism for the attack, whereby copper forms a complex with the acid, as cuprous formate or acetate.[58] This is then oxidized to give cuprous oxide and cupric formate or acetate. Because cuprous oxide has a larger volume, it will tend to crack and hence be less protective, permitting solution to contact the metallic copper beneath. The cupric formate, or acetate, can then be reduced to the cuprous form by reaction with copper dissolved from the metal. The pitting is fine because the cracks in the cuprous oxide permit attack at the pit base, but the oxide on the pit sides acts as the cathode.

While at BNF Metals Technology Centre, Campbell investigated the failures of copper pipes placed in troughs in a concrete floor in a new warehouse. The pipes leaked due to external corrosion after about a year. Only the cold water pipes were attacked, whereas the hot water pipes were free of corrosion. Both sets of pipes were insulated with a compound containing phenol formaldehyde. The cold water pipes were damp when the coating was applied because they already contained water from final pressure testing. The combination of moisture and air led to the oxidation of the formaldehyde to formic acid, which caused the formicary corrosion. The hot water pipes were unattacked because the hot water dried the pipes before the insulation was applied.

So far, all the failures discussed have been associated with copper tubes and pipes, but Olszewski and Corbett reported the failure of 90/10 copper-nickel tubes used in a heat exchanger to cool hydrogen, with seawater inside the tubes.[59] The tubes leaked after 8 months in service from the inside, and the attack had the typical appearance of formicary corrosion. It seems unlikely that significant quantities of carboxylic acids would be present in the seawater, and it seems more likely that the attack occurred during storage of the tubes. This may have involved the use of an insulation or packing compound that contained formaldehyde. Clearly formicary corrosion is not confined to copper; copper alloys can also be susceptible.

5.7 Rosette Corrosion

This type of attack is a relatively recently recognized phenomenon and is only found in the United Kingdom and other regions that use low-pressure, copper hot water cylinders.[60] Although most of the water in these cylinders is hot, the very bottom region never gets warm and can suffer Type I pitting in localities with a supply water having a high pitting propensity. To combat this, Lucey suggested fitting a small aluminum anode to the base of the cylinder, which protects the copper by depressing the potential during the early part of its life. Once the aluminum is consumed, the copper has formed a protective oxide film and is no longer susceptible to Type I pitting. Unfortunately, cylinders with aluminum protector rods have been fitted in many areas where the water does not support Type I pitting, in the mistaken belief that the aluminum protects against all corrosion.

In some waters, corrosion has occurred beneath the aluminum corrosion products on the cylinder base. The aluminum corrosion products vary in color from bluish white to light green. Underneath the deposit, it is a vivid brick red and magenta color, consisting of large platelets of cuprous oxide separated by deep trenches—hence the name *Rosette* attack. It was originally thought to be a type of pitting, but it is more properly accelerated general corrosion, due to the aggressive conditions beneath the deposits. In some cases, this can lead to perforation and leakage.[60] Figure 5.12 shows a typical example of Rosette corrosion.

One possible cause of the attack is that under some conditions, there is aluminum chloride ($AlCl_3$) as well as AlOOH in the corrosion product deposit. Aluminum chloride hydrolyses readily and could produce very low pH conditions beneath the deposit to exacerbate corrosion.

Another possibility is the fact that the galvanic current between the copper and aluminum is sufficiently high that it is possible to reduce nitrate to ammonia (NH_3) and sulfate to sulfide.[60] Both these anions are aggressive to copper and would increase attack. In some parts of the United Kingdom, there has been a general increase in the nitrate content of the water due to increased agricultural runoff. This increase, and the ability of the nitrate ion to be reduced to ammonia, may explain why it has taken over 30 years of the use of aluminum protector rods for the problem of Rosette corrosion to be noticed.[60]

5.8 Methods of Prevention

5.8.1 Type I Pitting

As described previously, one of the main causes of Type I pitting is carbon film from the manufacturing process. Carbon films are less of a problem nowadays, as most manufacturers use water-soluble drawing lubricants, and

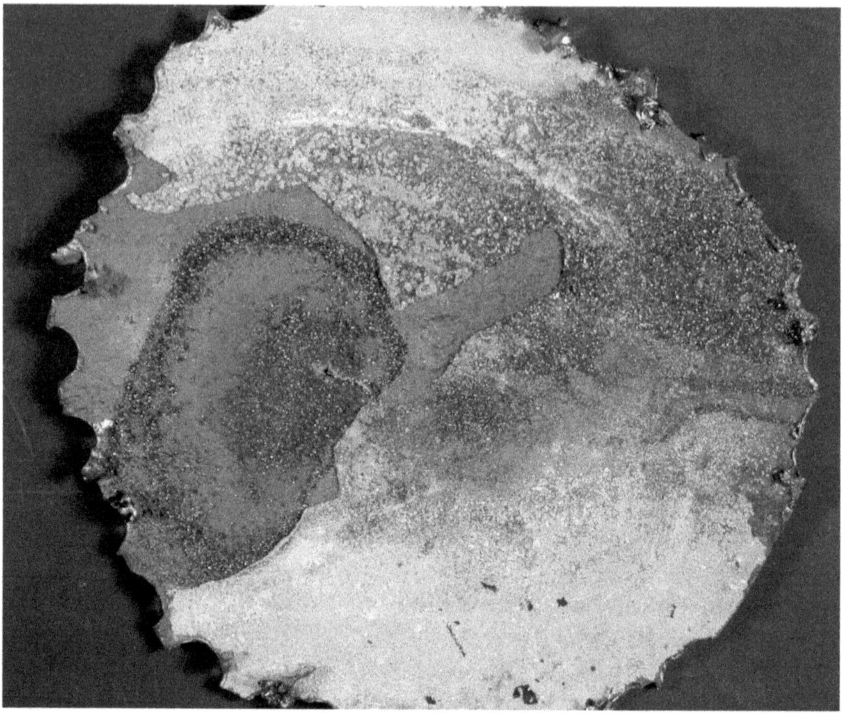

FIGURE 5.12 Typical appearance of Rosette corrosion. (Photograph courtesy of P. Munn.)

some also bright anneal in an atmosphere that is slightly oxidizing, rather than reducing. This has the effect of oxidizing any cracked drawing lubricant to carbon dioxide. Other manufacturers clean the tube bore by internal grit blasting or by pickling. However, not all manufacturers are so assiduous, and the European standard for copper tubes, EN 1057,[61] specifies that tubes must be free of deleterious films in the bore. The older British Standard for copper alloy heat exchanger tubes, BS 2871 Part 3,[62] also includes this stipulation. The standards do not say how to check for such films. Shone and Grimm[32] suggested the use of a simple electrochemical test, but it has been found that the results for tubes on the borderline are not very reproducible.[63] An alternative, which the author has used on several projects, is the detection of carbon by heating a sample such as in the LECO†-type test. In this test, a half-section of tube (abraded on the outside surface) is heated in a stream of oxygen, and the CO_2 produced at different temperatures is determined.

† Trade name.

By heating at increasing temperatures, the carbon due to various sources can be isolated:

200°C	solvents
650°C	greases
850–900°C	carbon films

Only the carbon measured at the highest temperature is used for assessing the tube quality. The pass-fail criterion that has been adopted for copper tubes for use in fresh waters is carbon ≤ 0.2 mg/dm^2. Tubes that fail this test can be cleaned by internal grit blasting or by pickling and then retested. The test should be applied to several samples from each batch of tubes. The author has also specified this test for copper alloy heat exchanger tubes carrying seawater. Although research has shown that higher levels of carbon are required to cause pitting, compared with fresh waters, the level of carbon of $\leq$0.2 mg/dm^2 was still applied, and the manufacturer had no trouble meeting this standard. Clearly the test is only required for annealed and half-hard tubes that are heat treated after drawing.

It is also clear that long periods of stagnation, after tubes are first filled with water, may lead to pitting, and this should be avoided. The number of long, rarely used legs should be kept to a minimum, and these should have a slow flow of water through them for 2 to 3 weeks to form protective films. This was found to work at a school in Torquay, United Kingdom, whose problems were investigated by BNF Metals Technology Centre.

Munn[16] has raised the question of phosphates interfering with protective film formation and possibly encouraging pitting. This is an area that clearly requires more research.

Where low-pressure hot water cylinders are used, it is best to use one fitted with an aluminum protector rod in waters that have a pitting propensity rating greater than 1.[64] Oliphant recommends the use of a modified design of cylinder, where the heating coil is much lower in the body.[60] This makes sure that even water in the very bottom of the cylinder is heated above 40°C so that Type I pitting does not occur and an aluminum protector rod is not required.

It is clearly important to know whether a particular water composition will support Type I pitting. The nomogram of Lucey, described in Section 5.1.2, was a good start, but it has deficiencies and is not widely available. The nomogram cannot cope with waters in which the chloride exceeds 57 mg/L, and it underestimates the aggressivity of waters containing less than 10 mg/L sulfate.[65]

In an attempt to get around these problems, Shaw and Lucey developed a simple test in which a sample of the water could be evaluated.[66] The test cell consists of an acrylic block with 12 copper anodes and a single copper cathode (Figure 5.13). A glass rod created a crevice over one anode, and a constant current was passed through the cell for 100 hours. The greater the current passed by the shielded electrode after 100 hours, the greater was the propensity to cause Type I pitting. Figure 5.14 shows the correlation between the current passed by the shielded anode and the PPR, calculated from Lucey's nomogram. There is a good correlation between the current and the PPR such that Shaw and Lucey suggested the following aggressivity rating:

Proportion of total current carried by shielded anode (%)	Aggressivity
<20	Nonpitting
20–50	Borderline (no failures expected)
50–70	Fairly aggressive (PPR = 1–2)
>70	Very aggressive (PPR > 2)

The BNF test cell is a little cumbersome to use in the field, and the anode block needs regrinding after every test. Ansuini developed a simpler test cell using an etched copper circuit board for the electrodes, designed such that a new board could be plugged into the cell for each test.[67] The board is connected to a battery-operated power supply that also provides a visual display of the PPR, as shown in Figure 5.15. The cell is easy to take to site and connect to a slow flow of water through the cell (Figure 5.15). The results with this compact unit correlated well with the BNF data, and the PPR was found to hardly change after 50 hours of testing.[65-66] The test equipment is still available and is recommended for assessing water aggressivity toward Type I pitting.

Work has also been carried out to look at water composition modification to avoid Type I pitting. The first consideration must be to avoid the use of activated charcoal or similar filters that can remove the naturally occurring inhibitor from surface-derived waters that might support Type I pitting. If water blending is possible, then mixing a surface-derived water (with the natural inhibitor) with an aggressive well water can prevent Type I pitting.

Depommier examined four water treatments to try to reduce the aggressivity of Brussels tap water.[68] These were as follows:

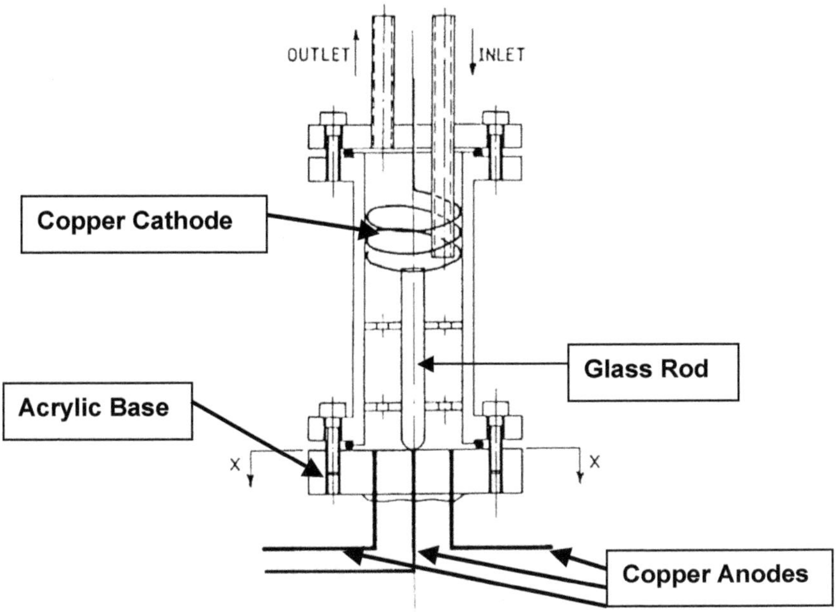

FIGURE 5.13 Schematic of BNF test cell for Type I pitting.

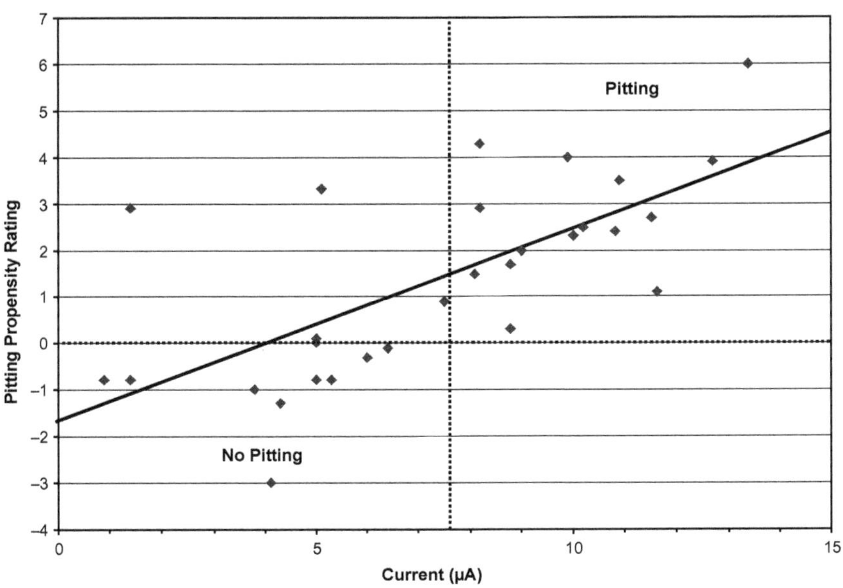

FIGURE 5.14 Relationship between current on shielded anode and the pitting propensity rating from Lucey's nomogram.[65]

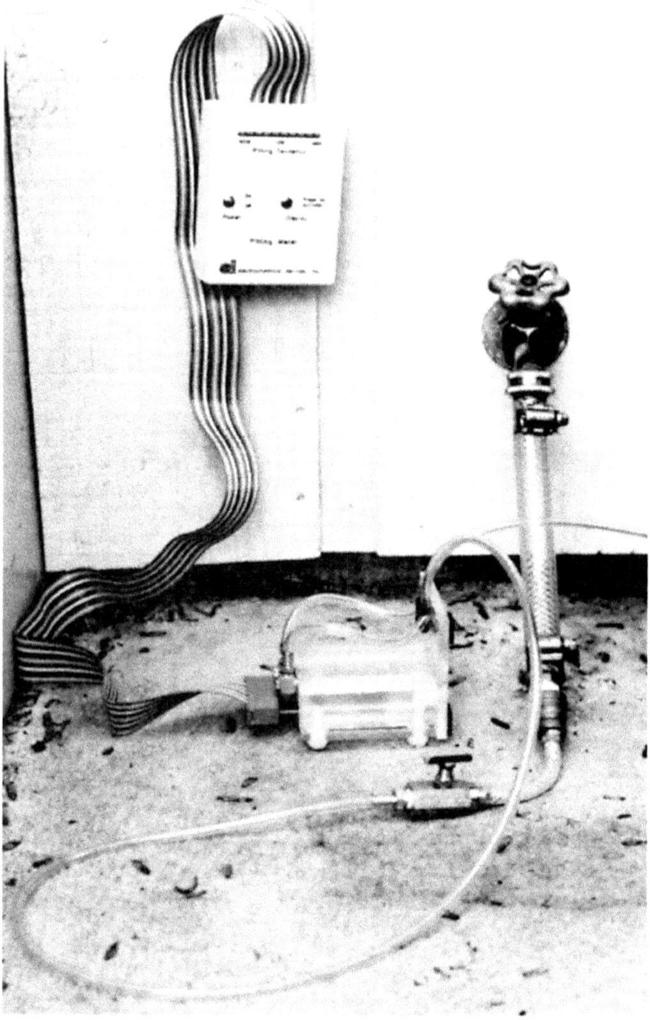

FIGURE 5.15 Picture of test cell by Ansuini with portable power source.[67] (Photograph courtesy of F. Ansuini.)

1. Add NaOH to increase pH to 8;
2. Add $Ca(OH)_2$ to increase pH to 8;
3. Partially soften and add NaOH to increase pH to 8; and
4. Partially soften and pass through a Magno-Dol filter (DIN 19621)

A test rig was constructed that contained copper tubes from a variety of manufacturers, including some with carbon films. Depommier found that neutralizing the CO_2 with NaOH after partially softening (treatment 3) was the only treatment that was successful. Even this treatment was not 100% successful, and some half-hard tubes with a heavy carbon film still pitted. He concluded that removal of the carbon films (by internal grit blasting or pickling) or annealing in a slightly oxidizing atmosphere offered a more reliable solution to problems of Type I pitting.

5.8.2 Type II Pitting

Mattson and Fredriksson recommended the following measures for preventing Type II pitting[35]:

1. Reduce the water temperature to $<60°C$;
2. Keep the water pH >7;
3. Increase the bicarbonate content >70 mg/L; or
4. Increase the bicarbonate-sulfate ratio to greater than 1

If the water temperature is reduced too much, there is an increasing risk of Type III pitting. In practice, a water temperature of $\sim 55°C$ is recommended as the best compromise; however, it is important to also consider regulations for the control of *Legionella*. In the United Kingdom, the Health and Safety Executive Document L8 requires all hot water systems to operate at 55°C or greater to prevent the growth of *Legionella*.

Bicarbonate can be added to water by acidifying a sidestream with carbon dioxide, passing the water through dolomite limestone, blending it back into the main supply, and adjusting the pH. In severe cases, all the water might require treatment through a limestone bed, which increases costs.

Where the manganese content of a water is high and is contributing to Type II attack, as described in Section 5.2, it may be necessary to include a manganese filter, as well, to significantly reduce the manganese concentration. Commercial units to do this are available.

Where recirculating hot water is being dosed with chlorine to prevent Legionnaires' disease, there is a clear risk of increasing the susceptibility to Type II pitting, as described in Section 5.2. Hamamoto et al. carried out tests and showed that the propensity for Type II pitting was not increased, provided the chlorine residual did not exceed 0.25 mg/L.[69]

Kasahara et al. investigated the use of ultraviolet irradiation of hot water circuits to prevent the growth of bacteria.[70] They found that bacteria could be controlled by ultraviolet irradiation and a chlorine residual of 0.2 mg/L, compared with a residual of 5 to 10 mg/L without ultraviolet treatment. At the low chlorine concentration with ultraviolet treatment, no pitting occurred in hot Tokyo water.

5.8.3 Type III Pitting

Although the large outbreaks of Type III pitting in the 1980s led to a great deal of research, there is less agreement on methods of prevention. Where bacterial action does not appear to be important, Linder and Lindman recommended increasing the bicarbonate content of the water to >70 mg/L.[45] This can be done as described in Section 5.8.2. In addition, increasing the pH above 8 is also beneficial as it removes most of the excess carbon dioxide.

Where microbial action contributes to the problem, Fischer et al. reviewed the methods adopted in a German hospital to reduce the incidence of Type III pitting.[71] With the hot recirculated water system, it was recommended that the water temperature be maintained in the range 55 to 60°C. The bacteria do not thrive at 55°C, but Type II pitting may occur if the temperature goes above 60°C. In addition, dead legs should be avoided wherever possible, and the water should be pumped to avoid stagnant, cooler zones, where bacteria might thrive.

Cleaning of the corroded pipes with dilute acid did not prevent reinitiation of pitting on refilling with water.[71] Irradiation with ultraviolet light did not prevent bacterial growth. Filtering of the water reduced bacterial growth, but did not prevent it, as was confirmed by Bond.[49]

Increasing the bicarbonate content of the water reduced the incidence of attack. The higher the bicarbonate content, the better, provided that the bicarbonate-sulfate ratio exceeded 1.[71]

5.8.4 Alloy 400

Alloy 400 is most susceptible to pitting in quiescent water, and if the flow rate can be kept to 1 m/s or more, the alloy will perform well. In the presence of chlorine, it is important to keep the chlorine concentration to around 0.5 mg/L to avoid initiating pitting. In the presence of sulfide, pitting may occur at any velocity if the sulfide concentration is sufficiently high; for such cases, an alternative material is recommended.

5.8.5 Formicary Corrosion

Most of the incidences of formicary corrosion have involved packing or insulation that contained formaldehyde and the presence of a moist, oxidizing atmosphere, which could oxidize the formaldehyde to formic acid. The problem can be avoided by the use of packing or insulation materials that do not contain formaldehyde. Similarly, compounds that can oxidize to form acetic acid should also be avoided. In addition, packing should be done so that the contents remain dry until required, by the use of desiccants or water-absorbing gels.

5.8.6 Rosette Corrosion

This problem is best avoided by the use of the new design of hot water cylinder that heats the water above 40°C even at the cylinder base.[60] Even in an aggressive water, an aluminum protector rod is not necessary with such a design.[60] Where this type of cylinder is not available, then one with an aluminum protector rod should only be used in waters known to cause Type I pitting. Although magnesium anodes have been used in some countries, aluminum is now the preferred material because of its lower cost and longer life.

References

1. H.S. Campbell, *J. Inst. Met.* 77 (1950): p. 345.
2. H.S. Campbell, *J. Appl. Chem.* 4 (1954): p. 633.
3. H.S. Campbell, *Proc. Soc. Water Treat. Exam.* 3 (1954): p. 100.
4. H.S. Campbell, "A Review: Pitting Corrosion of Copper and its Alloys," BNF Miscellaneous Publication, MP574 August 1972. (This is an expanded version of a paper presented at the U R Evans International Conference on Localized Corrosion, Williamsburg, VA, USA, December 1971.)
5. M. Edwards, J.F. Ferguson, S.H. Reiber, *J. Am. Water Works Assoc.* 87 (1994): p. 74.
6. P. Devroey, C. Depommier, unpublished report, Visseries et Trefileries Reunies S.A., 1965.
7. H.J. Boorsma, C.H.J. Elzenga, H. Nijholt, unpublished report, KIWA.
8. F.J. Cornwell, G. Wildsmith, P.T. Gilbert, *Brit. Corros. J.* 8 (1973): p. 202.
9. M. Pourbaix, J. van Muylder, P. van Laer, N. de Zoubov, CEBELCOR Rapport Technique, 125, 1965.
10. M. Pourbaix, J. van Muylder, P. van Laer, N. de Zoubov, CEBELCOR Rapport Technique, 126, 1965.
11. M. Pourbaix, J. van Muylder, P. van Laer, N. de Zoubov, CEBELCOR Rapport Technique, 127, 1965.

12. M. Pourbaix, J. van Muylder, P. van Laer, N. de Zoubov, CEBELCOR Rapport Technique, 128, 1965.
13. M. Pourbaix, J. van Muylder, P. van Laer, N. de Zoubov, CEBELCOR Rapport Technique, 133, 1965.
14. V.F. Lucey, Brit. Corros. J. 2 (1967): p. 175.
15. R. Riedl, J. Klimbacher, *Int. J. Mater. Product Tech.* 4 (1989): p. 159.
16. P. Munn, "Effect of Phosphates on the Corrosion of Copper in Domestic Hot and Cold Water Systems," U.K. Corrosion Conference (Sheffield, UK: Institute of Corrosion, 2007).
17. P. Devroey, C. Depommier, unpublished report, Visseries et Trefileries Reunies S.A., 1962.
18. M.F. Obrecht, W.E. Sartor, J.M. Keyes, 4th International Congress on Metallic Corrosion (Amsterdam, Holland, 1969), p. 576.
19. V.F. Lucey, BNF Metals Research Association Research Report, A1692, June 1968.
20. V.F. Lucey, BNF Metals Research Association Research Report, RRA1838, December 1972.
21. M. Edwards, J. Rehring, T. Meyer, *Corrosion* 50 (1994): p. 366.
22. S. Jacobs, S. Reiber, M. Edwards, *J. Am. Water Works Assoc.* 90 (1998): p. 62.
23. R.B. Diegle, W.E. Berry, INCRA Project Final Report, 272, April 1980.
24. P.D. Goodman, V.F. Lucey, C. Maselkowski, *The Use of Synthetic Environments for Corrosion Testing* STP 970 (Philadelphia, PA: ASTM International, 1988), p. 165.
25. R. May, *J. Inst. Met.* 32 (1953): p. 65.
26. M. Sosa, S. Patel, M. Edwards, *Corrosion* 55 (1999): p. 1069.
27. C. Breckon, J.R.T. Baines, *J. Inst. Marine Eng.* 67 (1955): p. 1.
28. R. Retief, *Brit. Corros. J.* 8 (1973): p. 264.
29. D.D. Marsden, *MP* 17 (1978): p. 9.
30. M. Grassiani, "The Effect of Carbon Residue Films on Localised Corrosion of Brass Condenser Tubes," International Congress on Metallic Corrosion, vol. 4 (Toronto, Canada, 1984), p. 206.
31. S. Sato, K. Nagata, S. Yamauchi, *Sumitomo Light Met. Tech. Rep.* 20 (1979): p. 107.
32. E.B. Shone, G.C. Grimm, *Trans. Inst. Marine Eng.* 98 (1985): Paper 11.
33. R. Francis, BNF Metals Technology Centre Research Report, R461/4, October 1985.

34. S. Sato, T. Minamoto, K. Seki, H. Yamamoto, T. Takizawa, S. Yamauci, Y. Hisamatsu, I. Suzuki, T. Fuji, T. Kodama, H. Baba, K. Nawata, "Case Studies on Pitting Corrosion Failures of Copper Tubes in Hot Waters," International Symposium on Corrosion of Copper and Copper Alloys in Buildings (Tokyo, Japan: CDA, 1982).

35. E. Mattson, A.M. Fredriksson, *Brit. Corros. J.* 3 (1968): p. 246.

36. K. Kasahara, S. Komukai, Y. Konishi, T. Fujiwara, "An Investigation of Cases of Pitting Corrosion in Copper Pipes of Centralised Hot Water Systems," Annual Report of the Tokyo Gas Research Institute, 1986, p. 213.

37. H. Baba, T. Kodama, T. Fuji, Y. Hisamatsu, Y. Ishikawa, *Boshoku Gijutsu* (*Corros. Eng.*) 30 (1981): p. 113.

38. H. Baba, T. Kodama, T. Fuji, Y. Hisamatsu, *Boshoku Gijutsu* (*Corros. Eng.*) 30 (1981): p. 161.

39. I. Suzuki, Y. Ishikawa, Y. Hisamatsu, *Corros. Sci.* 23 (1983): p. 1095.

40. H. Baba, T. Kodama, T. Fuji, *Trans. Natl. Res. Inst. Met.* 28 (1986): p. 42.

41. R. Francis, H. Saxton, BNF Metals Technology Centre Research Report, R658/6, January 1991.

42. T. Hamamoto, M. Kumagai, K. Kawano, S. Yamauchi, *Sumitomo Light Met. Tech. Rep.* 28 (1987): p. 16.

43. R. Francis, BNF Metals Technology Centre Research Proposal, PR963/6, January 1990.

44. O. von Franque et al., *Werkst. Korros.* 26 (1975): p. 4.

45. M. Linder, E.-K. Lindman, "Investigation of Type III Pitting Corrosion in Copper Pipes," Ninth Scandinavian Corrosion Congress (Stockholm, Sweden: Swedish Corrosion Institute, 1983).

46. H.S. Campbell, A.H.L. Chamberlain, P.J. Angell, "An Unusual Form of Microbially Induced Corrosion in Copper Water Pipes," in *Corrosion and Related Aspects of Materials for Potable Water Supplies,* P. McIntyre, A.D. Mercer, eds. (London: Institute of Materials, 1993), p. 222.

47. W. Fischer, I. Hanssel, H.H. Paradies, *Microbial Corrosion,* vol. 1, Conference held in Sintra, Portugal (New York, NY: Elsevier, 1998), p. 300.

48. J.L. Nuttall, "Copper," in *Corrosion and Related Aspects of Materials for Potable Water Supplies,* P. McIntyre, A.D. Mercer, eds. (London, U.K.: Institute of Materials, 1993), p. 65.

49. S. Bond, BNF Metals Technology Centre Research Report, R688/6, December 1991.

50. C.W. Keevil, *Water Sci. Technol.* 49 (2004): p. 91.

51. H. Siederlarek, D. Wagner, M. Kropp, B. Fussinger, I. Hanssel, W.R. Fischer, "Effects of Water Composition and Operating Conditions on the Corrosion

Behaviour of Copper in Potable Waters" in Corrosion and Related Aspects of Materials for Potable Water Supplies, P. McIntyre, A.D. Mercer, eds. (London, U.K.: Institute of Materials, 1993), p. 122.

52. P.A. Klein, R.A. Hays, R.J. Ferrara, R.M. Kain, CORROSION/91, paper no. 509 (Houston, TX: NACE International, 1991).

53. D.G. Tipton, R.M. Kain, CORROSION/80, paper no. 36 (Houston, TX: NACE International, 1980).

54. P. Gallagher, A. Nieuwhof, R.J.M. Tausk, "Experiences with Sea Water Chlorination on Copper Alloys and Stainless Steels," in *Marine Corrosion of Stainless Steels—Chlorination and Microbial Effects*, European Federation of Corrosion Publication 10 (London, U.K.: Institute of Materials, 1993), p. 73.

55. J.M. Keyes, "Pinhole Pitting of Copper Tube in Storage," International Copper Research Association Symposium on Pitting Corrosion of Water Tubes (Brussels, Belgium, 1965).

56. J.O. Edwards, R.I. Hamilton, J.B. Gilmour, *MP* 16 (1977): p. 18.

57. S. Yamauchi, K. Nagata, S. Sato, B. Shimono, *Sumitomo Light Met. Tech. Rep.* 25 (1984): p. 1.

58. T. Notoya, T. Hamamoto, K. Kawano, "Unusual Corrosion in Copper Tubes," Step into the 90's Conference (Queensland, Australia, 1989).

59. A. Olszewski, R. Corbett, *MP* 46 (2007): p. 52.

60. R. Oliphant, "Causes of Copper Corrosion in Plumbing Systems," Foundation for Water Research Review, FR/R0007, May 2003.

61. EN 1057, "Copper and Copper Alloys, Seamless, Round Copper Tubes for Water and Gas in Sanitary and Heating Applications" (Brussels, Belgium: EN).

62. BS 2871 Part 3, "Specification for Copper and Copper Alloys. Tubes. Tubes for Heat Exchangers" (London, U.K.: BSI).

63. S. Bond, R. Francis, BNF Metals Technology Centre Research Report, R660/8, November 1991.

64. V.F. Lucey, *Brit. Corros. J.* 7 (1972): p. 36.

65. D. Shaw, V.F. Lucey, BNF Metals Technology Centre Research Report, A1956, June 1979.

66. D. Shaw, V.F. Lucey, INCRA Project Final Report, 302, 1984.

67. F. Ansuini, ICA Project Final Report, 442, July 1991.

68. C. Depommier, *ATB Metall.* 14 (1984): p. 387.

69. T. Hamamoto, M. Kumagai, K. Kawano, S. Yamauchi, *Sumitomo Light Met. Tech. Rep.* 28 (1987): p. 70.

70. K. Kasahara, S. Komukai, T. Fujiwara, *Corros. Eng.* 37 (1988): p. 361.

71. W.R. Fischer, D. Wagner, H. Siedlarek, *MP* 34 (1995): p. 50.

CHAPTER 6

Crevice Corrosion

6.1 Mechanism

Crevice corrosion occurs when part of the metal surface is shielded from the bulk environment. This may be because of deposits, or the crevice may be engineered such as those on a flange joint or under a washer on a bolt. Unlike the oxygen concentration cell that is required to initiate crevice corrosion on stainless steels, crevice corrosion of copper alloys depends on a copper ion concentration cell. As the metal corrodes, copper ions are released into solution, and in the bulk environment, they are either swept away or they precipitate as a copper compound such as copper oxide (Cu_2O) or copper hydroxychloride ($Cu_2Cl[OH]_3 \cdot 4H_2O$). In the creviced area, there is a shortage of reactants to form copper compounds, and the copper ion concentration increases, making the metal in the creviced region cathodic to the metal outside the crevice. Corrosion occurs at the anodic region just outside the crevice, often along much of its length.

Figure 6.1A provides an example of crevice corrosion on 70/30 copper-nickel after exposure in seawater, followed by cleansing. Its appearance is similar to that of most copper alloys experiencing crevice corrosion. Apart from the location just outside the crevice, the appearance is very similar to pitting corrosion in seawater (see Chapter 5). In the example, most of the crevice sites are where barnacles were attached to the plate during the 60-day exposure to slowly moving seawater. Figure 6.1B shows two plates of Alloy 400 that had been exposed for 6 months in quiescent seawater with nonmetallic annular crevice formers, as described by Oldfield.[1] Like 70/30 copper-nickel, the attack is just outside the crevice, but pitting has also occurred on openly exposed surfaces. Alloy 400 is susceptible to this type of attack in stagnant or slowly moving seawater, as described in Chapter 5.

(A)

(B)

FIGURE 6.1 Appearance of crevice corrosion on (A) 70/30 copper-nickel and (B) Alloy 400, exposed to quiescent seawater. (Photographs courtesy of R.M. Kain.)

The concentration of copper ions in the crevice provides the cathodic reaction, and redeposited, metallic copper is often seen in the shielded area when crevice corrosion is occurring, that is,

$$Cu^{+} + e \rightarrow Cu. \tag{6.1}$$

The supply of copper ions for this reaction is limited and will soon become exhausted. However, once crevice corrosion has initiated, it is self-sustaining, and if the crevice is removed, the corrosion will continue under the accumulated corrosion products. The corrosion now looks somewhat similar to a Type I pit, as described in Chapter 5, with a mound of green, basic copper salts on top of the pit and the cathode above the pit. The copper ion concentration that was required to initiate attack is no longer necessary. Although crevice corrosion is seen quite often in corrosion tests, it is rarely a cause of service failure (see Section 6.4).

6.2 Fresh Water

Copper pipe is commonly used for fresh water distribution, particularly in domestic premises. Upland waters and some river waters have a high organic content and/or solids burden, and this can settle in stagnant or slow-moving water regions.[2] Soft waters and waters of low pH are also more likely to cause corrosion of cast iron water mains, and thus deposits may also contain iron corrosion products. These deposits create a shielded area, and crevice corrosion can initiate on copper surfaces at the edge of such deposits. However, propagation rates are low, and penetration of a 1-mm-thick copper tube usually takes 10 years or more if the attack is not stifled.[2]

Flux residues from soldering may also cause crevice corrosion, but this can be much more rapid.[2] This is because older-style fluxes contain ammonium (NH_4Cl) and/or zinc chloride ($NaCl_2$), and this means that conditions beneath a flux deposit can be very acidic, hence the high rate of corrosion. Modern fluxes for potable water pipes must be free of chlorides and water-soluble and so are less corrosive; however, some flux residues dissolve only slowly, and crevice corrosion like that seen with other deposits may occur.

6.3 Seawater

There is more data on crevice corrosion of copper alloys in seawater and modified seawater than in any other medium. Kain reviewed the test

TABLE 6.1 Depth of crevice corrosion after 180 days in natural seawater[4]

Alloy	Condition	Crevice former	Finish	Flow(A)	Max. depth (mm) A	Max. depth (mm) B
M	Cast	Gasket	Phonograph	Q	0.34	0.32
Bronze	Cast	Gasket	Phonograph	F	0	0
NAB	Cast	Gasket	Phonograph	Q	0.60	<0.01
	Cast	Gasket	Ground	F	<0.01	<0.01
70/30	Cast	Gasket	Phonograph	Q	<0.01	0.25
Cu-Ni	Cast	Gasket	Phonograph	F	0.05	<0.01
70/30	Wrought	PTFE	Ground	Q	0.10	0.04
Cu-Ni	Wrought	PTFE	Ground	F	0.17	0.11
90/10 Cu-Ni	Wrought	Gasket	Phonograph	Q	1.05	1.28
Alloy	Wrought	PTFE	Ground	Q	0.67	0.69
400	Wrought	PTFE	Ground	F	0.25	0.27

(A) Q, quiescent; F, flowing (1.6 m/s).

methods available to study crevice corrosion, including the parameters that should be considered and the various methods of evaluating the results.[3]

The U.S. Navy conducted a study of the crevice corrosion resistance of newer alloys, and as part of that study, they included the copper alloys currently used in U.S. naval vessels:[4] M bronze (C92200), cast nickel aluminum bronze (NAB), 90/10 and 70/30 copper-nickels, and Alloys 400 and K-500. Tests were conducted in both flowing seawater and quiescent conditions using crevice assemblies consisting of rubber gasket material under metal washers and also polytetrafluorethylene (PTFE) crevice formers. Surface finish of the test samples was also a variable that was examined. Some of the results are presented in Table 6.1; they show that crevice corrosion was much deeper under quiescent conditions and was negligible for M bronze, NAB, and 70/30 copper-nickel under flowing conditions. Note that the crevice geometry and, in particular, the cathode to anode ratio, were different in flowing and quiescent conditions. The cathode area on the crevice assemblies exposed to quiescent seawater was much larger than that used on the samples in flowing conditions, and this would be expected to produce more severe crevice corrosion, as seen in Table 6.1. Although there was deep attack on 90/10 copper-nickel under quiescent conditions, no data were generated under flowing conditions. The results for 70/30 copper-nickel show that a PTFE crevice former creates a more severe crevice than one with a rubber gasket. This is also the case for stainless steels, but the reason for this, a smaller crevice width, does not explain the reactions on copper alloys, where the crevice region is the cathode.[5] It is possible that the greater depth of attack with a PTFE crevice former was due to an earlier transition to a mechanism in which the reduction of dissolved oxygen (DO) was the cathodic reaction. The depth of attack was generally more severe for Alloy 400 than the copper-rich alloys, with the exception of 90/10 copper-nickel.

The author conducted a series of tests of copper alloy condenser tubes utilizing the Campbell condenser tube test rig, which contains both crevices exposed to cold water and crevices exposed to warm water (see Section B.2 in Appendix B).[6-8] The alloys under test were aluminum brass, 90/10 copper-nickel, 70/30 copper-nickel, 66/30/2/2 copper-nickel-iron-manganese, and Alloy 722. In tests investigating the effects of chlorine and ferrous sulfate additions to seawater, there was no significant crevice corrosion in either region at all chlorine concentrations up to 4 mg/L continuous.[6] When ammonia (NH_3) was present, corrosion occurred in the warm water crevice.[7] As this was more akin to dealloying than crevice corrosion, it is described in detail in Chapter 9.

When sulfide was added to simulate polluted waters, the effects were variable.[8] At low levels of sulfide (0.01 mg/L), there was an increased tendency for crevice attack to occur outside the crevice on all alloys, presumably because the sulfide hindered the formation of protective films outside the crevice. At higher concentrations of sulfide (0.03 and 0.1 mg/L), there was no classical crevice corrosion, but general corrosion and pitting occurred in the crevice. It appears that the sulfide in the seawater prevented the establishment of a copper ion concentration cell by causing copper sulfide (CuS) to precipitate.

In an investigation of the variability of seawater on corrosion, Phull et al. reported the results of corrosion tests of 90/10 copper-nickel exposed at 14 seawater sites around the world.[9] The test panels were bolted to the supporting frame with a nylon bolt and washer, which created four crevices on each panel. The results are summarized in Table 6.2. It should be noted that samples from seven of the sites showed no crevice corrosion at any exposure time. Also, the 3- and 5-year results from California could not be quantified because erosion corrosion had occurred in addition to crevice corrosion. The depth of crevice corrosion was very variable, with half the sites reporting none. While pitting away from the crevices was seen on panels exposed at all the sites, it was often quite shallow. The variability of the attack shows that it is influenced not so much by seawater conditions (temperature, DO, etc.) as by local conditions such as sulfides, deposits, and so on.

TABLE 6.2 Maximum depth of crevice corrosion on 90/10 copper-nickel after up to 5 years' exposure in natural seawater[9]

Site[(A)]	Exposure time (years)	Max. depth (mm)	
		A	B
Wrightsville Beach, NC, East Coast, USA	0.5	0	0
	1.0	0	0
	3.0	0	0
	5.0	0.05	0.01
Freeport, Texas, Gulf Coast, USA	0.5	0	–
	1.0	0.02	0.07
	3.0	0.62	0.52
	5.0	0.16	0.71
Port Hueneme, California, West Coast, USA	0.5	0	0
	2.0	1.3	0.8
	3.0	(B)	(B)
	5.0	(B)	(B)
Peru	5.0	0.10	0.33
Australia	1.0	0	0.26
	3.0	0.33	2.06
	5.0	1.01	0.10
Denmark	0.5	0.11	0.05
	1.5	0.20	0.30
	3.1	0.31	0.30
	5.0	0.36	0.49
United Kingdom	0.5	0	0
	1.0	0	0
	3.0	0.40	0.24
	5.0	0.60	0.55

(A) No crevice corrosion found on panels exposed at Ocean City, Maryland; Key West, Florida; Hawaii; Japan; Italy; and Sweden (two sites).
(B) Depth not reported because of erosion corrosion damage at crevice site.

Klein et al. examined the effect of chlorine additions to seawater on the crevice corrosion of 70/30 copper-nickel and Alloy 400 using various crevice formers on tube and plate under both flowing and quiescent conditions.[10] Replicate samples were exposed to natural seawater, seawater with 1 mg/L chlorine (to simulate a normal antifouling regime), or seawater with 90 mg/L chlorine (to simulate a concentrated sodium hypochlorite [NaOCl] line). The results for 70/30 copper-nickel and Alloy 400 under quiescent conditions are shown in Tables 6.3 and 6.4, respectively. Note that there were cathode to anode area ratio differences between samples exposed to quiescent seawater and flowing seawater, as described previously,[4] which explains the more severe attack under quiescent conditions.

TABLE 6.3 Maximum depth of crevice corrosion of 70/30 copper-nickel under quiescent conditions with different crevice formers[10]

Material form	Crevice former	Chlorine (mg/L)	Depth of attack (mm)	
			45/60 days[(A)]	180 days
Tube	Vinyl sleeve	0	0.01	0.07
		1	0.01	0.03
		90	0.05	0.05
	Buna N sleeve	0	0.01	0.08
		1	0.01	0.01
		90	0.08	0.13
	O-ring	0	0.01	0.07
		1	0.01	0.03
		90	0.03	0.07
Plate	Polyacetal annulus	0	0.01	0.02
		1	0.01	0.02
		90	0.15	0.07
	PTFE annulus	0	0.01	0.02
		1	0.01	0.01
		90	0.06	0.06
	Nylon annulus	0	0.01	0.05
		1	0	0.02
		90	0.09	0.10
	PTFE O-ring	0	0.01	0.01
		1	0.01	0
		90	0.01	0.06
	Rubber O-ring	0	0.01	0
		1	0.01	0
		90	0.01	0

[(A)] Forty-five days for plate and 60 days for tube samples.

The results for 70/30 copper-nickel show that very little crevice corrosion occurred in natural seawater, and this was reduced even more in the presence of 1 mg/L chlorine. However, 90 mg/L chlorine increased the attack to the same as that seen in natural seawater or more. Despite the differences in product forms and crevice geometry, there was no discernible difference between the results for any of the crevice formers on tube or plate.

With Alloy 400, there was deep attack in natural seawater, which was reduced in the presence of 1 mg/L chlorine, but there was still a significant depth after 180 days' exposure. The addition of 90 mg/L chlorine made the conditions more aggressive—about the same as seen in natural seawater. This is presumably because small additions of chlorine reduce the potential below the pitting potential, while high levels of chlorine take it back above

TABLE 6.4 Maximum depth of crevice corrosion of Alloy 400 under quiescent conditions with different crevice formers[10]

Material form	Crevice former	Chlorine (mg/L)	Depth of attack (mm)	
			45/60 days[A]	180 days
Tube	Vinyl sleeve	0	0.27	0.56
		1	0.01	0.20
		90	0.20	0.24
	Buna N sleeve	0	0.23	0.08
		1	0.01	0.11
		90	0.25	0.26
	O-ring	0	0.27	0.49
		1	0.02	0.25
		90	0.31	0.38
Plate	Polyacetal annulus	0	0.41	0.76
		1	0.08	0.34
		90	0.36	0.67
	PTFE annulus	0	0.41	0.68
		1	0.07	0.30
		90	0.35	0.70
	Nylon annulus	0	0.38	0.68
		1	0.07	0.31
		90	0.33	0.65
	PTFE O-ring	0	0.10	0.23
		1	0.08	0.19
		90	0.12	0.20
	Rubber O-ring	0	0.08	0.19
		1	0.01	0.05
		90	0.06	0.15

[A] Forty-five days for plate and 60 days for tube samples.

this potential.[11] There was no real difference between the different crevice formers on tube samples, but the O-rings on plate samples were a little less severe than the annular crevice formers. One observation for Alloy 400 was that the rate of propagation of crevice corrosion in natural seawater decreased with time, while it was more or less constant in chlorinated seawater. Another observation was that deep pitting occurred on Alloy 400 in natural seawater, but there was hardly any in chlorinated seawater. Deep pitting also occurred with 90 mg/L chlorine. The effect of chlorine on the pitting behavior of Alloy 400 is discussed in Chapter 5.

Under flowing conditions, the results were a little different, as shown in Table 6.5. For 70/30 copper-nickel, there was deeper attack in flowing, natural seawater than under quiescent conditions, presumably because there

TABLE 6.5 Maximum depth of crevice corrosion of 70/30 copper-nickel and Alloy 400 tubes under flowing (1.6 m/s) conditions[10]

Alloy	Crevice former	Chlorine (mg/L)	Depth of attack (mm)		
			60 days	180 days	360 days
70/30 Cu-Ni	Vinyl sleeve	0	0.1	–	0.36
		1	<0.01	–	0.02
		90	0.48	–	Perforated
	Buna N sleeve	0	0.01	0.21	–
		1	<0.01	0.02	–
		90	0.7	1.5	–
	Nylon compression fitting	0	<0.01	–	0.28
		1	<0.01	–	0.03
		90	0.27	–	1.01
Alloy 400	Vinyl sleeve	0	0.37	–	0.63
		1	0.1	–	0.45
		90	1.7	–	Perforated
	Buna N sleeve	0	0.36	1.4	–
		1	0.11	0.27	–
		90	1.7	1.95	–
	Nylon compression fitting	0	0.05	–	0.43
		1	0.02	–	0.56
		90	0.04	–	0.17

was a more steady supply of DO to the cathode to prevent polarization. In seawater with 1 mg/L chlorine, the attack was reduced to very low levels. This suggests that the reduction of chlorine to chloride is a less efficient cathodic reaction than the reduction of DO in the presence of a biofilm.[12] With 90 mg/L chlorine, the crevice corrosion was much deeper than in natural seawater, with perforation, or near perforation, of the tubes. Again, there were no discernible differences between the different crevice formers for 70/30 copper-nickel.

The results followed the same trend for Alloy 400 as for 70/30 copper-nickel, but with much deeper attack. Even though the depth of attack with 1 mg/L chlorine was reduced, the depth of attack was still significant. The nylon compression fitting appeared to be a less aggressive crevice former for this alloy compared with the sleeves.

Kain et al. examined the possible use of 70/30 copper-nickel for reverse osmosis (RO) desalination plants.[13] Samples were exposed under three separate conditions. The first was filtered seawater (~19,000 mg/L chloride), as found in the high-pressure section of an RO pilot plant. The second was water with 200 to 2,000 mg/L chloride, as found after partial

TABLE 6.6 Maximum depth of crevice corrosion on 70/30 copper-nickel exposed in various parts of a reverse osmosis plant[13]

Medium	Chloride (mg/L)	Attacked region	Depth of attack (mm)
Filtered seawater	19,000	All sites	<0.01
Second-pass	200–2,000	PCA	0.04
brine		Rubber gland	0.04
		20.6-mm O-ring	<0.01
		23.8-mm O-ring	0.08
Third-pass	<0.02	PCA	<0.01
permeate		23.8-mm O-ring	<0.01
		Rubber gland	0
		20.6-mm O-ring	0

treatment of the seawater. The third condition was water with <0.2 mg/L chloride, as found in the final stage of water treatment. The tests lasted for 90 days, with O-ring crevice formers on tubes and acrylic crevice assemblies (PCA) for plate samples, as described by Oldfield.[1] All the samples were exposed in empty RO membrane housings.

The results, in Table 6.6, show the depths of attack at the various crevice sites. There was negligible crevice corrosion at all the sites, in both filtered seawater and the final permeate. In the second-pass permeate, with moderate chloride concentrations, there was deeper attack at most of the crevice sites. It is not clear why this intermediate concentration of chlorides should be more aggressive in terms of crevice corrosion initiation. Another fact that comes out of Table 6.6 is that the looser O-ring produced deeper attack than the tighter O-ring. This is probably because the cathode under an O-ring is much smaller for a tight O-ring and so cannot produce an ion concentration cell to supply much current to create deep attack.

Kain and Weber examined the effects of lay-up conditions on the crevice corrosion resistance of 70/30 copper-nickel and Alloy 400 under RO conditions.[14] The samples were tubes and plates creviced as described earlier.[13] The exposure cycle was 500 hours of flowing seawater, followed by 500 hours in either stagnant, filtered seawater or stagnant lay-up solution, which consisted of 1% sodium metabisulfite in water containing 1,000 mg/L chloride. The samples were examined after one and three cycles of exposure, and Figure 6.2 shows the appearance of the 70/30 copper-nickel samples after one and three cycles. After the first cycle, the tube was largely covered in a black sulfide film, and the crevice corrosion where the O-rings were located is clearly visible. After three cycles, there were also some green and brown corrosion products, with the crevice attack at the O-rings again still clearly visible. The depths of attack are shown in Table 6.7.

FIGURE 6.2 Appearance of 70/30 copper-nickel tubes with O-ring crevices after one and three cycles of 500 hours in flowing seawater followed by 500 hours of stagnant seawater. (Photographs courtesy of R.M. Kain.)

TABLE 6.7 Depth of crevice corrosion after exposure to alternating flowing and stagnant conditions[14]

Alloy	No. of cycles	Crevice former[(A)]	Depth of attack (mm)	
			Bisulfite lay-up	Seawater lay-up
70/30 Cu-Ni	1	PCA[(B)]	<0.01	<0.01
		Rubber gland	<0.01	<0.01–0.02
		O-ring	<0.01	<0.01–0.01
	3	PCA[(B)]	0.05	<0.01
		Rubber gland	<0.01–0.27	<0.01–0.08
		O-ring	0.02–0.15	<0.01–0.03
Alloy 400	1	PCA[(B)]	0.14	0.28
		Rubber gland	<0.01–0.32	<0.01–0.15
		O-ring	0.01–0.30	0.05–0.30
	3	PCA[(B)]	0.22	0.24
		Rubber gland	0.01–0.46	0.01–0.29
		O-ring	0.15–0.56	0.04–0.40

(A) PCA is on plate; gland and O-ring are on tubes.
(B) PCA, perspex crevice assembly.[1]

The attack on the 70/30 copper-nickel was generally shallow, but some deeper pits developed over three cycles with the bisulfite solution during lay-up. The attack was deeper on Alloy 400, but there was no significant difference between depths after one and three exposure cycles. This suggests that the bulk of the attack occurs early during exposure and the rate of further propagation is low. There was no particular difference in the severity of attack on Alloy 400 with either lay-up solution. There was also no particular difference between the different crevice formers for both alloy 400 and 70/30 copper-nickel. The stagnant seawater lay-up did not produce much sulfide, which might be expected to have a deleterious effect on the resistance of these alloys to crevice corrosion.[14]

Wharton and Stokes investigated crevice corrosion of NAB in both 3.5% sodium chloride (NaCl) solution and natural seawater.[15] In sodium chloride, they found conventional crevice corrosion outside the crevice to a modest depth, but much more severe attack along the kappa III phase in the acidic region within the crevice. This is discussed in more detail in Chapter 9. Wharton and Stokes also found selective attack of the kappa III phase at the base of the crevice corrosion pit, where the acidic conditions make the kappa III anodic to the alpha phase (Figure 6.3).[15] The attack in seawater was similar, but much deeper, with the crevice attack spreading horizontally away from the crevice for some distance. This is because, once established, the crevice attack is similar to Type I pitting, with the cathode above the

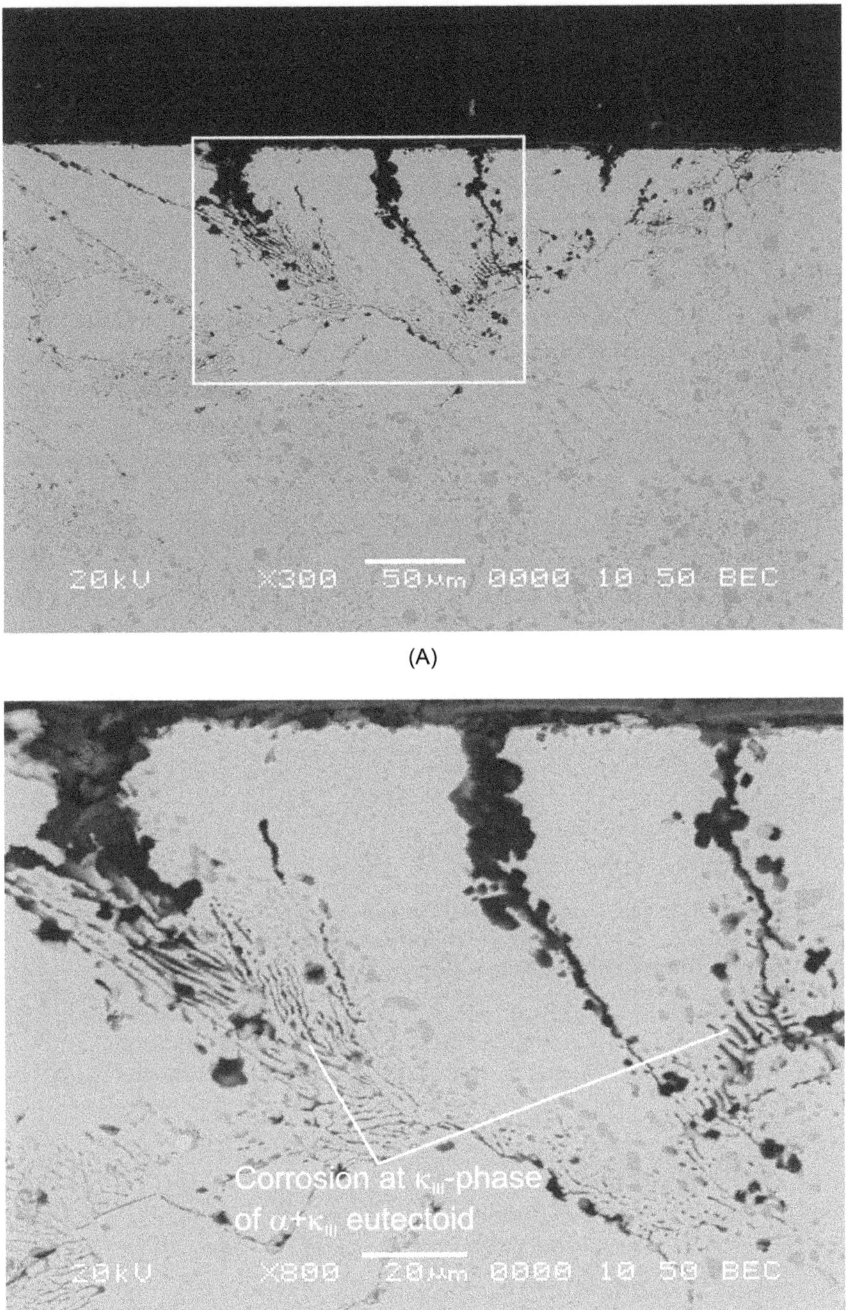

(A)

(B)

FIGURE 6.3 Cross sections of NAB after 6.5 months' immersion in 3.5% sodium chloride solution: (A) attack at the edge of the crevice; (B) detail from Figure 6.3A showing attack of kappa III phase at base of pit. (Photographs courtesy of J.A. Wharton.)

mound. This favors sideways propagation in high-conductivity media like seawater.

6.4 Avoidance of Crevice Corrosion

Although crevice corrosion of copper alloys is a well-recognized phenomenon, it rarely causes failures. In over 30 years as a corrosion engineer, the author has never seen a failure of a copper alloy due to crevice corrosion, although crevice corrosion has been observed on some components that failed for other reasons. Where silt or deposits may cause problems, the fitting of an adequate filtration system is recommended. The exception to this is NAB, but the failures are due to selective phase attack of the kappa III phase (see Chapter 9), rather than true crevice corrosion.

When thin material is being employed in a critical application, it is best to avoid a design that creates a crevice. Where a crevice is unavoidable, the evidence is that very tight crevices are less likely to cause crevice corrosion with copper alloys. Hence, greasing screwed joints or using mastic or gaskets containing a copper corrosion inhibitor (e.g., benzotriazole) can be considered for critical applications.

References

1. J.W. Oldfield, *Int. Met. Rev.* 32 (1987): p. 153.
2. J.L. Nuttall, "Copper," in *Corrosion and Related Aspects of Materials for Potable Water Supplies*, P. McIntyre, A.D. Mercer, eds. (London, U.K.: Institute of Materials, Minerals, and Mining, 1993): p. 65.
3. R.M. Kain, "Seawater Crevice Corrosion Testing of Stainless Steels, Ni-Base and Cu-Ni Alloys: Perspectives on Methodologies and Interpretation of Results," EUROCORR 2006 (Maastricht, Holland: NCC/European Federation of Corrosion, 2006).
4. D.M. Aylor, R.A. Hays, R.M. Kain, R.J. Ferrara, CORROSION/99, paper no. 329 (Houston, TX: NACE International, 1999).
5. R. Francis, *The Selection of Materials for Seawater Cooling Systems—A Practical Guide for Engineers* (Houston, TX: NACE International, 2006).
6. R. Francis, *MP* 21 (1982): p. 44.
7. R. Francis, *Brit. Corros. J.* 20 (1985): p. 167.
8. R. Francis, *Brit. Corros. J.* 20 (1985): p. 175.
9. B. Phull, S. Pikul, R.M. Kain, "Seawater Corrosivity Around the World; Results from Five Years of Testing," ASTM STP 1300 (1997): p. 34.

10. P.A. Klein, R.A. Hays, R.J. Ferrara, R.M. Kain, CORROSION/91, paper no. 509 (Houston, TX: NACE International, 1991).

11. P. Gallagher, A. Nieuwhof, R.J.M. Tausk, "Experiences with Seawater Chlorination on Copper Alloys and Stainless Steels," in *Marine Corrosion of Stainless Steels—Chlorination and Microbial Effects*, European Federation of Corrosion Publication 10 (London, U.K.: Institute of Materials, Minerals, and Mining, 1993).

12. R. Francis, *Galvanic Corrosion—A Practical Guide for Engineers* (Houston, TX: NACE International, 2001).

13. R.M. Kain, W.L. Adamson, B. Weber, "Corrosion Coupon Testing in Natural Waters: A Case History Dealing with Reverse Osmosis Desalination of Seawater," ASTM STP 1300 (1997): p. 122.

14. R.M. Kain, B.E. Weber, CORROSION/97, paper no. 422 (Houston, TX: NACE International, 1997).

15. J.A. Wharton, K.R. Stokes, *Electrochim. Acta* 53 (2007): p. 2463.

CHAPTER 7

Erosion Corrosion and Erosion

Copper and its alloys have good resistance to corrosion at moderate water flows. However, when the local velocity is high enough, the local shear stress becomes sufficient to remove the corrosion products. Metal dissolution occurs, followed by film formation, which is again disrupted by the water flow. This is termed *erosion corrosion*, or *impingement attack*, because it often occurs where water flow impinges on the metal surface. This is an electrochemical process because the application of cathodic protection prevents attack.

The attack usually takes the form of pits that are often undercut on the downstream side. Where the attack in a tube is severe, the attack often has the appearance of horseshoe imprints, with the horse walking upstream, as shown in Figure 7.1. The leading edge of the attack is frequently undercut due to the swirling action of the water.

When solids are present in the water, then erosion can occur at velocities at which the metal loss would be low in the absence of solids. Section 7.1 discusses erosion corrosion of copper and its alloys, whereas Section 7.2 deals with erosion in the presence of solids. Section 7.3 offers some advice on mitigating attack due to both of these corrosion mechanisms. The data presented come from a variety of sources, using several different test methods, from flow loops to high-velocity jet test rigs. The different methods of carrying out such tests and their relevance to service performance are discussed in Appendix B.

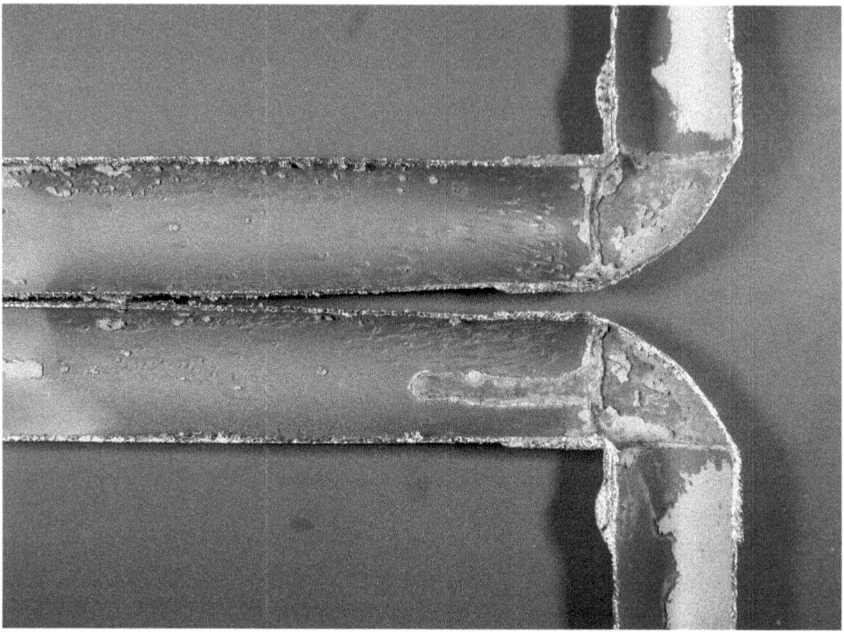

FIGURE 7.1 Erosion corrosion of 15-mm copper pipe after a 90-degree elbow. (Photograph courtesy of P. Munn.)

7.1 Erosion Corrosion

7.1.1 Fresh Water

In fresh waters, the most commonly used copper alloys are copper and brass (both single-phase and duplex), with gunmetal being used mostly for pumps and valves and, occasionally, for fittings. Other copper alloys have only limited use.

There has been little published data on erosion corrosion in fresh waters, but this type of attack is sometimes thought to be more common in soft waters, where calcium carbonate ($CaCO_3$) cannot form. Myers and Obrecht carried out extensive studies in a sodium zeolite–softened well water, whose composition is shown in Table 7.1.[1] They tested tubes made of copper, red brass (C23000), admiralty brass, and 90/10 copper-nickel over a range of temperatures and velocities. Tests were conducted in three waters: fully softened, a blended water with intermediate hardness, and softened water with reduced dissolved oxygen (DO) and increased carbon dioxide (CO_2) concentrations. There is insufficient space to present all their findings here, but some of their data and the main conclusions are included.

TABLE 7.1 Composition of raw well water used for erosion corrosion tests[1]

		Concentration (mg/L)
Calcium	As $CaCO_3$	210–250
Magnesium	As $CaCO_3$	100–120
Sodium	As Na	6–9
Bicarbonate	As $CaCO_3$	300–325
Sulphate	As SO_4	14–24
Chloride	As Cl	6–8
Free carbon dioxide	As CO_2	10–40
Oxygen	As O_2	6–12
Silica	As SiO_2	8–12
Total hardness	As $CaCO_3$	310–370
Total alkalinity	As $CaCO_3$	300–325
pH		6.8–7.5

Figures 7.2 and 7.3 show the maximum annual metal loss as a function of velocity and temperature, respectively, for copper tube in fully softened water. The data show that there is an increasing loss of metal with an increase of velocity, but this is not excessive up to 60°C. At higher temperatures there is a much greater metal loss with increasing velocity. At 93°C, the metal

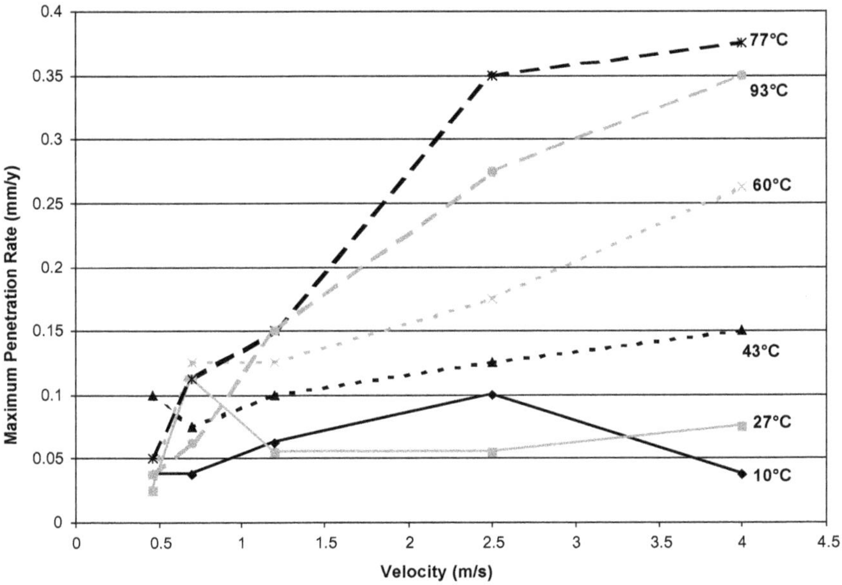

FIGURE 7.2 Maximum penetration rate vs. velocity for copper tube in softened water.[1]

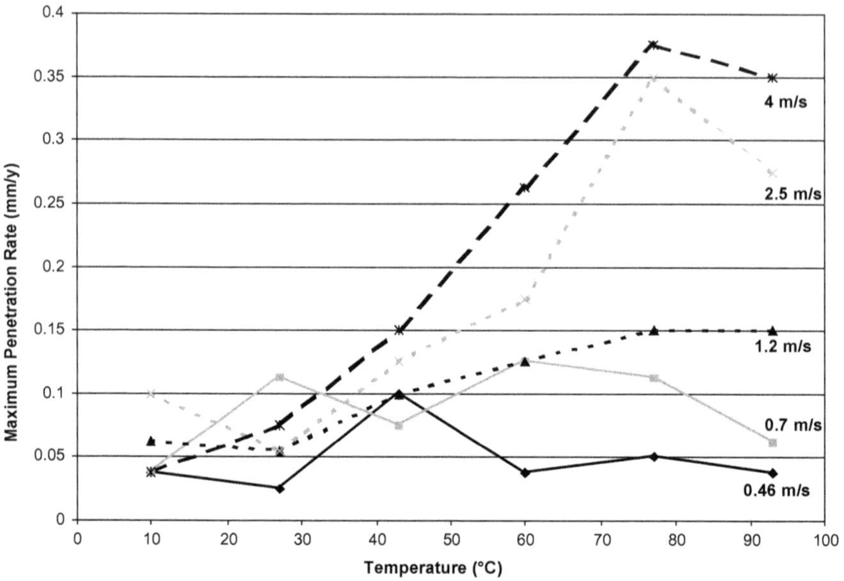

FIGURE 7.3 Maximum penetration rate vs. temperature for copper tube in softened water.[1]

loss is less than at 77°C, which is probably due to the low level of DO in the water at the higher temperature. The rate of metal loss is then controlled by the availability of DO to sustain the cathodic reaction, the reduction of DO.

Figures 7.4 and 7.5 show the effect of velocity on red brass, admiralty brass, and 90/10 copper-nickel at 10°C and 60°C, respectively. These graphs show that red brass behaves similarly to copper, while admiralty brass suffers less attack at higher temperatures, and 90/10 copper-nickel suffers even less. On the basis of these test data, Myers and Obrecht recommended safe operating velocities for the alloys that they tested.[1] A summary of these is shown in Table 7.2, which suggests that copper and red brass are suitable

TABLE 7.2 Recommended safe maximum velocities for copper and its alloys in fresh water[1]

Alloy	UNS no.	Temperature range (°C)	Max. velocity (m/s)
Copper	C12200	0–60	1.2
Red brass	C23000	0–60	1.2
Admiralty brass	C44300	0–77	3.8
90/10 copper-nickel	C70600	0–93	3.8

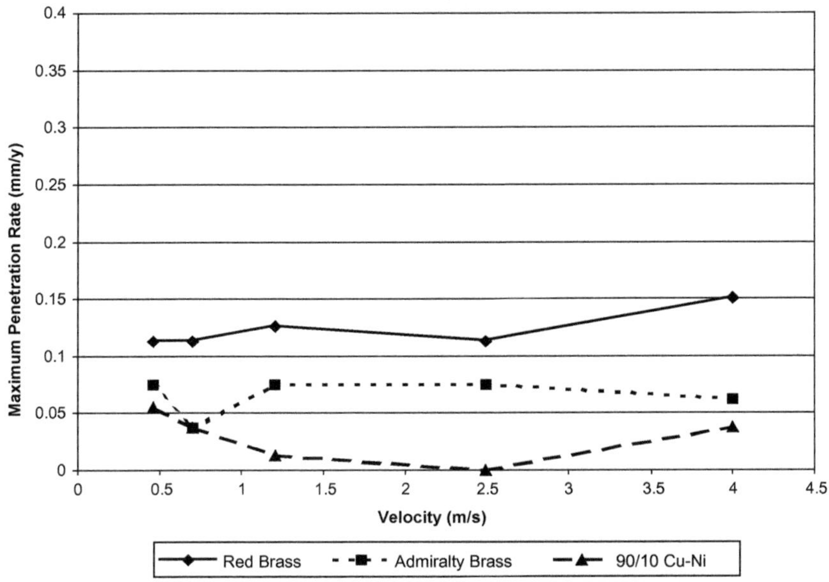

FIGURE 7.4 Effect of velocity on the erosion corrosion rate in soft water at 10°C.[1]

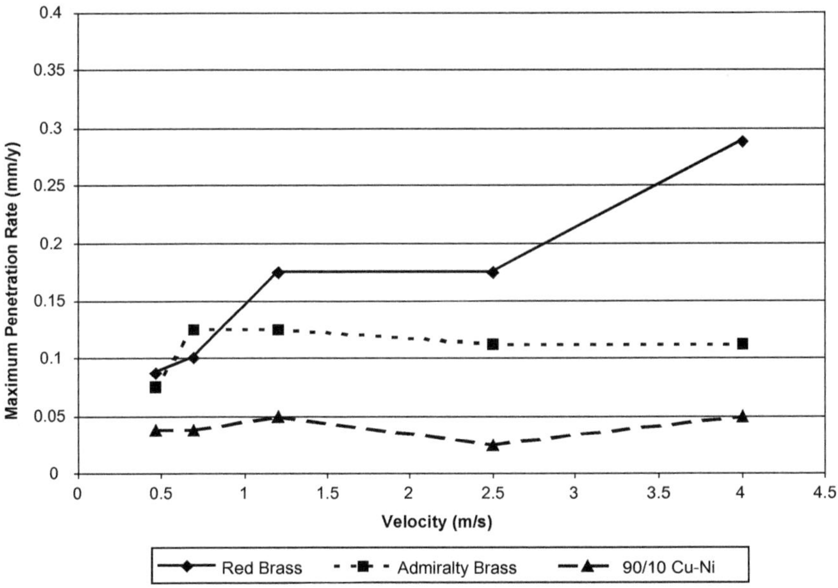

FIGURE 7.5 Effect of velocity on the erosion corrosion rate in soft water at 60°C.[1]

TABLE 7.3 The corrosion rates of some common alloys as a function of velocity in seawater[5]

		Corrosion rate (mm/y)			
		0–0.6 m/s			
Alloy	UNS no.	General	Deepest pit	8 m/s	40 m/s
Carbon steel	–	0.075	2.0	1.0	4.5
Cast iron	–	0.550	4.9	4.4	13.3
Gunmetal	C83600	0.018	0.325	1.80	1.33
Admiralty gunmetal	C90500	0.028	0.25	0.875	1.08
Ni-Al-bronze	C96320	0.055	1.125	0.225	0.975
90/10 Cu-Ni	C70600	<0.025	0.200	0.45	0.825
70/30 Cu-Ni	C71500	<0.025	0.250	0.125	1.48
316L S/S	S31603	<0.025	1.80	<0.025	0.005
Alloy 400	N04400	<0.025	1.30	0.015	0.010

at moderate velocities up to 60°C. Oliphant made similar recommendations, but suggested that whereas the maximum water velocity in hot water should be 1.0 m/s, in cold water, it can be increased to 2.0 m/s.[2] In pipes that experience only intermittent flow, these maximum velocities can be relaxed, provided that the off time is sufficiently long to enable film repassivation to occur.

One location where erosion corrosion is commonly seen in copper piping systems carrying fresh water is immediately after sharp bends or elbows. If the fitting has been soldered into place, sometimes a protruding solder bead increases the local turbulence. The erosion corrosion is always on the copper and never on the solder, although the potential of solder is significantly more electronegative than that of copper.[3] Francis and Campbell believed that this is because tin oxide is more stable and mechanically strong than copper oxide (Cu_2O) in fresh waters. In unpublished work carried out at BNF Metals Technology Centre, they exposed samples of copper coated with solder in the May jet test at 10 m/s with 3% added air in a hard borehole water containing 100 mg/L chloride.[4] No measurable erosion corrosion had occurred on the solder after 30 days at 15°C. These results demonstrated the high resistance of solders to erosion corrosion in fresh water.

7.1.2 Seawater

Copper and its alloys are more susceptible to erosion corrosion in seawater compared with fresh water. This means that alloys that are more resistant than copper and brass are usually chosen. Table 7.3 shows the effect of

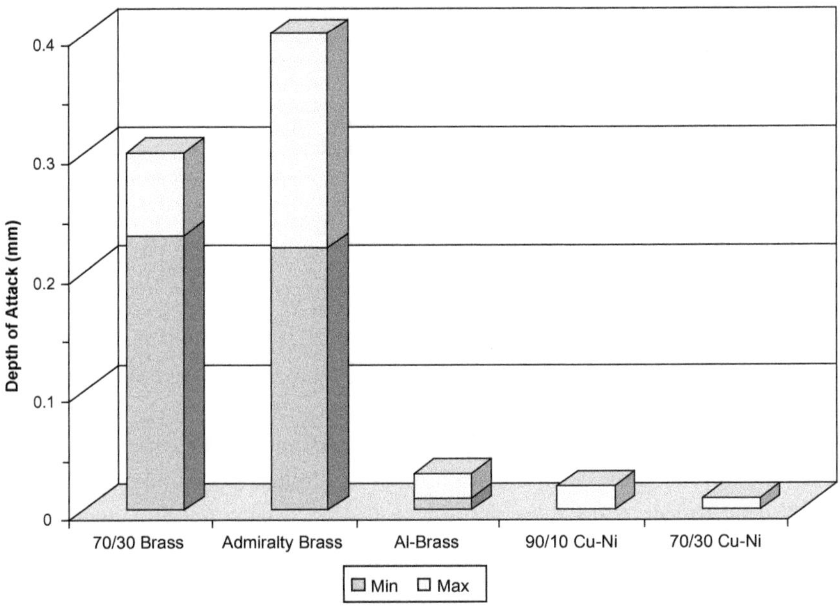

FIGURE 7.6 Depth of attack of some wrought copper alloys in the jet test at 9.3 m/s and 18°C.[6]

velocity on the corrosion rate of some common copper alloys, with the rates for carbon steel, cast iron, Alloy 400, and Type 316L stainless steel (SS) (UNS 531603) provided for comparison.[5] The rate of attack of the copper alloys increases with velocity, whereas those of Alloy 400 and 316L SS remain low. However, both these alloys are susceptible to localized pitting and/or crevice corrosion at low velocities. The corrosion rate of NAB in Table 7.3 is not as low as might be expected. However, the composition tested (C96320) contains only 2.5% iron, whereas the more common C95800 contains 4.0% to 4.5% iron and is more resistant to erosion corrosion (see the following discussion).

Bog[6] carried out jet impingement tests[4] on a range of copper alloys at a jet velocity of 9.6 m/s using recirculated seawater. The results for a selection of common, wrought copper alloys are shown in Figure 7.6. The results show the poor resistance of 70/30 brass and admiralty brass to erosion corrosion and the excellent resistance of aluminum brass and the copper-nickel alloys.

In later work, Bog evaluated the resistance of some cast copper alloys to erosion corrosion by the same method at 20°C.[7] The results, given in Table 7.4, show the benefit of increasing tin content on the performance of gunmetals and bronzes and the poor performance of high-tensile brass

TABLE 7.4 Depth of attack in jet impingement tests of some cast alloys at 9.6 m/s jet velocity and 20°C[7]

Alloy	UNS no.	Mean depth(mm)
Mn-bronze	C86400	0.5
Gunmetal (LG2) (5% Sn)	C83600	0.23
Gunmetal (LG4) (7% Sn)	C92410	0.05
Gunmetal (G1) (10% Sn)	C90500	0.02
Leaded tin bronze (3% Pb)	C92900	0.02
NAB	C95800[(A)]	0.0

[(A)] Aluminum slightly below specification minimum.

(manganese bronze). The best performance came from nickel aluminum bronze (NAB; C95800).

The copper-nickel alloys have good resistance to erosion corrosion, and variations in their composition have been studied to optimize it. Bailey, in his description of the development of 90/10 copper-nickel, showed that the maximum resistance to erosion corrosion is found when the iron content is 1.5 to 2.5 wt% (Figure 7.7).[8] Efird looked at the synergistic effect of nickel and iron on the susceptibility to erosion corrosion.[9] He obtained similar results to Bailey for 90/10 copper-nickel, and he found the optimum iron content for 70/30 copper-nickel to be 0.8 to 1.2 wt%.

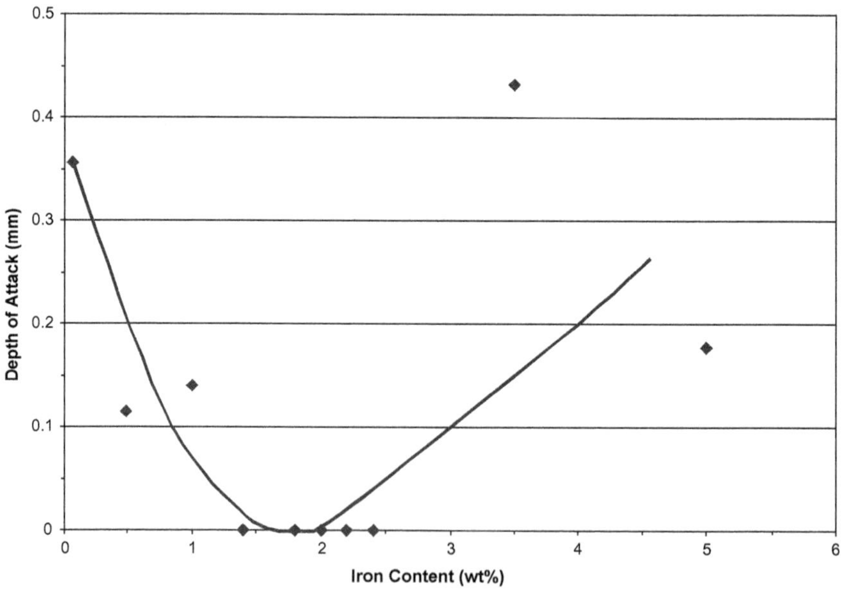

FIGURE 7.7 Effect of iron on the erosion corrosion resistance of 90/10 copper-nickel in the jet test at 5 m/s.[8]

These figures are interesting because the specifications (for example, ASTM B111[10]) for 90/10 and 70/30 copper-nickel (C70600 and C71500, respectively) are 1.0% to 2.0% and 0.4% to 1.0%, respectively. Where resistance to erosion corrosion is required, it is recommended that 90/10 copper-nickel have a minimum iron content of 1.5 wt% and that 70/30 copper-nickel have a minimum iron content of 0.7 wt%.

The iron content of 90/10 copper-nickel exceeds the solubility limit at room temperature. Bailey demonstrated that the iron must be in solid solution to confer the best erosion corrosion resistance.[8] This is achieved commercially by annealing in the range 850 to 900°C and then cooling rapidly to room temperature (for example, water quenching).[11] If 90/10 copper-nickel is heated such that the iron comes out of solution, then the erosion corrosion resistance decreases.[8]

In unpublished work at BNF, Guyoncourt and Harper aged copper-nickel alloys for 16 hours at 645°C and also investigated cold rolling, either before or after aging.[12] In jet impingement tests at 10°C and 9.5 m/s jet velocity, the depth of attack was 0.11 mm in both the solution-annealed and aged conditions for 90/10 copper-nickel. However, the area of attack was much greater on all the aged samples. The most interesting effect of aging was a fourfold increase in the general corrosion rate. With Alloy 722, there was no attack in the solution-annealed or aged conditions and only very slight attack (0.01 mm) on the cold-rolled samples. Alloy 722 contains more nickel and less iron than 90/10 copper-nickel, and so the fact that it does not lose corrosion resistance after thermal aging is not surprising.

Other research has also showed that aged 90/10 copper-nickel does not have a significantly different erosion corrosion resistance compared with solution-annealed material. Appel et al. conducted tests on pipes with seawater flows of 3.3–7.5 m/s.[13] The pipes had varying magnetic permeabilities from 1.0 (solution annealed) to 1.6 (fully aged). Not only was there no significant difference in erosion corrosion resistance between solution-annealed and aged pipes, but also, there was no real difference in the general corrosion rate. This is different from the results of Guyoncourt and Harper and suggests that the nature or distribution of the precipitates is critical to the subsequent performance.[12] Guyoncourt and Harper showed electron micrographs with the size and distribution of the precipitates, but Appel et al.'s paper showed only optical micrographs that do not properly characterize the precipitates.

Although laboratory tests on aged 90/10 copper-nickel do not show a large increase in susceptibility to erosion corrosion, service failures have occurred. Frick et al. reported failures of a 90/10 copper-nickel piping system in the North Sea after less than 12 months in service.[14] The water

FIGURE 7.8 Selective corrosion of parent 90/10 copper-nickel pipe because most of the iron was not in solution.

velocity did not exceed 2.2 m/s, well within the design limits for this alloy, and the iron content of the failed samples was 1.5% to 1.7%, within the specification for 90/10 copper-nickel. The high rates of corrosion that were seen in the vicinity of welds were attributed to much of the iron not being in solution. This was confirmed by magnetic permeability measurements. This is the reverse of what would be expected in welding, as the weld metal and heat-affected zone (HAZ) cool quickly and tend to hold the iron in solution. Corrosion of the welds suggests that they were welded with incorrect filler metal.

The author reported a similar series of failures from an oil platform in the North Sea.[15] The system was suffering a very high rate of general corrosion plus erosion corrosion after the bends, even though the water velocity was only ~3 m/s. The total iron content of the pipe was 1.7%. The clue to the cause of the problem was at the welds: not only was the 70/30 copper-nickel weld metal unattacked, but neither was the parent metal in the HAZ. Figure 7.8 shows a half-section of pipe, and the cut edge clearly shows the thicker metal in the weld and HAZ. This was because the iron was not in solution in the pipe, but the weld thermal cycles in the HAZ had re-solutionized the iron and restored the corrosion resistance. The weld metal was not affected because iron is more soluble in the 70/30 copper-nickel filler metal. Determination of the magnetic permeability with a simple device, such as the Severn† magnetic gauge, showed the parent metal to have a value of >1.2, whereas the weld metal and HAZ were <1.05. It is suspected that the parent metal suffered severe general corrosion because the water velocity was close to the recommended limit, and as the weld and HAZ became increasingly proud, they increased turbulence downstream sufficiently to cause erosion corrosion. It was recommended by Yorkshire Imperial Metals that 90/10 copper-nickel should have a magnetic permeability of ≤1.1 to give optimum corrosion resistance.

Kirk carried out an extensive review of the effect of alloying additions to copper-nickels on their corrosion resistance.[16] His review includes a large quantity of data from the LaQue laboratory that were previously unpublished and includes other forms of attack in addition to erosion corrosion. Only the highlights are discussed here; anyone requiring additional data is referred to the original report.

The work of Efird, on the effect of nickel and iron and the way that they are interrelated, was described previously.[9] Kirk pointed out that several workers have shown that manganese acts in a similar way to iron, but at only about 20% of the efficiency; that is, 1% Mn is equivalent to 0.2% Fe.[16] This means that manganese supplements the benefit of iron additions, particularly when the iron is on the low side. Tests have shown that 70/30 copper-nickel alloys with high iron contents (>2%) have good resistance to erosion corrosion but increased susceptibility to localized attack such as crevice corrosion. Where components are to be used almost continuously with high water velocities, this can be of benefit. Kirk cites the use of a cast 70/30 pump impeller with ~5% iron that gave excellent performance for many years.[16]

Figure 7.9 shows the effects of various alloying additions on 90/10 copper-nickel at 2.4 and 4.6 m/s velocity. The tests lasted 6 years, which gives an indication of long-term performance. Figure 7.10 shows similar data for 70/30 copper-nickel alloys at the same velocities.

Figure 7.9 shows that increasing the iron content in 90/10 copper-nickel to 2% had no significant effect at 2.4 m/s and was detrimental at 4.6 m/s. This is surprising, and one wonders if all the iron was in solution in the 2% iron alloy. The addition of 2% aluminum was beneficial, particularly at the lower velocity, but only in the aged condition. Additions of niobium gave a small improvement at the lower velocity, but only in the solution-annealed condition. The addition of 0.5% molybdenum was detrimental at both velocities.

† Trade name.

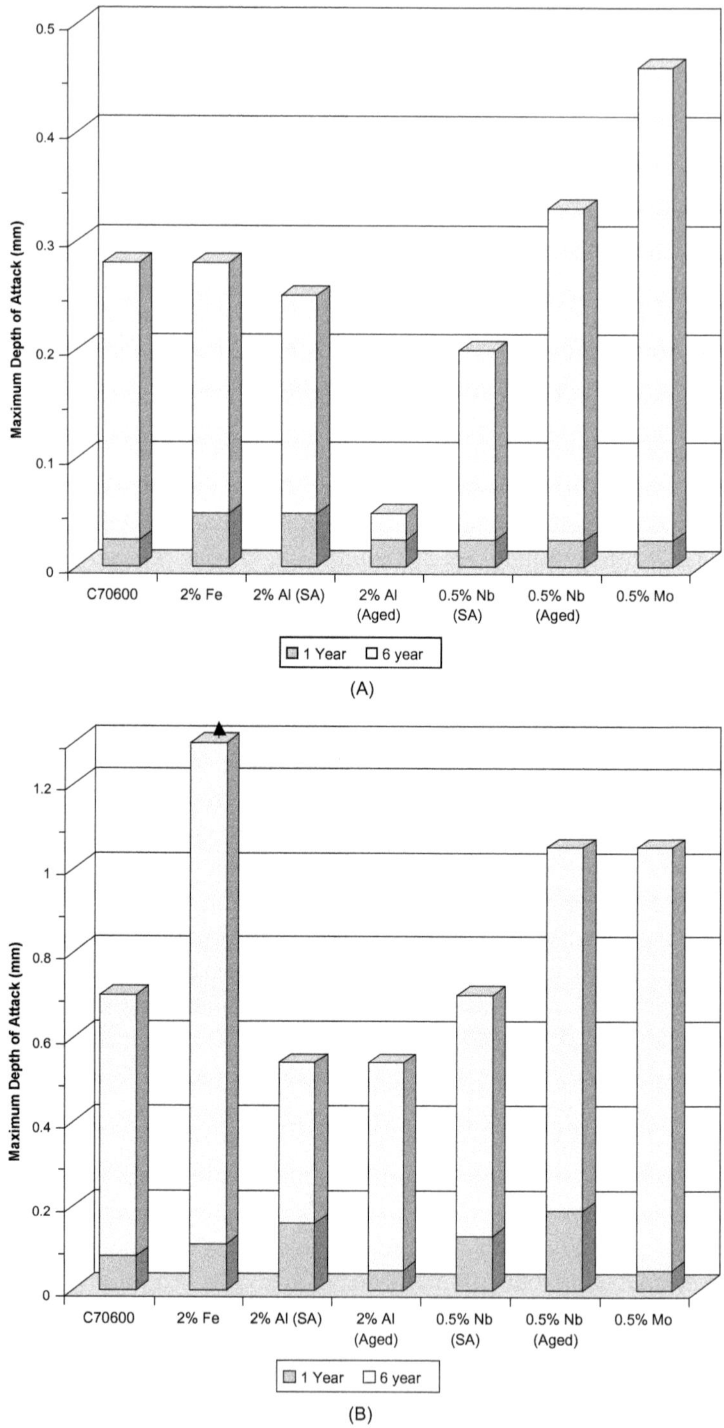

FIGURE 7.9 Effect of alloying additions on the erosion corrosion of 90/10 copper-nickel: (A) 2.4 m/s; (B) 4.6 m/s.[16]

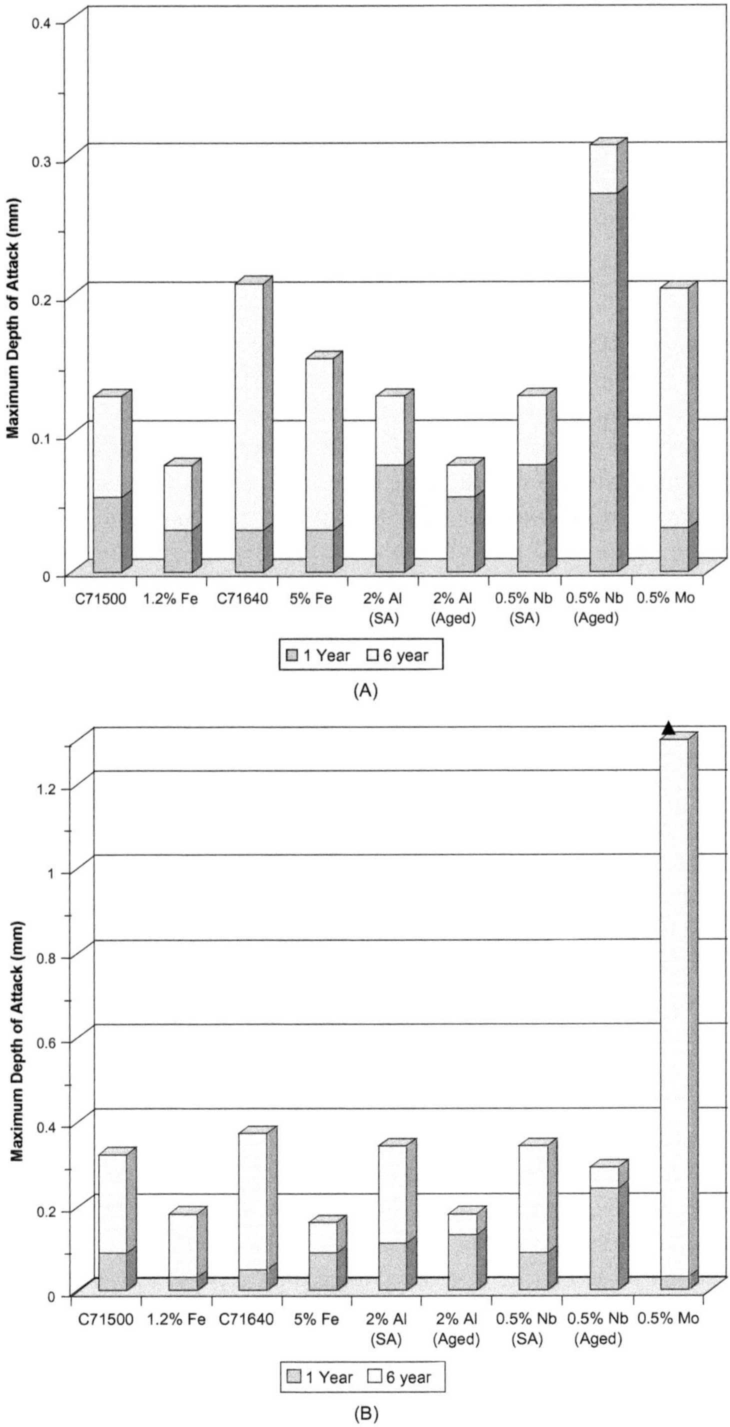

FIGURE 7.10 Effect of alloying additions on the erosion corrosion of 70/30 copper-nickel: (A) 2.4 m/s; (B) 4.6 m/s.[16]

TABLE 7.5 Strength of 70/30 copper-nickel with niobium and silicon additions[18–19]

Alloy	0.2% Proof stress (MPa)	Tensile strength (MPa)	Elongation (%)
C71500 (sol. ann.)	145	393	43
0.5% Si + 1.5% Nb (aged)	398	706	24

Figure 7.10 shows the effect of increasing iron content (0.5% to 5%) on the performance of 70/30 copper-nickel. At 2.4 m/s, the best performance was achieved with 1.2% iron, but at 4.6 m/s, the 5% iron alloy offered the best performance. As for 90/10 copper-nickel, the addition of 2% aluminum gave a small increase in performance, particularly at 4.6 m/s, but only in the aged condition. The addition of 0.5% niobium produced no significant change in corrosion resistance in either the solution-annealed or aged condition. The addition of 0.5% molybdenum decreased the erosion corrosion resistance at both velocities.

Additions to 70/30 copper-nickel to improve both its strength and hot ductility were investigated at the BNF Metals Research Association during the 1960s.[17–18] The most attractive additions were niobium and silicon. Additions of silicon alone gave poor ductility and weldability; the further addition of niobium gave high strength, toughness, and weldability. The optimum composition was an alloy of Cu/30Ni/0.5Si/1.3Nb/1.2Fe/1Mn. Figure 7.11 shows the depth of attack in jet tests at 10.7 m/s, with small additions of hydrogen sulfide (H_2S) to increase the aggressivity. The results show that the alloy with niobium and silicon additions in the aged condition gave similar performance compared with the standard alloy (C71500). In clean seawater, there was no difference between the two alloys. In seawater polluted with large quantities of H_2S, the modified alloy was slightly inferior. The mechanical properties of the wrought alloy in the aged condition compared with those of C71500 are shown in Table 7.5. These indicate the much greater strength of the modified alloy, similar to that of NAB.

Townsend also investigated the effect of additions of niobium and silicon to 90/10 copper-nickel.[19] The addition of 0.5% silicon and 0.3% niobium gave a large increase in strength in the aged condition, while not significantly affecting the resistance to erosion corrosion. He also investigated the effect of adding 2% aluminum and found, like Kirk, that it gave good results in the aged condition but increased the susceptibility to crevice corrosion.[16]

Townsend also investigated the effects of zirconium additions to 70/30 copper-nickel.[17] With 0.25% zirconium, the hot ductility improved, and

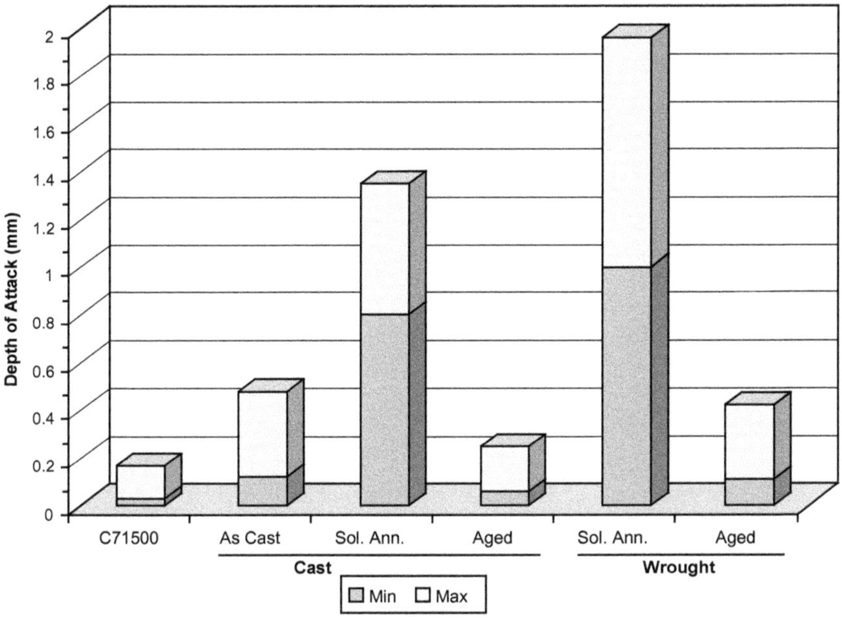

FIGURE 7.11 Depth of attack on 70/30 copper-nickel with niobium and silicon in jet test at 10.7 m/s additions.[18]

there was no change in the resistance to erosion corrosion in clean seawater. However, in polluted seawater, the attack was much worse on the alloy with zirconium additions compared with conventional 70/30 copper-nickel.

One element that has a big effect on erosion corrosion when it is added to copper-nickel alloys is chromium. Kirk summarized the test results from a number of studies.[16] The results from all the studies were two alloys, one a 70/30 copper-nickel with 3% chromium (C71900) and a Cu/15Ni/0.8Fe/0.7Cr/0.6Mn alloy (C72200). Both these alloys can be cold worked to give a higher strength than C71500, while the cast version of the 70/30 alloy with chromium additions is also stronger in the as-cast condition. In jet impingement tests, Anderson and Efird compared some chromium-containing alloys with conventional copper-nickel alloys, as shown in Figure 7.12.[20] The results show the significant improvement in resistance to erosion corrosion with chromium additions, particularly to the 70/30 copper-nickel alloys. IN768 is a cast copper-nickel alloy with the nominal composition Cu/30Ni/1.6Cr/0.7Fe/0.6Mn.

A particular problem can occur in heat exchanger tubes when a foreign body becomes lodged, causing local high turbulence. This can increase local flow velocities by up to a factor of 5. Polan et al. reported the results

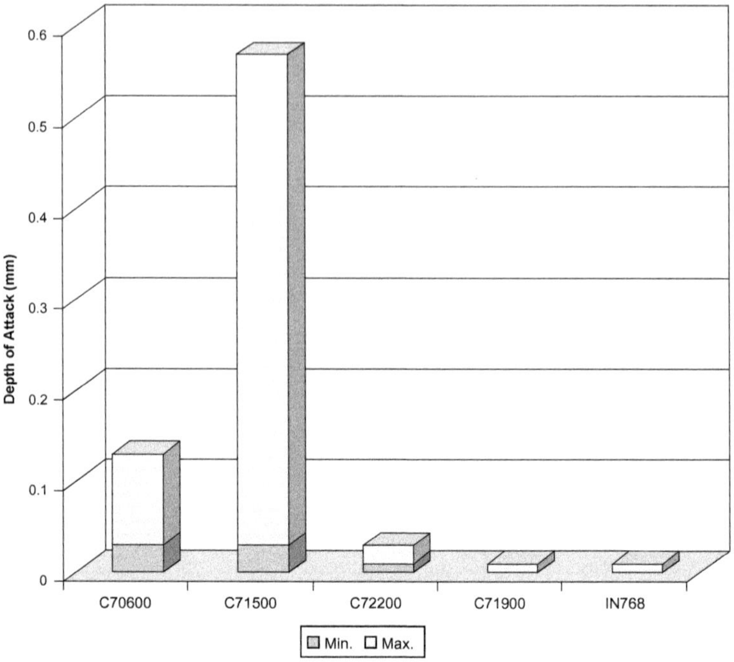

(A)

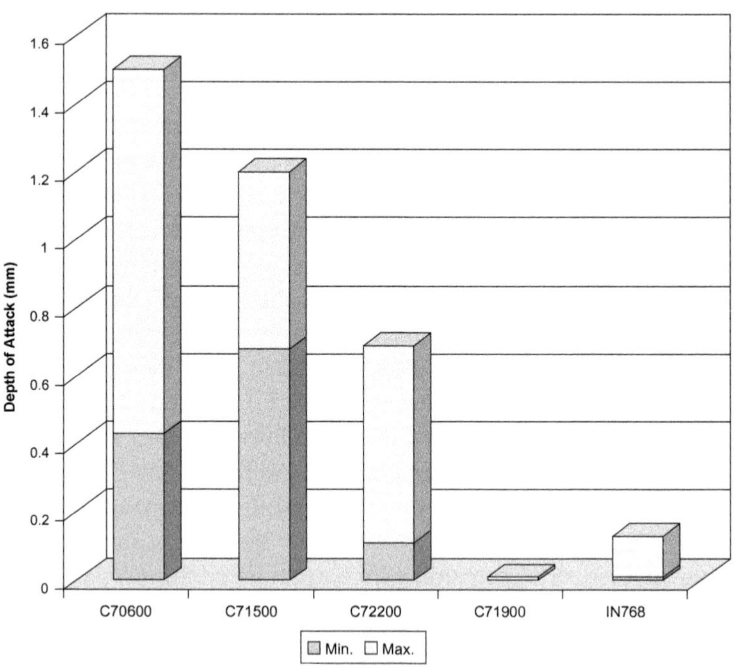

(B)

FIGURE 7.12 Depth of attack of some copper-nickel alloys containing chromium at a jet velocity of (A) 6.8 m/s and (B) 15.3 m/s.[20]

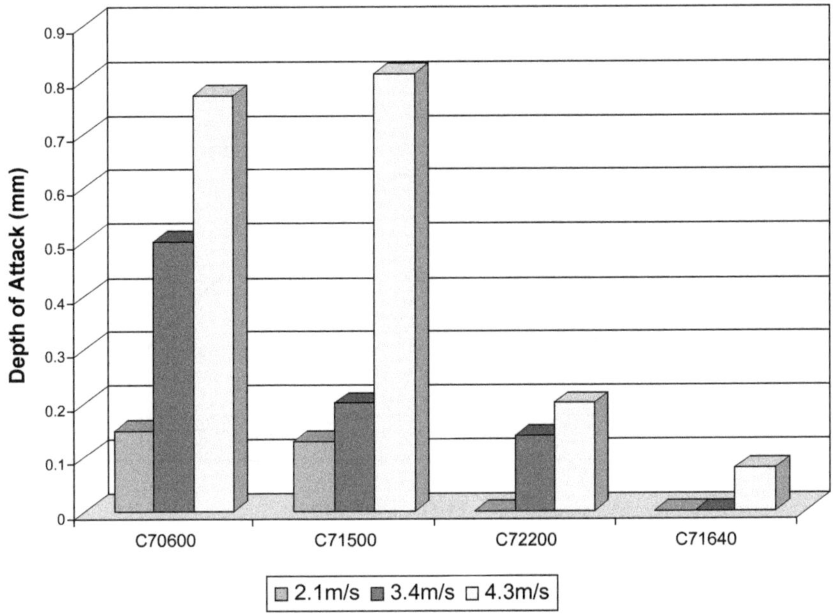

FIGURE 7.13 Maximum depth of attack around a blockage of various copper-nickel alloys after 12 months in seawater.[21]

of tests with artificial blockages.[21] Figure 7.13 shows the results at three different nominal flow velocities. Although Alloy 722 with chromium was better than 90/10 or 70/30 copper-nickel, the best performance came from C71640 (66/30/2/2 Cu-Ni-Fe-Mn). Note, however, that there was some crevice corrosion at the blockage with both Alloy 722 and C71640, while there was none with 90/10 or 70/30 copper-nickel.

Kirk reported the results of tests of the 70/30 copper-nickel alloy with 3% chromium and additions of beryllium.[16] At a level of 0.15%, the rate of attack increased, but at levels of 0.2% and 0.5%, it decreased. However, beryllium additions also made welds susceptible to stress corrosion cracking.

When zinc is added to copper-nickel in substantial quantities, the nickel silvers are produced. These alloys have very poor resistance to erosion corrosion in seawater compared with 90/10 and 70/30 copper-nickel.[16] Additions of 0.9% aluminum to nickel silvers improved the resistance to erosion corrosion, whereas manganese additions (3% to 7%) lowered the resistance and also made the alloy susceptible to stress corrosion cracking.

In Japan, during the 1960s, there were problems in power stations, with accelerated corrosion of copper alloy condenser tubes due to polluted water containing sulfides. To combat this, AP bronze was developed, a

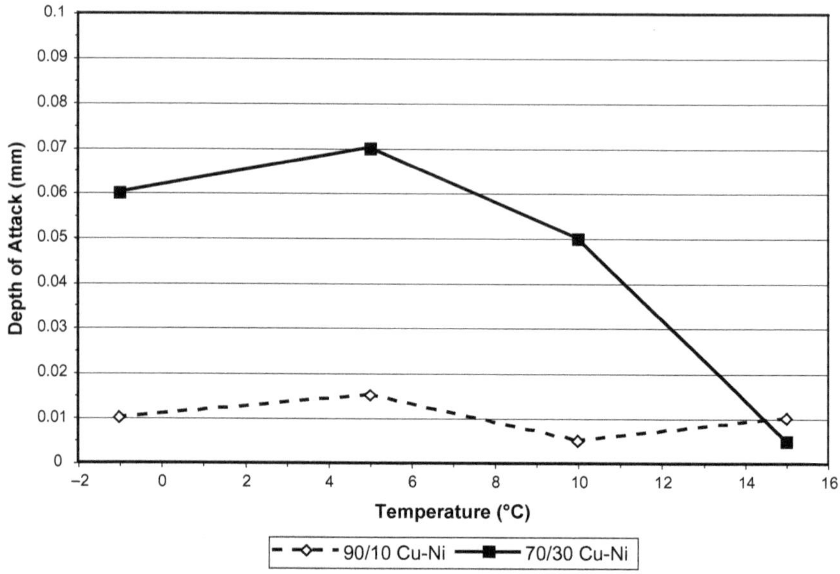

FIGURE 7.14 Depth of attack in jet test at various seawater temperatures.[24]

Cu/8Sn/1Al/0.1Si alloy, which also has good resistance to corrosion in clean seawater. Nosetani et al. described the results of model condenser tests, and in clean seawater, at normal flow velocities (2.0–2.7m/s), the corrosion rates of aluminum brass and AP bronze were similar.[22]

In an investigation of the microstructure of NAB, Michels and Kain carried out jet impingement tests on cast NAB (C95800) with aluminum contents from 8.1% to 9.4% and in a range of heat treatment conditions.[23] Despite the range of microstructures produced, the depth of attack at both 4.6 m/s and 9.1 m/s jet velocity was very similar in all conditions. This was in contrast to the effect of microstructure on localized corrosion, as discussed in Chapter 9.

Besides composition, there are a number of physical factors that can affect the susceptibility of copper alloys to erosion corrosion. One of these is temperature. Mills reported the effect of seawater temperature from −1 to 15°C in the jet test at 9 m/s.[24] The results, shown in Figure 7.14, show an increase of attack on 70/30 copper-nickel below 10°C, whereas 90/10 copper-nickel was unaffected. This has been attributed to the slower formation of iron-rich films at lower temperatures and was discussed in detail in Chapter 3.

Campbell reported jet test results for 70/30 copper-nickel and NAB (Table 7.6), which show deeper attack at 10°C compared with 15°C for both alloys.[25]

TABLE 7.6 Erosion corrosion in jet impingement test for NAB and 70/30 copper-nickel vs. seawater temperature[25]

Temperature (°C)	Alloy	Attack at jet Diameter (mm)	Depth (mm)
10	Cast NAB (C95800)	10	0.12
	70/30 Cu-Ni (C71500)	14	0.10
15	Cast NAB (C95800)	0	0
	70/30 Cu-Ni (C71500)	2.1	0.05

The potential of 90/10 copper-nickel in natural seawater is about −200 mV SCE (saturated calomel electrode). However, coupling to more noble metals can increase the potential. Rowlands and Angell investigated depth of attack as a function of the distance from the center of an impinging jet.[26] They found that at 20 m/s, little attack occurred at −100 mV Ag/AgCl (silver/silver chloride), but deep attack occurred as the potential increased up to 0 mV (Figure 7.15).

The effect of coupling 90/10 copper-nickel to other copper alloys when it is undergoing erosion corrosion was investigated by LaQue

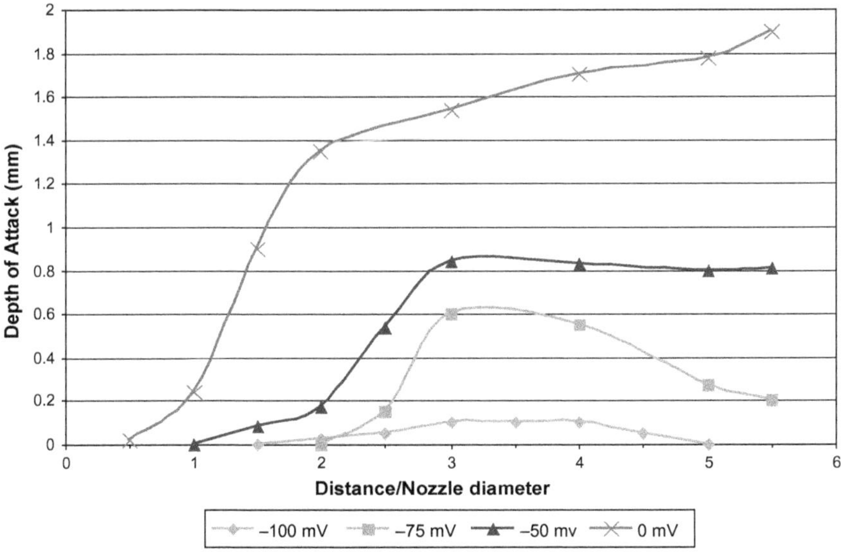

FIGURE 7.15 Depth of attack of 90/10 copper-nickel as a function of potential and distance from the center of an impinging jet after 1,000 hours.[26]

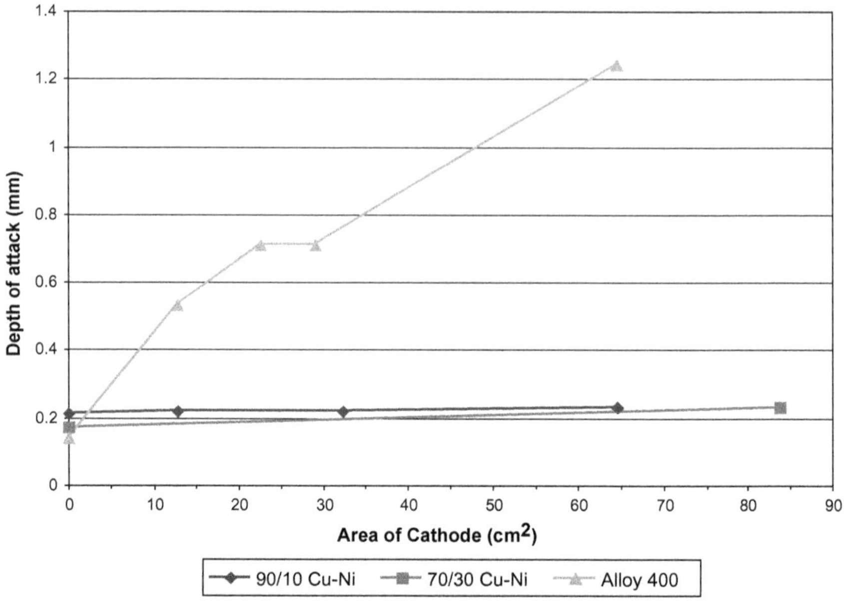

FIGURE 7.16 Effect of copper alloy cathodes on the depth of erosion corrosion of 90/10 copper-nickel.[27]

(Figure 7.16).[27] The results show that coupling a 90/10 copper-nickel anode (high velocity) to large cathodic areas of 90/10 copper-nickel (low velocity) had no effect, and coupling to 70/30 copper-nickel only increased the depth of attack slightly. However, coupling 90/10 copper-nickel to Alloy 400, which has a significantly more electropositive potential and is also an efficient cathode, produced a significant increase in attack as the cathodic area was increased.

Campbell and Carter investigated the erosion corrosion of cast manganese bronze propellers on marine vessels.[28] They found that a high velocity jet of seawater alone did not induce much erosion corrosion. By the use of dyes and high-speed photography, they demonstrated that air bubbles within a certain size range induced additional turbulence into the water jet. The increased turbulence produced much greater attack on manganese bronze, typical of that seen in service.

Besides external effects, sometimes corrosion products themselves can influence erosion corrosion. When copper alloys are first immersed, they often corrode quite rapidly until a protective film forms (see Chapter 3). Cheung and Thomas reported erosion corrosion studies of 90/10 copper-nickel in recirculating, synthetic seawater.[29] In the first tests, the corrosion

TABLE 7.7 Surface shear stress and critical velocity for erosion corrosion of some copper alloys at 20°C[31]

Alloy	Critical shear stress (MPa)	Critical velocity (m/s)
Copper	9.6	1.3
Al-brass	19.2	2.4
90/10 Cu-Ni	43.1	4.2
70/30 Cu-Ni	47.9	4.4
Alloy 722	296.9	10.9

rates were very low, much lower than any published data. When filters were fitted to remove the high levels of copper corrosion products in the seawater, the corrosion rates increased to more normal levels. Such filters were standard equipment on the BNF jet test when used with recirculated seawater.

The effect of initial surface condition was examined by Sato et al., who tested aluminum brass condenser tubes finished by bright annealing (the standard finish), annealing in air, bright annealing after pickling, and pickling after bright annealing.[30] Although electrochemical tests showed differences in the films on these surfaces, after a few days in seawater at either 2 m/s or 7.7 m/s, there was no discernable difference between the corrosion rates in any of the tubes. Instead, corrosion rates were greatly affected by both entrained solids in the water and sulfide pollution.

Much of the testing of copper alloys for resistance to erosion corrosion has been conducted using the jet test, where a high-velocity water jet impinges on the sample at 90 degrees.[4] However, in many applications, the flow is more or less parallel to the surface such as in a pipe. Efird conducted parallel flow tests on several copper alloys to determine the critical shear stress at which the protective film broke down and accelerated metal loss occurred.[31] Efird found that the shear stress was not only a function of the bulk velocity, but also of the water temperature.[31] Table 7.7 shows the critical velocities and shear stresses for several common copper alloys, normalized to 20°C. The results show that the critical velocity and shear stress increase with alloy content. Most notable is the substantial increase in critical shear stress for Alloy 722. It should be noted that this only applies to parallel flow and takes no account of local turbulence produced by bends or surface irregularities, which can increase local velocities.

Efird's tests were conducted in small-diameter pipes, and it is often argued that flow is smoother and more uniform in larger-diameter pipes.[31] Kirk carried out tests with NPS 4 pipe in 90/10 copper-nickel at velocities from 2.4 to 7.3 m/s.[32] The system included straight pipes and long radius bends, all welded together. After 16 weeks, there was no significant erosion

TABLE 7.8 Erosion corrosion rates of various alloys in seawater at very high jet velocities.

Alloy	Average temperature (°C)	Velocity (m/s)	Mean corrosion rate (mm/y)
Grey cast iron[(A)]	28	42	13.3
Aust. cast iron	19	39	2.19
90/10 Cu-Ni	9	39.6	0.56
	16	40	0.66
	27	39	0.84
70/30 Cu-Ni	9	42	1.17
	16	40.5	1.83
	27	40.5	2.67
66/30/2/2 Cu-Ni-Fe-Mn	16	40.5	2.34
70/30 Cu-Ni (5% Fe)	11	40.5	0.36
LG2 gunmetal	25	40.5	1.35
Admiralty gunmetal	20	38	0.76

[(A)] Ten days only.

corrosion at any velocity, except downstream of welds. The maximum depth of attack in these regions was 0.23–0.25 mm at all velocities. Clearly the extra turbulence caused by the weld bead was sufficient to rupture the protective film. The results showed that the critical shear stress was not passed in a larger-diameter pipe, except where local turbulence was increased.

Some studies have looked at very high water velocities because even when the film has been damaged, metal loss rates may not be very high. The LaQue Laboratory carried out very high-velocity jet tests for 30 days on a range of alloys. The jet velocity was mostly around 40 m/s, although a few tests were conducted at lower velocities. Table 7.8 shows a summary of the data for copper alloys compared with austenitic cast iron and gray cast iron. The results for the copper-nickel alloys show increasing metal loss with temperature. The corrosion rates for 90/10 copper-nickel were significantly lower than those for 70/30 copper-nickel, suggesting that at very high velocities, the iron content is more important than the nickel content. Certainly the corrosion rate for 70/30 copper-nickel with 5% iron was very low. The rate for C71640 (Cu/30Ni/2Fe/2Mn) was higher than that for the other copper alloys. This is unusual as this alloy has exceptionally good resistance to erosion corrosion (see the previous discussion). It is possible that the iron was not all in solution in the samples used for these tests. The results for the gunmetals show decreasing metal loss as the tin

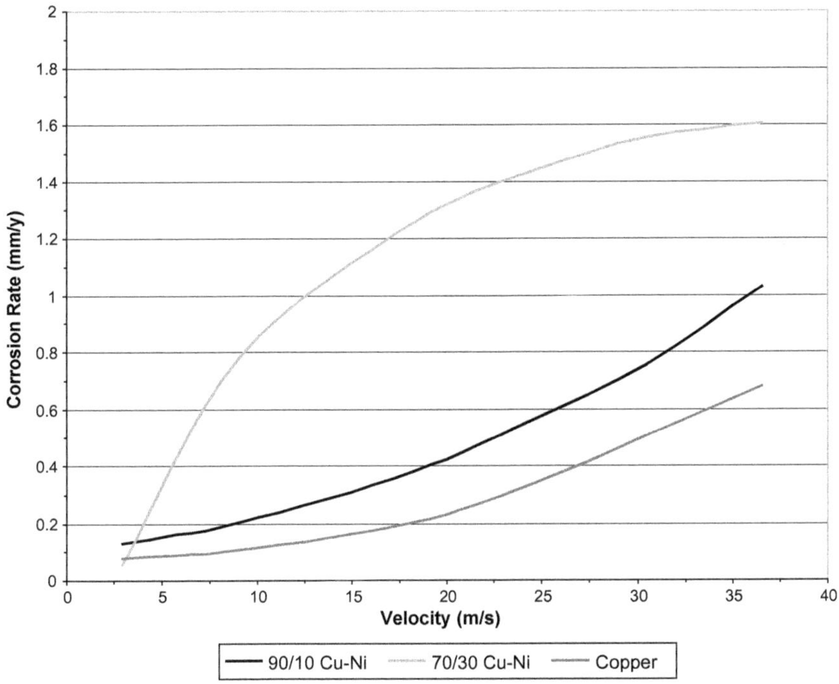

FIGURE 7.17 Corrosion rate of some copper alloys in high-velocity tests in seawater.[32]

content of the alloy was increased, as seen at lower velocities (see the preceding discussion).

A study was carried out comparing high-velocity erosion corrosion test results using a multivelocity parallel flow rig, a rotating disc test, and the BNF jet test.[33] In the multivelocity rig, velocities of 2.9–36.6 m/s were investigated. Some of the results are shown in Figure 7.17. The surprising result is the low weight loss of copper compared with the copper-nickel alloys. The reason for this is the degree of protection conferred by the film. The film on pure copper was poorly protective, and the attack was spread more or less uniformly over the sample, whereas the film on 70/30 copper-nickel was very protective and only broke down in a few places, where attack was deep. The results for 90/10 copper-nickel were intermediate between these two extremes. The importance of the protectiveness of the film, as well as the velocity, was brought out by Cotton in his discussion of the preceding work.[34] The author also described the change from narrow, deep attack to shallow, broad attack, as the film becomes less protective.[35]

In some unpublished work, the U.S. Navy looked at high flows in an NPS 2 piping system with 90/10 copper-nickel piping and bends and bronze valves and flanges. The flow velocities were 2.4, 4.6, and 9.1 m/s. Where orifice plates produced cavitation downstream, there was severe attack, particularly at the highest velocity. Where noncavitating orifices were fitted, more normal erosion corrosion was seen downstream. At 9.1 m/s, there was heavy roughening in turbulent areas and light to moderate roughening elsewhere. Leakage occurred in the most turbulent area within 6 months. At 4.6 m/s, there were no failures over the 8 months of testing, but there was heavy roughening in the most turbulent areas. Attack was worst on or after bends. There was also attack of the bronze valves, but this did not lead to failure because of the higher wall thickness. However, the depth of attack was greater on the bronze than on the 90/10 copper-nickel. The attack on the piping at 2.4 m/s was more normal, with no significant attack even in the turbulent zones. The conclusion was that high water velocities cannot be sustained for extended periods in copper alloy piping systems without leaks due to erosion corrosion.

However, there are occasions when copper alloys may experience intermittent high velocities. This was investigated by Melton using a small anode experiencing a high-velocity water jet coupled to a larger cathode in gently moving seawater, both of 90/10 copper-nickel.[36] The currents between the anode and cathode, measured on freshly abraded samples over a range of velocities from 4 to 10 m/s, stabilized at levels equivalent to a metal loss of 1.0–1.2 mm/y. However, for samples that were prefilmed in slow-moving water, the current between the anode and cathode was effectively zero, even with a jet velocity of 19 m/s. These tests only lasted a few days, but the results show that once a protective film forms on 90/10 copper-nickel, it can survive high seawater velocities, at least for a few days.

This is borne out by the experience with 90/10 copper-nickel in offshore firewater systems. These generally operate at velocities up to 10 m/s, but they are only tested for about 1 hour per week and remain stagnant for the rest of the time. There have been no failures due to erosion corrosion reported, either from cold waters, such as in the North Sea, or from tropical waters, where pipe wall temperatures may reach 60 or 70°C under stagnant conditions when exposed to direct sunlight.

7.1.3 Iron/ Ferrous Sulfate

Throughout the first half of the 20th century, admiralty brass (C44300) and aluminum brass (C68700) were the most commonly used alloys for

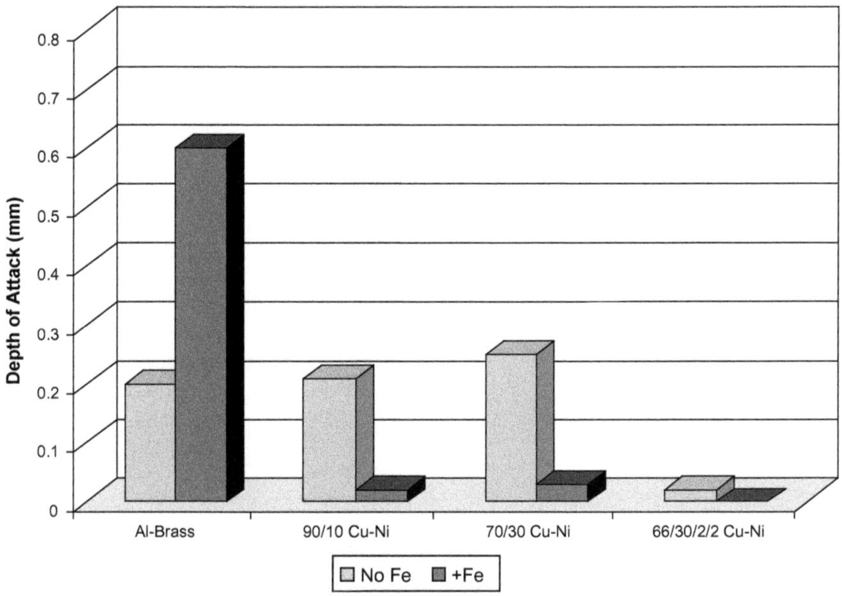

FIGURE 7.18 Effect of iron dosing (1 mg/L for 1 h/d) on the depth of attack at 9 m/s.[35]

tubes in seawater-cooled heat exchangers. In the 1950s, there was a move to replace bare carbon steel and cast iron components with more corrosion-resistant alloys or to apply coatings to increase reliability. Following this, there was an increase in the number of heat exchanger tube failures due to erosion corrosion. In 1961, Bostwick described how daily additions of ferrous sulfate to the cooling water replaced the iron-rich films that had previously been formed by corroding steel, with a large reduction in the number of aluminum brass tube failures.[37] Shortly after, Lockhart described a similar success story at a U.K. power station.[38] Since that time, additions of ferrous sulfate solution (typically 1 mg/L Fe^{2+} for 1 hour/day) have been made at many plants to solve heat exchanger corrosion problems.

The use of iron additions to combat corrosion of copper alloy heat exchanger tubes was surveyed among U.K. power stations as well as the published literature to determine when ferrous sulfate dosing would be effective and for which alloys.[39] Ferrous sulfate was clearly efficacious on aluminum brass at normal flow velocities (2 m/s), and a number of laboratories conducted tests at higher velocities and included other alloys. Figure 7.18 shows the effect of ferrous sulfate dosing (1 mg/L Fe^{2+} for 1 hour/day) on a number of copper alloys tested in the Campbell condenser tube apparatus at a jet velocity of 9 m/s.[35] The results show that even at this high

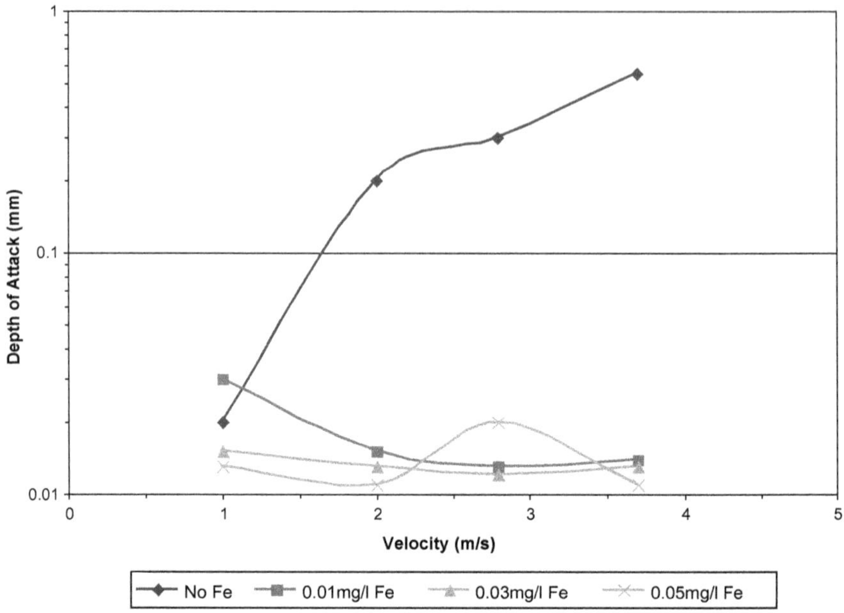

FIGURE 7.19 Effect of continuous iron dosing rate and velocity on the corrosion of aluminum brass.[40]

velocity, ferrous sulfate dosing significantly reduced the depth of attack on the copper-nickel alloys, but not on aluminum brass. The reason for the anomalous result with aluminum brass is the nature of the protective film. At 9 m/s, ferrous sulfate dosing produces a more protective film, but it cannot prevent attack under the jet. Hence, the attack is narrow and deep. Without ferrous sulfate dosing, the film is not so protective, and the attack is more widespread and shallow.[34]

Initially, iron was added as ferrous sulfate solution on an intermittent basis. However, iron can also be added continuously by the use of electrically stimulated soft iron or mild steel anodes. Iron added continuously at 0.04 mg/L Fe^{2+} is roughly equivalent to 1 mg/L Fe^{2+} for 1 hour/day. Sato et al. investigated the effect of a range of continuous iron dosing levels on the corrosion of aluminum brass tubes at various velocities.[40] The results (Figure 7.19) show that iron dosing levels as low as 0.03 mg/L Fe^{2+}reduced the depth of attack at velocities up to 3.7 m/s. Sato also showed that ferrous sulfate dosing reduced the attack around lodged obstructions in aluminum brass heat exchanger tubes, but it did not totally stop it.[41]

Nosetani et al. showed that the duration of the test also affected the perceived benefit of ferrous sulfate additions.[42] Figure 7.20 shows the

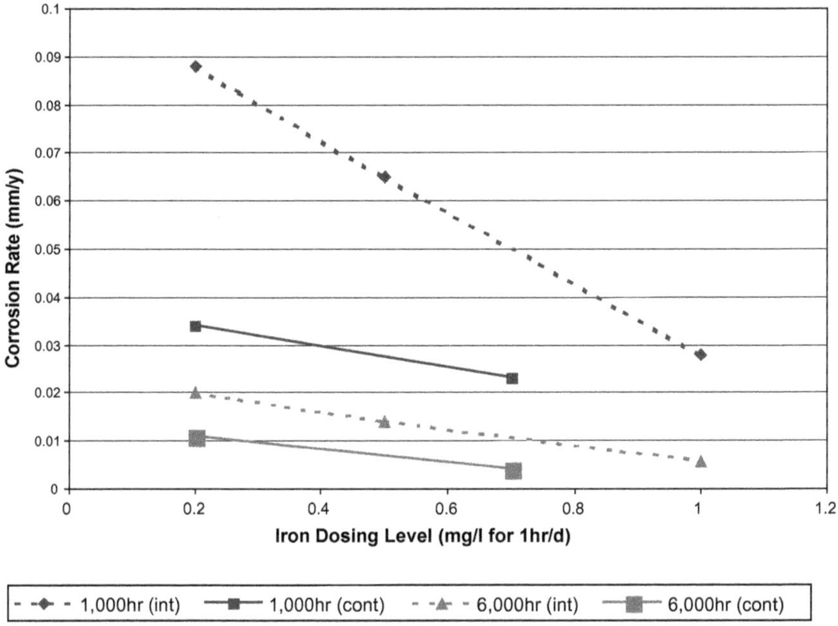

FIGURE 7.20 Effect of time and manner of iron dosing on corrosion of aluminum brass at 2 m/s.[41]

corrosion rate of aluminum brass tubes at a flow rate of 2 m/s with both continuous and intermittent iron dosing. The continuous iron dosing levels have been converted to the equivalent amount of iron dosed for 1 hour/day. The results show that the corrosion rate was much lower at all dosing levels after 6,000 hours, compared with 1,000 hours' exposure. This is because the protective films become more efficient as they grow and growth rates are slow. The results also show that continuous low-level dosing with iron is more protective than the equivalent dosed intermittently. The same effect as 1 mg/L Fe^{2+} for 1 hour/day can be achieved with approximately half that equivalent level dosed continuously.

Although intended originally for brass tubes, ferrous sulfate has occasionally been used successfully to solve corrosion problems with 90/10 and 70/30 copper-nickel. The original corrosion problem that iron additions were used to overcome was erosion corrosion (impingement attack), but they have also been used to try to solve other corrosion problems with varying degrees of success. In Japan, ferrous sulfate dosing solved problems of localized attack of aluminum brass tubes that had a cathodic film of manganese dioxide (MnO_2) on them.[43] This film was produced by the oxidation of manganese in solution in the seawater by chlorine, added to control fouling.

Ferrous sulfate dosing has also solved corrosion problems caused by ammonia (NH_3) pollution. The author used ferrous sulfate dosing for 2 months at the start-up of a multistage flash (MSF) desalination plant to form more protective films on 90/10 copper-nickel tubes. The water was intermittently polluted with ammonia, which had caused low-temperature hot spot attack in the heat exchangers when the first unit was started. These were retubed, and ferrous sulfate dosing was started from new at a level of 1 ppm Fe^{2+} for 1 hour every day. The plant has now been in operation since 1994 with no further failures. At another MSF plant in the Middle East, the ammonia pollution was almost continuous because the water intake was near the outfall of a fertilizer plant. Daily ferrous sulfate dosing throughout its life was recommended for this plant as it was tubed with aluminum brass. There had been failures of 90/10 copper-nickel tubes in an earlier plant on the same site. The plant has now been in operation since 1998 with no problems.

Ferrous sulfate does not work on aluminum brass tubes in water polluted with H_2S, unless protective films can first be formed in clean water.[44] Hack, Gudas, and their colleagues investigated the effect of iron additions on the corrosion of 90/10 and 70/30 copper-nickel in polluted seawater.[45-46] They found that intermittent ferrous sulfate dosing had little effect, but continuous iron additions were more efficient, particularly with 70/30 copper-nickel. The iron additions worked best when the sulfide pollution was at a low continuous level. A U.K. power station, operating in a tidal estuary, used ferrous sulfate dosing to prevent corrosion of 70/30 copper-nickel tubes, where H_2S was an intermittent pollutant at certain states of the tide.[39]

It is well documented that chlorine/hypochlorite dosing to control fouling interferes with the efficacy of ferrous sulfate dosing.[35,47] This is because of the way in which ferrous sulfate acts. When added to water, it forms a colloid of FeOOH, whose charge increases the smaller the particle size. The colloid is attracted to the tube wall by the local electric field, so the greater the charge on the colloid, the more will be attracted to the surface. Chlorine causes very rapid growth of the colloid so that the charge is much weaker and is less likely to be attracted to the tube wall. Figure 7.21 shows the increasing pit depth with chlorine concentration for 90/10 copper-nickel under turbulent, high-velocity water conditions.[35] It also shows that with no chlorine, ferrous sulfate additions suppressed attack, but this effect was cancelled when there was more than 0.5 mg/L chlorine in the water. The recommendation is that if the chlorination is continuous, it should be

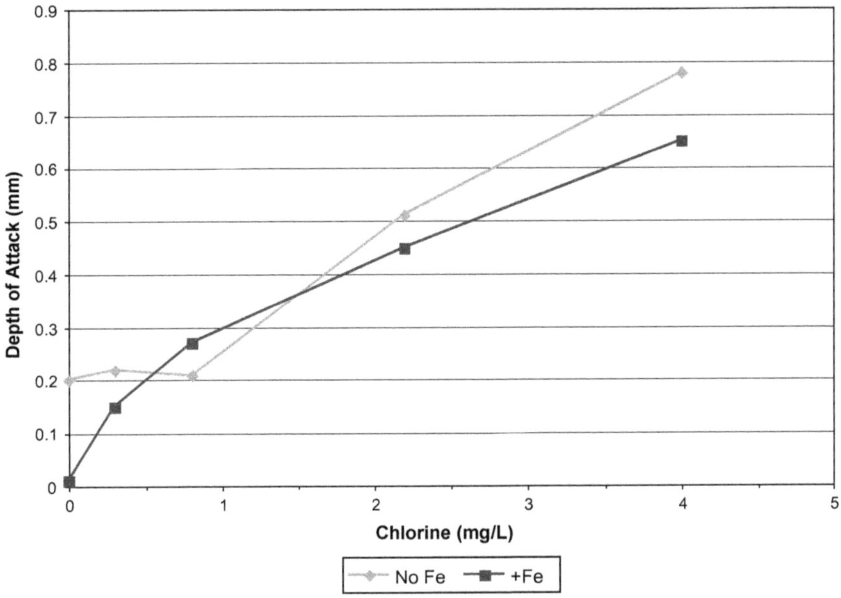

FIGURE 7.21 Effect of ferrous sulfate and chlorine on the depth of attack on 90/10 Cu-Ni at 9 m/s.[35]

turned off while ferrous sulfate dosing is taking place. If the chlorination is intermittent, then it should be staggered so that it does not occur during ferrous sulfate dosing.

7.1.4 Effect of Chlorine

Most cooling waters contain living organisms that can colonize metals. In a cooling system, this can lead to blockage if macrofouling (weed and shellfish) is not controlled. Copper and its alloys are very resistant to macrofouling (see Chapter 13), but microfouling by slimes and bacteria can impede the efficiency of heat exchangers. One of the most common methods of fouling control is the use of chlorine additions. These may be added as hypochlorite solution, or chlorine may be generated electrolytically from brine or seawater. Chlorine is a powerful oxidizing agent and could have a detrimental effect on copper alloys.

Sato carried out tests on a range of copper alloy heat exchanger tube materials in a Japanese power station.[48] The water flow rate was nominally 2 m/s, and a range of chlorine levels were investigated over several months. The results, given in Figure 7.22, show that the corrosion rate increased as

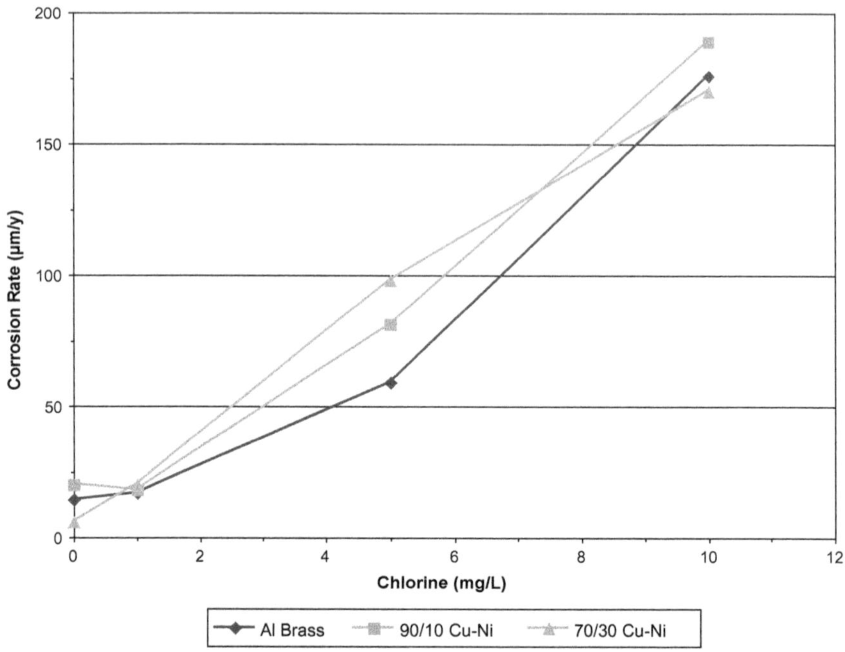

FIGURE 7.22 Effect of chlorine on copper alloy condenser alloys at 2 m/s.[48]

the chlorine concentration increased. However, what is of more concern is localized corrosion caused by chlorine.

In the 1970s, failures began occurring in the aluminum brass heat exchanger tubes of several oil supertankers. The attack took the form of erosion corrosion at scratch lines (Figure 7.23). All the tankers with failures had continuous chlorine dosing equipment as well as ferrous sulfate dosing. Tankers with only ferrous sulfate dosing were not experiencing problems. BNF undertook a program of research to investigate the problem.[35] The initial results showed that chlorine could prevent the formation of the normal protective film, particularly after local damage, leading to accelerated local attack (see Chapter 3). Figure 7.21 shows the results of tests on 90/10 copper-nickel, whereas Figure 7.24 shows the results for aluminum brass and 66/30/2/2 Cu-Ni-Fe-Mn. High levels of chlorine accelerated attack, particularly for aluminum brass, whereas the 66/30/2/2 Cu-Ni-Fe-Mn alloy was hardly affected. Figure 7.24A shows that the attack on tubes also dosed with ferrous sulfate was deeper than on those with chlorine alone. This effect is the same as that discussed for tests on copper alloys at very high velocities, where attack with a more protective film is localized, but very

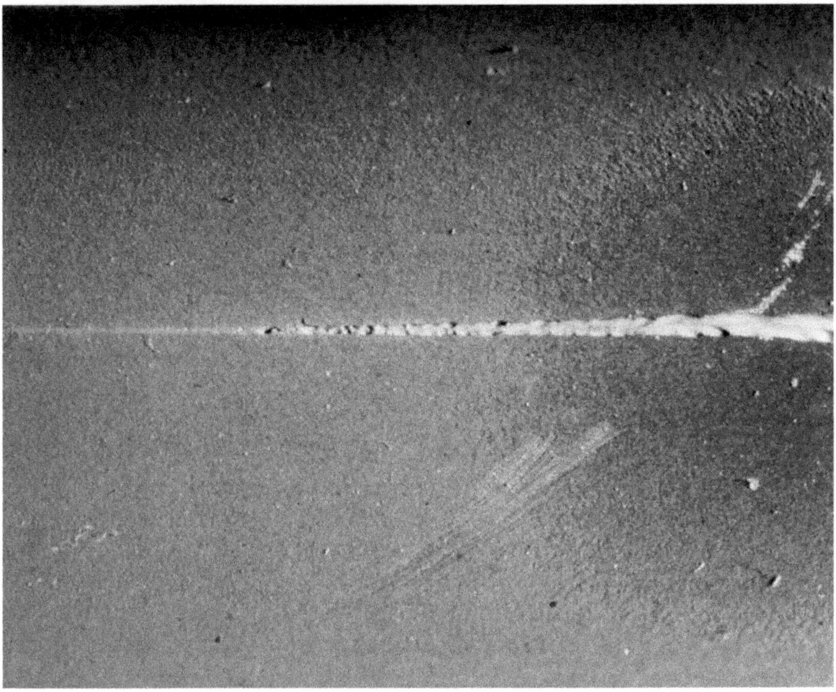

FIGURE 7.23 Erosion corrosion at a scratch line on an aluminum brass tube, where repassivation did not occur due to overchlorination.

deep, while the attack with a poorly protective film is more widespread and shallow. Further testing was conducted to identify safe chlorine levels for the common copper alloys.[35] These were as follows:

All alloys	When ferrous sulfate dosing is being used, switch off continuous chlorination during the dosing period
Aluminum brass	Intermittent chlorination preferred (e.g., 1 mg/L for 1 hour every 12 hours), but up to 0.5 mg/L continuous with ferrous sulfate dosing
90/10 Cu-Ni	Up to 0.5 mg/L continuous
70/30 Cu-Ni	Up to 1.0 mg/L continuous
66/30/2/2 Cu-Ni-Fe-Mn	Up to 2.0 mg/L continuous

Note: These chlorine concentrations are measured in the feed piping or inlet water box.

When the water is polluted with sulfide, the additional presence of chlorine can produce a large acceleration of attack at sites experiencing high water velocities. This is discussed in detail in the next section.

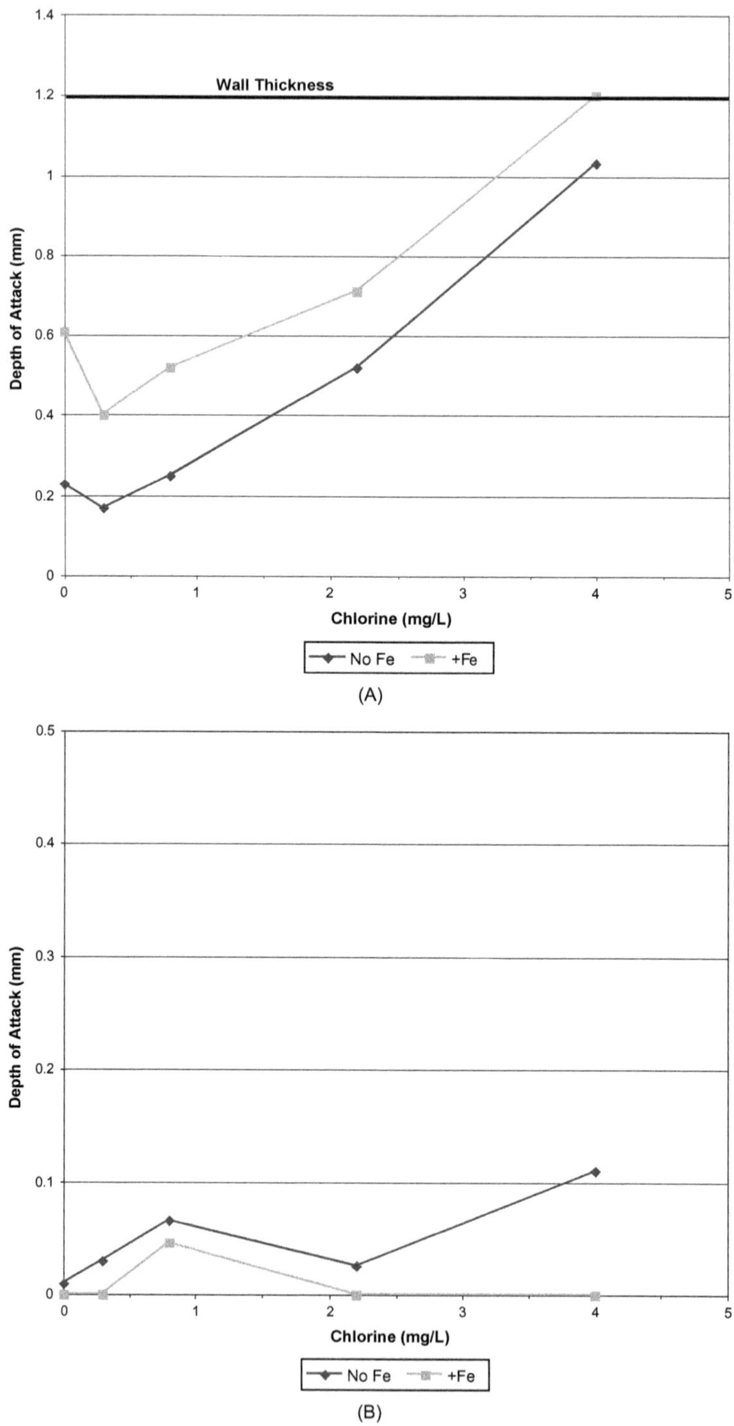

FIGURE 7.24 Depth of attack in Campbell condenser tube test rig vs. chlorine level at 9 m/s: (A) aluminum brass; (B) 66/30/2/2 Cu-Ni-Fe-Mn.[35]

The effect of chlorine and flow rate on the corrosion of NAB has also been investigated.[49] Cast NAB (C95800) was exposed in quiescent seawater (~0.1 m/s) and also in parallel flow rigs at 2.5 m/s in both natural seawater and seawater with a chlorine residual of 0.5 mg/L. All the tests were at 10°C. In the first 4 weeks of exposure, the corrosion rates in natural seawater at 0.1 and 2.5 m/s were ~80 and ~155 μm/y, respectively. In chlorinated seawater, the corrosion rates were ~30 and ~75 μm/y. Although the corrosion rate increased with velocity, the rates in chlorinated seawater were significantly lower. However, after 12 weeks' exposure, there were no significant differences between the corrosion rate in natural and chlorinated seawater at 0.1 m/s (~25 μm/y). At 2.5 m/s, the corrosion rates in natural and chlorinated seawater were ~10 and ~40 μm/y, respectively. The corrosion rates had decreased with exposure time, but more so for the samples in natural seawater. It should be noted that even in chlorinated seawater, the corrosion rate is not excessive.

When the chlorine levels are higher (e.g., ~5 mg/L) and velocities are also high, then accelerated erosion corrosion of NAB can occur. The author has seen failure by erosion corrosion of the first stage of an NAB multistage seawater lift pump on an offshore platform. The pump was situated inside a steel caisson, and strong hypochlorite solution was injected close to the pump intake. Failure occurred in less than 12 months.

7.1.5 Sulfides

When hydrogen sulfide is present in seawater, it is almost always produced by the action of sulfate-reducing bacteria (SRB). These bacteria need anaerobic conditions to thrive and are normally inactive in aerated seawater. When the water is stagnant or the access of oxygen is restricted, for example, in bottom mud, the aerobic bacteria gradually consume the DO. Once the oxygen is gone, the anaerobic bacteria, including SRB, become active. They reduce the sulfate in seawater to H_2S as part of their metabolic process. In deaerated seawater containing H_2S, the corrosion rate of most metals is low.

During the 1940s and 1950s, there was a marked increase in the number of failures of copper alloy heat exchanger tubes in both power stations and ships operating with polluted cooling water. BNF began an extended test program utilizing the jet test.[4] H_2S was added either continuously or intermittently to the tests using recirculated seawater. Bem and Gilbert reported the results of tests on a range of alloys under a variety of operating conditions.[50] Both low-level, continuous dosing of sulfide and intermittent, high-level dosing with H_2S produced deep attack on all the common copper alloys, similar to that seen in service. The results for 90/10 copper-nickel showed that an alloy with 1% iron performed better than one with 2% iron. Similarly, 70/30 copper-nickel with 0.8% iron suffered less attack than 66/30/2/2 Cu-Ni-Fe-Mn. Aluminum brass suffered severe attack under all conditions. Bem and Gilbert also conducted tests in which the sulfide was all oxidized to elemental sulfur by the DO in the seawater, that is,

$$2H_2S + O_2 \rightarrow 2H_2O + 2S. \quad (7.1)$$

All the alloys were still attacked more than in clean seawater, but there was less attack than in water containing H_2S, particularly for the copper-nickel alloys. As before, those with the lower iron content suffered less attack.

In later tests, Bem showed that if the H_2S was present in deaerated seawater, there was no erosion corrosion on any copper alloy, and the general corrosion rates were low.[51] This demonstrated the importance of the presence of oxygen with the H_2S to cause accelerated attack.

The work at BNF was not published in the open literature, and further work was undertaken in the 1970s in the United States, following the premature failure of 90/10 copper-nickel piping on a U.S. Navy vessel.[52] Gudas et al. showed that accelerated corrosion occurred on copper-nickel alloys that were first exposed to deaerated polluted water and then clean seawater, but not to the same extent as seen in the service failure.[52] When they investigated the effect of low levels of sulfide in aerated seawater, the severe attack seen in service was reproduced.[53] The tests also showed that velocity was a critical factor and the higher the velocity, the deeper the attack. On aluminum brass, this sometimes takes the form of accelerated pitting corrosion, rather than the smooth water-swept pits more normally associated with erosion corrosion. The mechanism of sulfide attack was discussed in Chapter 3.

Lee et al. carried out further studies on several heat exchanger alloys as a function of sulfide concentration and velocity.[54] The results, given in Figure 7.25, show the isocorrosion curves (0.1 mm/y) as a function of velocity and sulfide. This is quite a high corrosion rate for a condenser tube alloy in that an 18 SWG tube would have lost half its thickness in 6 years—a relatively short life. The results show that 90/10 copper-nickel (in this case, with 1.1% iron) performed the best. At sulfide levels below 0.01 mg/L, 70/30 copper-nickel was next best, with aluminum brass being the worst. However, at higher sulfide concentrations, aluminum brass was slightly better than 70/30 copper-nickel. At a typical heat exchanger velocity of 2 m/s, 90/10 copper-nickel was satisfactory up to 0.01 mg/L sulfide. These

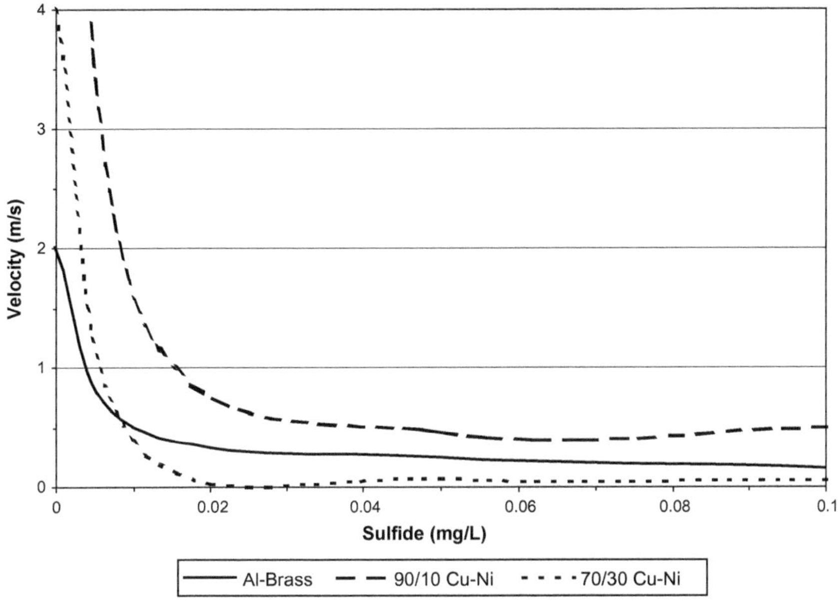

FIGURE 7.25 Isocorrosion curves (0.1 mm/y) as a function of velocity and sulfide level.[54]

sulfide levels appear low, but they are continuous in aerated water. Under intermittent pollution, the alloys can tolerate more sulfide, especially when a protective film is first formed in clean seawater.

A number of methods have been attempted to ameliorate the effects of sulfide attack. Rowlands recommended a pretreatment of copper-nickel alloys with a hot solution of sodium dimethyldithiocarbamate.[55] Christensen conducted trials on aluminum brass as well as both 90/10 copper-nickel and 70/30 copper-nickel tubes in both the treated and untreated conditions.[56] The tubes were first exposed for 5 weeks in a heavily polluted Danish harbor, followed by exposure in a heat exchanger at 1.8 m/s with clean seawater for several weeks. The overall conclusion was that the pretreatment, far from being beneficial, resulted in increased corrosion of all three alloys compared with untreated tubes.

The work of Gudas, Hack, and their colleagues,[45-46] who investigated the effects of ferrous sulfate dosing in polluted water, was described in Section 7.1.3. They found that continuous iron dosing was better than intermittent dosing, and it worked better on 70/30 copper-nickel than on 90/10 copper-nickel. In no case did iron/ferrous sulfate dosing completely suppress sulfide attack; rather, it only reduced it.

TABLE 7.9 Depth of attack in jet impingement test at 20°C for some cast copper alloys[7]

		Depth of attack (mm)	
Alloy	UNS no.	Clean Seawater	Seawater with sulfide
Mn-bronze	C86400	0.5	0.42
Gunmetal (LG2) (5% Sn)	C83600	0.23	0.07
Gunmetal (LG4) (7% Sn)	C92410	0.05	0.04
Gunmetal (G1) (10% Sn)	C90500	0.02	0.02
Leaded tin bronze (3% Pb)	C92900	0.02	0.03
NAB	C95800[(A)]	0.0	0.14

[(A)] Aluminum slightly below specification minimum.

INCRA sponsored work to determine whether ferrous sulfate dosing combined with sponge ball cleaning could mitigate the effects of heavy sulfide pollution.[57] The experimenters found that sponge ball cleaning plus 0.18 mg/L Fe^{2+} continuously (equivalent to 4.3 mg/L Fe^{2+} for 1 hour/day) minimized the corrosion of 90/10 copper-nickel in the presence of 0.08 mg/L sulfide, but it did not prevent all accelerated attack. Joseph and Hammond also found that excessive sponge ball cleaning accelerated attack and recommended that the rate of sponge ball cleaning and the iron dose be optimized by measuring the heat transfer efficiency of the heat exchanger.[57]

Many attempts have been made to reduce the attack in polluted waters by alloying. Bog carried out jet impingement tests on a range of cast copper alloys, the results of which are shown in Table 7.9.[7] The results show the poor performance of NAB in polluted water as well as the poor performance of manganese bronze. It is clear that the resistance to sulfides increases as the tin content of the bronzes increases, with the 10% tin alloys performing best. The benefits of tin additions to copper alloys were explored at BNF in the 1950s. Bem reported on trials with the tin brasses containing about 17% zinc and 4% to 6% tin.[58] The best performance in polluted water was obtained with a brass containing 6% tin and from a straight tin bronze containing 7.5% tin. These alloys were never developed commercially in the United Kingdom, but Sumitomo Light Metals experienced similar corrosion problems with copper alloy heat exchanger tubes in polluted waters around Japan and introduced AP bronze in the 1970s. This alloy has a nominal composition of Cu/8Sn/1Al/0.1Si, and test data in polluted water were reported by Nosetani et al. (Table 7.10).[22] The results show similar performance of aluminum brass and AP bronze in clean seawater and a big increase in corrosion rate for aluminum brass in polluted water, while the corrosion rate of AP bronze is virtually unchanged. The laboratory data

TABLE 7.10 Corrosion rates of aluminum brass and AP bronze heat exchanger tubes at 2 m/s[22]

Alloy	Corrosion rate (mm/y)	
	Clean seawater	0.05 mg/L sulfide
Aluminum brass	0.08	0.23
AP bronze	0.07	0.08

and over 10 years of successful operating experience with AP bronze in Japanese power stations were reviewed by Atsumi et al.[59]

The effects of sulfide and chlorine together were investigated by the author.[60] The results for aluminum brass, 90/10 copper-nickel, and 66/30/2/2 Cu-Ni-Fe-Mn are shown in Figures 7.26, 7.27, and 7.28, respectively. It is clear that low levels of chlorine reduce the effects of sulfide on the corrosion of aluminum brass and 90/10 copper-nickel, but higher chlorine levels resulted in accelerated attack compared with sulfide alone, leading to tube perforation at the higher sulfide levels. The 66/30/2/2 Cu-Ni-Fe-Mn alloy was severely attacked by sulfide, and any dosing of chlorine further increased the corrosion. Alloy 722 performed similarly to the 66/30/2/2

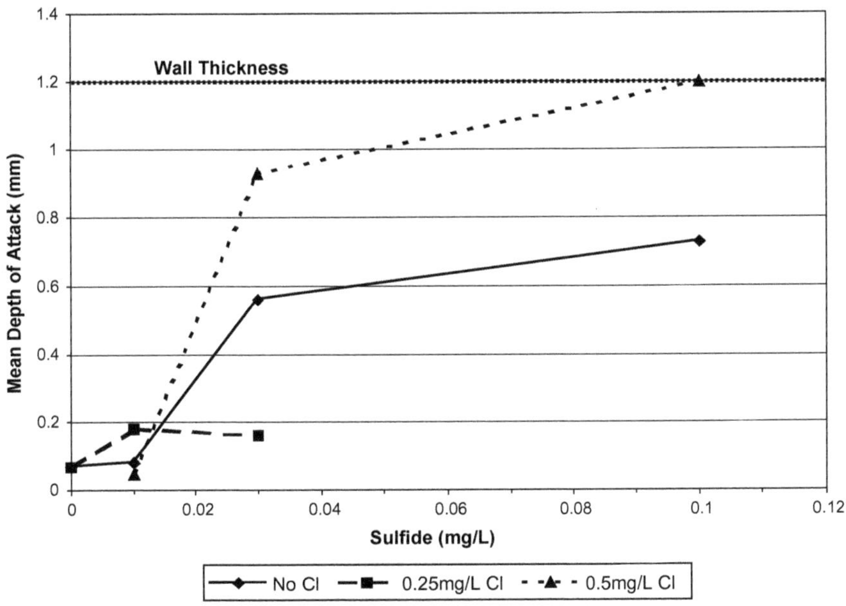

FIGURE 7.26 Effect of chlorine and sulfide on the erosion corrosion of aluminum brass at 7 m/s.[60]

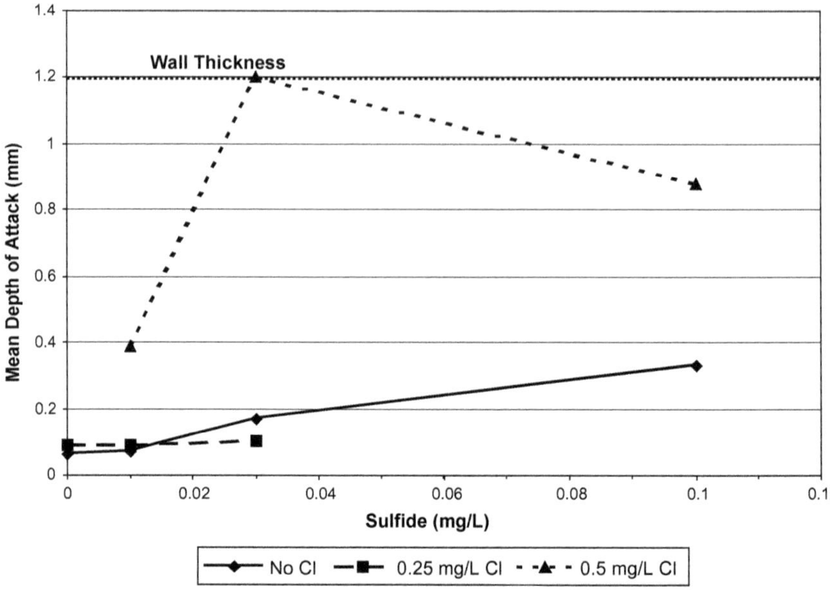

FIGURE 7.27 Effect of chlorine and sulfide on the erosion corrosion of 9/10 copper-nickel at 7 m/s.[60]

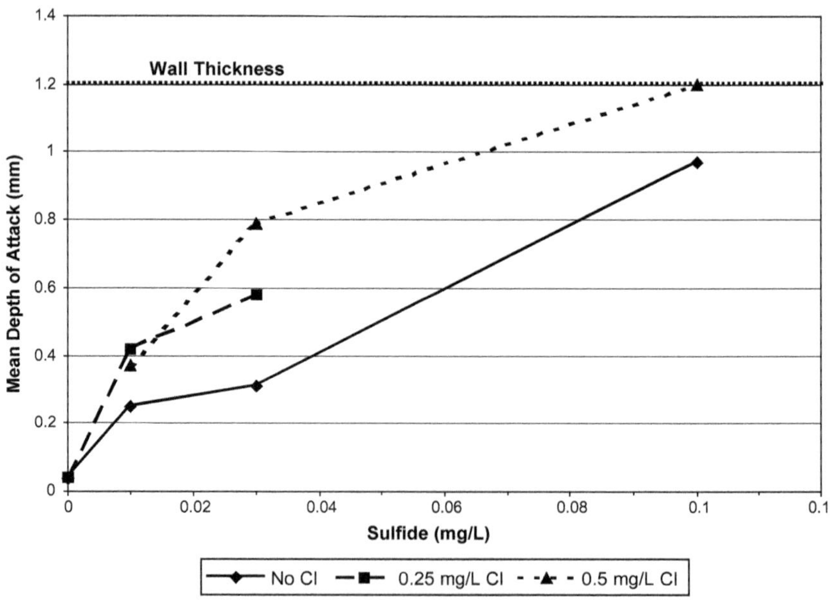

FIGURE 7.28 Effect of chlorine and sulfide on the erosion corrosion of 66/30/2/2 Cu-Ni-Fe-Mn at 7 m/s.[60]

Cu-Ni-Fe-Mn alloy, although the attack was not quite so severe; 70/30 copper-nickel was similar in performance to 90/10 copper-nickel, except that the corrosive effects of sulfide were reduced even with a 0.5 mg/L chlorine addition.

In the 1950s, a power station in Havana, Cuba, was experiencing severe corrosion of both aluminum brass and copper-nickel heat exchanger tubes due to polluted water. Fouling was also a problem, and chlorine dosing was carried out. When the chlorine dosing was stopped, the rate of tube failure decreased. These experiences mirror the test results with 0.5 mg/L chlorine.

The results of Francis demonstrate not only the severe effect of sulfide, but also the possibility of reducing its effects with low-level chlorine dosing.[60] This is discussed in more detail in Section 7.3.

7.2 Erosion

Erosion is the mechanical removal of the protective film and, usually, the underlying metal as well, most commonly by solids suspended in the liquid. In appearance, it is somewhat similar to erosion corrosion, often with a scored effect. Figure 7.29 shows a severely eroded gunmetal impeller operated in water containing high levels of suspended solids. With erosion corrosion, it can be suppressed by applying cathodic protection, whereas the mechanical effect of erosion cannot. However, it has been shown that where an alloy is corroding as well as eroding in a fluid, there is also a synergistic effect, which increases the metal loss.[61] The more the alloy is corroding, the greater the synergistic effect.

In erosion, the angle of incidence of the solids is important; the greater the angle of incidence, the more severe the attack. Most test rigs use a 90-degree angle of incidence, whereas there are many service applications in which the angle can be much lower, for example, 30 to 45 degree.

The suspended solids are usually present as silt or sand. These are usually introduced into cooling water when the area around the suction intake is relatively shallow and the bottom is stirred up by storms or tidal action. It can also occur when settling ponds are not dredged on a regular basis. Silt is fine material up to ~50 μm diameter, whereas sand usually varies from 50 to 250 μm diameter, or more.

Silt can be present in large quantities (hundreds of milligrams per liter), but it is less erosive than sand because of its fine nature. Sand is more erosive the larger its particles, and a small amount of coarse sand may do

FIGURE 7.29 Severe erosion of a gunmetal impeller due to a high level of suspended solids.

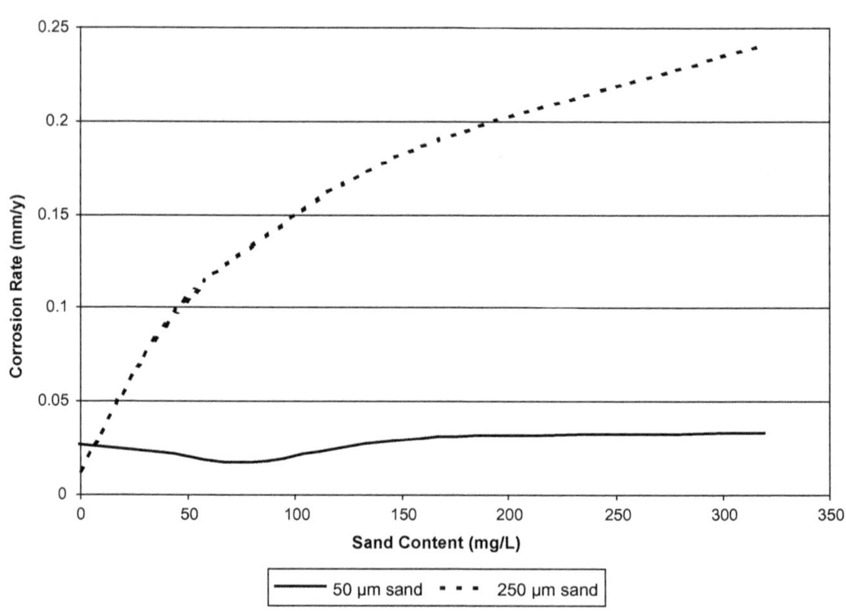

FIGURE 7.30 Erosion of aluminum brass after pretreating in seawater with ferrous sulfate.[62]

more damage than a larger quantity of silt. Sand and silt incursions are often intermittent, and so tubes that have had time to form protective films could be more resistant to erosion by sand. Sato et al. showed that aluminum brass tubes that were prefilmed by ferrous sulfate dosing could resist attack by 50-μm-diameter sand, but not 250-μm-diameter sand, as shown in Figure 7.30.[62] This is clearly of value where the solids loading is intermittent and there is time to form a protective film.

Corrosion by entrained sand gives rise to four distinct phases of attack, as shown in Figure 7.31 for aluminum brass.[41] The flow rate was 2 m/s with 50- to 100-μm-diameter sand particles:

- Stage I: Low sand content has no significant effect on corrosion rate.
- Stage II: The rate of attack increases as the sand content increases; here the protective film is being increasingly removed as the sand content rises.
- Stage III: The rate of attack is more or less constant with sand content; here the rate of film formation is much lower than the rate of film removal.
- Stage IV: The rate of attack rises sharply with an increase in sand content; electrochemical effects are minor, and the main damage is by mechanical removal of metal (pure erosion). This phase is rarely seen in practice.

Copper-nickel alloys formulated to increase the resistance to sand erosion, such as 66/30/2/2 Cu-Ni-Fe-Mn, were superior to aluminum brass, whereas 90/10 copper-nickel differed little from aluminum brass.[41]

Parker and Roscow reported the effect of varying silt quantities on the erosion rate of condenser alloys exposed at power stations along the Severn estuary in the United Kingdom.[63] The strong tidal action in the estuary stirs up silt from the bottom, but the quantity at the water intake varies from 50 mg/L near the ocean to ~5,000 mg/L high up the estuary. The mean silt content vs. erosion rate was plotted for a number of common condenser alloys, all exposed at ~2 m/s velocity. The results, in Figure 7.32, show that aluminum brass and 70/30 copper-nickel suffered severe attack as the silt level increased. The 70/30 copper-nickel alloy with 2% iron and 2% manganese (C71640), for increased erosion resistance, was better. A special 70/30 copper-nickel alloy with 3% iron and 3% manganese exhibited further improvement, but this alloy was more susceptible to localized attack during shutdowns, when stagnant conditions prevailed. Only titanium gave good results at the highest silt levels.

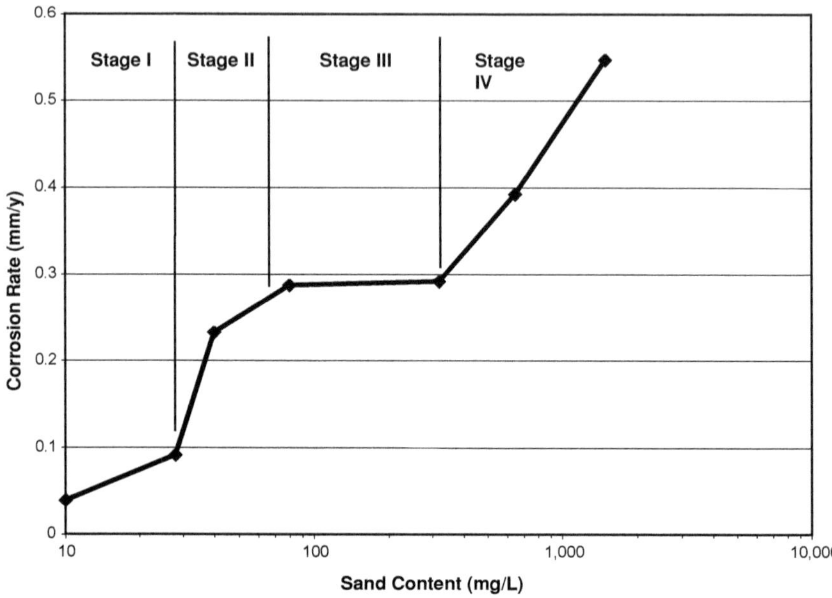

FIGURE 7.31 Effect of sand content on erosion of aluminum brass at 2 m/s by sand up to 0.1 mm diameter.[62]

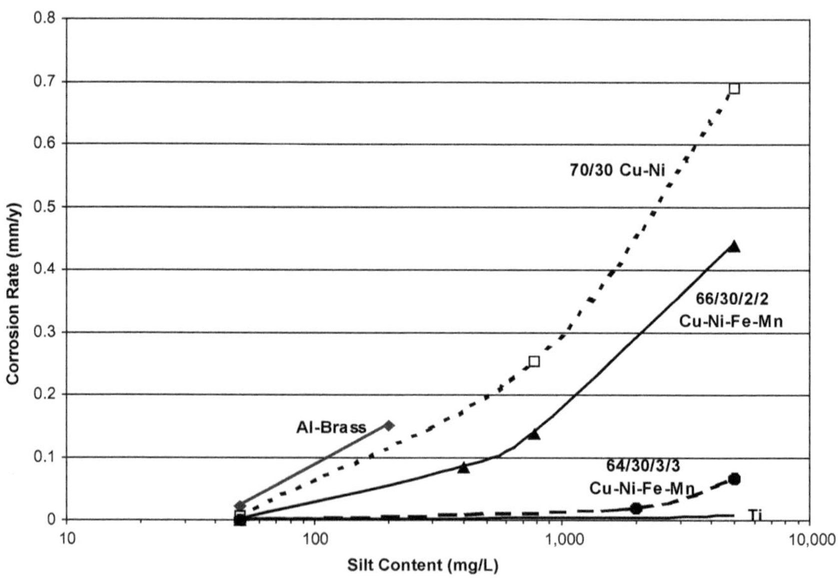

FIGURE 7.32 Metal loss of some heat exchanger alloys in silt-laden seawater.[63]

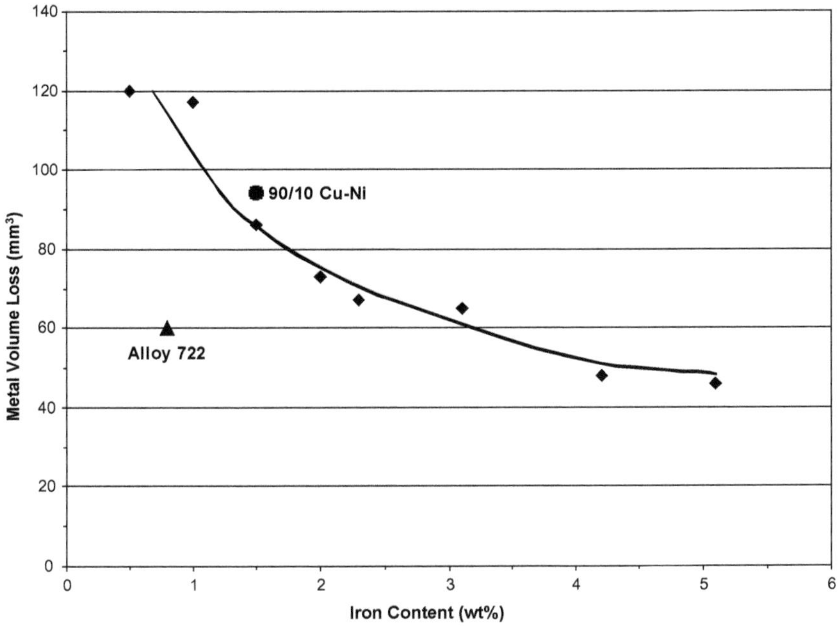

FIGURE 7.33 Effect of iron in a Cu/15Ni alloy on erosion by sand.[62]

The effect of alloying on erosion resistance has been studied by several workers. Sato et al. showed the effect on erosion of iron additions to a Cu–15% Ni alloy.[41] High iron contents significantly reduced the metal loss, as shown in Figure 7.33. This graph also shows data for 90/10 copper-nickel and Alloy 722. These points show that the addition of 5% nickel to 90/10 copper-nickel had only a small effect on erosion resistance. However, the 0.5% chromium addition present in Alloy 722 significantly improved the erosion resistance, compared with a chromium-free alloy with the same iron content.

Gaffoglio reported Sumitomo data for several copper heat exchanger alloys as a function of the silt or sand particle diameter.[64] The results showed a steady increase in erosion rate, with a particle diameter up to 100 μm for all the alloys, except aluminum brass. This followed the same trend as the copper-nickel alloys up to 40 μm diameter, followed by a sudden increase in erosion rate with larger-diameter particles (Figure 7.34). This suggests that a transition from stage I to stage II, as shown in Figure 7.31, had occurred. The higher-alloyed copper-nickels showed the best performance, but more rapid erosion rates would be expected with these alloys with sand particles of 250 μm and larger.

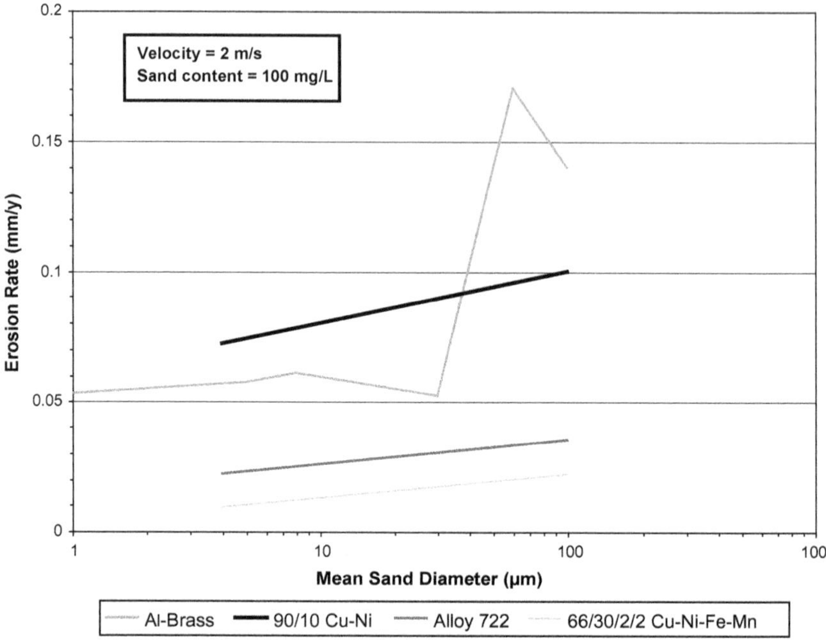

FIGURE 7.34 Erosion rate of various alloys as a function of sand diameter.[64]

A recent study[65] has investigated the interaction of velocity, temperature, and sand content in synthetic seawater, using a recirculating erosion rig of the type described by Dallas et al.[66] The results for cast NAB (C95800), shown in Figure 7.35, show that an increase in sand content from 50 to 500 mg/L had less effect than an increase in seawater temperature from 18 to 50°C. The effect of temperature on erosion corrosion in seawater has not previously been reported.

Another recent study, in the United States, investigated the effect of sand and sulfide on the corrosion of copper alloy heat exchanger tubes.[67] The flow rate was 2 m/s, sand was added at 200 mg/L (particle diameter 50–250 μm), and sulfide was added at 0.2 mg/L. The results, given in Figure 7.36, show that sand increased the corrosion rate by a factor of 2 to 3, but sulfide increased the corrosion rate by much more: 45 times for aluminum brass, 19 times for 70/30 copper-nickel, and nine times for 90/10 copper-nickel. When sand and sulfide were present together, the corrosion rate decreased dramatically for aluminum brass, but there were no significant changes to the rates for the copper-nickel alloys. This suggests that the corrosion products on aluminum brass in polluted seawater are cathodic to the bare

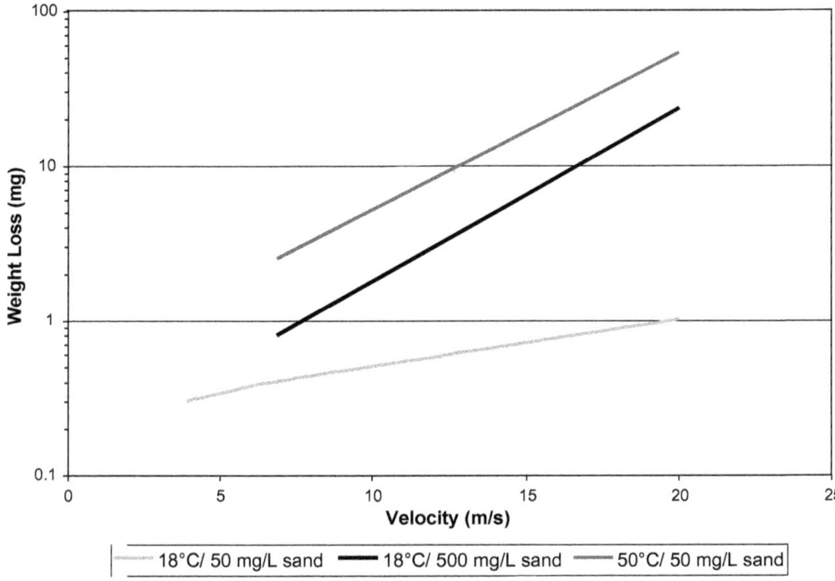

FIGURE 7.35 Erosion loss of NAB as a function of velocity, sand content, and temperature.[65]

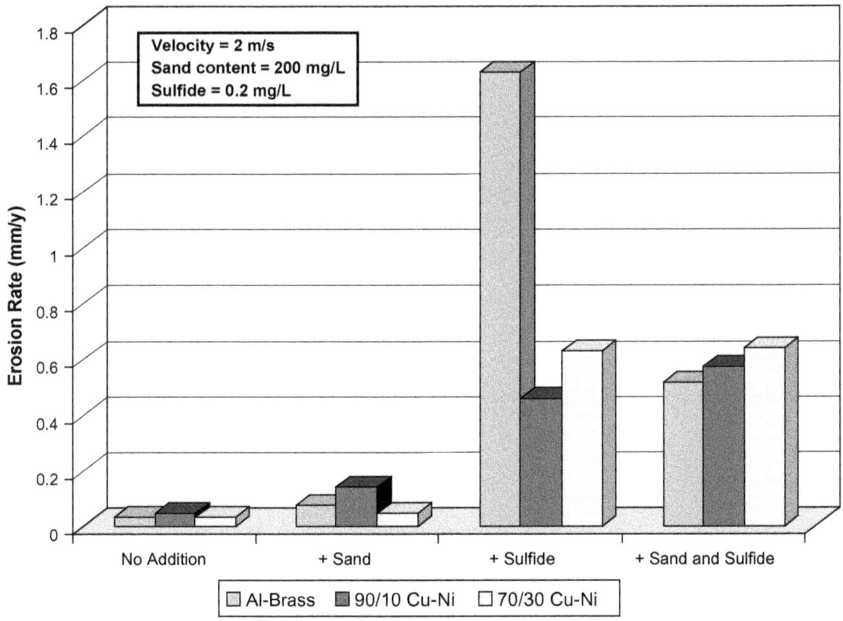

FIGURE 7.36 Effect of sand and sulfide on the erosion of some common condenser alloys.[67]

TABLE 7.11 Condenser tube plate corrosion rates with and without sand injection in seawater at 2 m/s[67]

Water	Alloy (plate/tube)	Corrosion rate (mm/y)	
		No sand	+200 mg/L sand
Clean seawater	Al-Bronze/Ti	0.820	0.423
	Muntz/Ti	2.830	1.540
	Muntz/Al-Brass	0.364	0.295
	Muntz/90/10 Cu-Ni	0.460	0.460
	Muntz/70/30 Cu-Ni	0.762	0.671
Sulfide polluted seawater	Al-bronze/Ti	1.570	0.824
	Muntz/Ti	6.080	4.530
	Muntz /Al-brass	2.410	2.300
	Muntz/ 90/10 Cu-Ni	2.930	2.020
	Muntz/70/30 Cu-Ni	3.340	3.570

metal, exacerbating attack. The removal of some of this film by the sand is probably the reason for the reduced corrosion rate (see Chapter 3). Ault et al. also observed that the corrosion rate decreased with time to a steady value in clean seawater, but when sand was present, there was a slow but steady increase in the corrosion rate over a period of 90 days.[67] The corrosion rate approximately doubled between day 10 and day 80.

Besides investigating condenser tubes, the study also examined tube plate corrosion.[67] The authors investigated aluminum bronze, Alloy D (C61400), coupled to titanium tubes, and muntz metal coupled to titanium and the three copper condenser tube alloys. The results, given in Table 7.11, show that sand had little effect on the corrosion of muntz metal coupled to any of the copper alloy tubes. However, the corrosion rate of both muntz metal and aluminum bronze decreased in both clean and polluted water when sand was added (reduction of 20% to 50%). The sand appeared to be having its biggest effect on the current from the tubes. This suggests that the sand was removing, or partially removing, a highly cathodic film.

7.3 Prevention of Erosion Corrosion and Erosion

The following sections discuss the best ways to avoid corrosion at high water velocities or in turbulent water. Recommendations for the avoidance of erosion corrosion and erosion are separated into two sections, for clarity.

7.3.1 Erosion Corrosion

The best way to avoid erosion corrosion of copper and its alloys in fresh water is to design to the velocities given in Section 7.1.1. In large buildings, such as hospitals and office buildings, it is necessary to use a pumped system for the hot water so that the feed to any tap from the loop is relatively short. Some engineers specify a smaller-diameter pipe for the return leg of the loop, assuming that much less water will be returning. Campbell pointed out that failures have occurred in some installations because hot water use was not that great, and hence, water velocities in the return leg were too high, causing failures by erosion corrosion.[68]

The best erosion corrosion resistance for the copper-nickel alloys is achieved with a minimum of 1.5% iron for 90/10 copper-nickel and 0.7% iron for 70/30 copper-nickel, as described in Section 7.1.2.

Previously, the nature of erosion corrosion and the limitations of copper alloys were discussed. However, what are important to the engineer are the safe design velocities for components in a flowing system.

The work of Efird was described previously, in which he determined the critical velocity at which the shear stress was sufficient to rupture the film.[31] However, his results take no account of any turbulence in the system, which can increase local velocities.

In long lengths of larger-diameter pipe (NPS 3 or 4 and greater), BS MA18 recommends velocities as follows[69]:

Aluminum brass	3m/s
90/10 copper-nickel	3.5m/s
70/30 copper-nickel	4m/s

This takes no account of restrictions or sharp bends, where local velocities may increase by a factor of 5 or 10. BS MA18 does recommend that the safe velocity be decreased as the pipe diameter decreases below NPS 3 because of increased turbulence. The document recommends velocities in the range 1.0–1.3 m/s for 19-mm outside diameter heat exchanger tube. This is clearly too low as there is excellent experience at around 2 m/s with copper alloys.

Table 7.12 shows some typical design velocities that have been used for copper alloys in clean seawater. Note that impeller tip speed is *not* a guide to water velocities in a pump because the water is also moving, and it is relative velocity that is important. In piping systems with sharp bends, the lower velocity in Table 7.12 is preferred. It is more important to keep velocities low in thin-walled components, such as heat exchanger tubes (for

TABLE 7.12 Typical velocity limits for copper alloys in clean seawater

	Maximum velocity (m/s)	
Alloy	Heat exchanger	Pipe (>NPS 3)
Admiralty brass	1.0–1.5	–
Aluminum brass	1.5–2.0	2.0–2.5
90/10 copper-nickel	2.0–3.0	3.0–3.5
70/30 copper-nickel	2.5–3.5	3.0–4.0
NAB	–	3.5–4.5

which the typical wall thickness is 1.2 mm), compared with castings for pumps and valves (with a typical thickness >15 mm), where a little erosion corrosion can sometimes be tolerated.

The velocities in Table 7.12 are meant as guidelines and must be used in conjunction with good design practice. The important thing to note is that there is a definite ranking order with copper alloys. The copper-nickel alloys have superior resistance to erosion corrosion compared with the brasses, and NAB has similar resistance to 70/30 copper-nickel. Table 7.12 does not include Alloy 722 or 66/30/2/2 Cu-Ni-Fe-Mn, which can be ranked equivalent to or better than NAB. However, these two alloys are more often selected when there are suspended solids in the water (see Section 7.3.2).

The successful use of copper alloys and the avoidance of erosion corrosion depend on the quality of the design. Utilization of the velocities in Table 7.12 alone is not sufficient. It is important to avoid designs that increase local turbulence. Hence 5-D bends are preferred to 3-D bends and 3-D bends are preferred to elbows. Traditional 90 degree elbows produce a lot of turbulence just downstream of the bend, which can lead to erosion corrosion. Smoothing out the flow with a radiused bend minimizes this effect.

Similarly, in heat exchangers, it is important not only to use a suitable velocity for the tubes, but also to use a water box design that distributes water evenly to all the tubes, without raising turbulence. Shone and Grimm warned against just increasing water box dimensions proportionately when designing a large heat exchanger.[47] To avoid turbulence, water boxes need to be deep, particularly if sacrificial anodes must be accommodated. Also, the water inlet should not be immediately opposite the tube plate because this tends to force most of the water down the central core of tubes. An angled inlet, or one at right angles to the tube plate, is preferred, with internal baffles, if necessary, to ensure uniform water flow to all tubes.[70]

Esso demonstrated the consequences of poor design in the 1970s, when it built a model of a heat exchanger designed for a very large crude carrier. With the inlet facing the tube plate and a shallow water box, there was good water flow in the central tubes, but in the outer ring of tubes, there was a steady flow of water in the *opposite* direction. The importance of water box design was clearly demonstrated.

Caruso discussed the design of water boxes, including the problems of a water box that is too deep, multipass heat exchangers, and feeding two heat exchangers from a common water box.[70] In heat exchangers, it is also important to avoid entrained air, which can lead to turbulence.[70] If there is inlet end erosion corrosion due to excessive turbulence, it can be mitigated in some cases by nylon tube inserts, some five or six tube diameters long. It is important to taper the trailing edge of the insert to avoid it raising turbulence there. If this is not done, the erosion corrosion is merely moved from the tube inlet to the end of the insert.[68] Erosion corrosion can also occur on the steam side of a condenser if the inlet velocity of the steam is too high. This is usually prevented by inserting a baffle at the inlet to spread the steam flow.[70]

Page described the problems that can arise from the positioning of condensers in power stations.[71] When the final condenser is high up and the waste cooling water flows down from the discharge water box, it is possible to create pressure conditions such that symphonic air release occurs in the condenser. The release of air increases turbulence and can cause erosion corrosion.

The ability of ferrous sulfate dosing to prevent erosion corrosion was discussed in Section 7.1.3. It is almost always required with brass tubes, particularly aluminum brass, and its use with copper-nickel alloys should be considered, depending on water composition.

Where it is important to protect aluminum brass tubes immediately on exposure to seawater, Sumitomo developed APF-coated tubes.[72-73] APF is a coating of hydrated iron oxide held in place with a polymeric binder. In unpublished work, it was shown that the coating conferred good resistance to erosion corrosion in turbulent water, but the coating was slowly eroded away and had to be reinforced with regular ferrous sulfate dosing.

Where chlorination is carried out, to prevent fouling, it is important to switch off the chlorine/hypochlorite when ferrous sulfate dosing is ongoing. Safe chlorination levels to prevent acceleration of erosion corrosion for the common condenser alloys are given in Section 7.1.4.

When sulfides are present in the water, either intermittently or continuously, it is important to select alloys carefully. Both 90/10 and 70/30 copper-nickel are more resistant than aluminum brass and 66/30/2/2 Cu-Ni-Fe-Mn, and ferrous sulfate dosing will help if the pollution is intermittent.

For more heavily polluted waters, tin bronze tubes, such as AP bronze, are better. Similarly, the high tin cast copper alloys are preferred to NAB for pumps and valves. The 10% tin bronzes are best, but it can be difficult to produce pressure-tight castings in these alloys. Hence, the 10% tin bronzes are used for pump impellers and valve internals, while LG4 gunmetal is used for pump cases and valve bodies. Even in clean seawater, if the pump or valve is to spend significant time full of stagnant seawater, where H_2S may be produced, the high-tin bronzes are preferred to NAB.

Chlorine reacts with pollutants in seawater, such as H_2S, with the following reaction:

$$H_2S + Cl_2 \rightarrow 2HCl + S. \tag{7.2}$$

Elemental sulfur is less corrosive toward copper alloys than H_2S, and hence, corrosion will be reduced.[74] However, if there is excess chlorine, then the corrosion rate can be greatly accelerated with some copper alloys such as 90/10 copper-nickel, as described in Section 7.1.5. Cigna et al. used the fact that oxidation of H_2S reduces the aggressivity of seawater toward copper alloys to solve corrosion problems at a Belgian power station.[75] The tidal, brackish cooling water to four power stations caused severe corrosion of the aluminum brass heat exchanger tubes after 1–2 years at two of the stations. The corrosion was so severe that the stations were retubed in 70/30 copper-nickel. The third station had its aluminum brass tubes acid cleaned, and measures were installed to prevent further accelerated attack. Water analysis showed that H_2S was present in the aerated water, and its presence reduced the redox potential. By controlling the redox potential with small doses of chlorine, Cigna et al. were able to prevent corrosion by sulfides without accelerating corrosion by overdosing with chlorine.

7.3.2 Erosion

In waters containing suspended solids that may cause erosion, it is important to keep velocities low and to avoid turbulence by good design, as described in the previous section. Where the main problem is silt (particles up to 50 μm diameter), the data of Parker and Roscow suggest that aluminum brass, 90/10 copper-nickel, and 70/30 copper-nickel may be used up to ~100 mg/L silt content without undue acceleration of corrosion.[63] Alloy 722 and 66/30/2/2 Cu-Ni-Fe-Mn are more resistant and may be used up to ~400 mg/L silt content.

Where the problem is coarser particles of sand (≥100 μm diameter), the data of Ault et al. suggest that levels up to 100–200 mg/L sand will roughly double the corrosion rate.[67] This is surprising as it suggests that sand is less of a problem than silt, whereas the reverse is actually correct. Note that the data of Ault et al. were obtained with a recirculating test loop, whereas those of Parker et al. were obtained from power station experiences in the Severn estuary, United Kingdom. Hence, it would be prudent to limit the use of 90/10 and 70/30 copper-nickel to a maximum of around 50 mg/L sand, whereas Alloy 722 and 66/30/2/2 Cu-Ni-Fe-Mn could be used up to approximately 100 mg/L sand. Note that any source of local turbulence will increase local erosion rates.

Any methods of reducing sand and silt contents are to be preferred. Filters and screens are effective, provided that there is sufficient pumping pressure to maintain flow as these block up. It is important to ensure that filters are cleaned regularly and are not just bypassed when the filters get blocked.

A common alternative is to use settling tanks. When the first fills up, a second is used while the first is dredged to remove accumulated solids. It is important that this is maintained. The author investigated erosion problems with copper alloy piping at a plant that were caused by the postponing of settling tank dredging to save money. The costs of shutdowns and pipe replacements, due to the ingress of sand, were more than the savings realized by not dredging the ponds.

References

1. J.R. Myers, M.F. Obrecht, *MP* 12 (1973): p. 17.
2. R.J. Oliphant, "Causes of Copper Corrosion in Plumbing System," Foundation for Water Research Report, FR/R0007, May 2003.
3. R. Francis, *Galvanic Corrosion—A Practical Guide for Engineers* (Houston; TX: NACE International, 2001).
4. R. May, R.W. de Vere Stacpoole, *J. Inst. Met.* 77 (1950): p. 331.
5. B. Todd, *Trans. Inst. Marine Eng.* 80 (1968): p. 161.
6. S. Bog, "The Resistance to Impingement Attack by Seawater of Copper Alloy Tubes," Fifth Scandinavian Corrosion Congress (Copenhagen, Denmark: 1968).
7. S. Bog, "The Resistance of Copper Alloys to Different Types of Corrosion in Seawater," Seventh Scandinavian Corrosion Congress (Trondheim, Norway: 1975).
8. G.L. Bailey, *J. Inst. Met.* 79 (1951): p. 243.
9. K.D. Efird, *Corrosion* 33 (1977): p. 347.
10. ASTM BILL, "Standard Specification for Copper and Copper-Alloy Seamless Condenser Tubes and Ferrule Stock" (West Conshohocken, PA: ASTM).

11. K. Nagata, S. Sato, *Sumitomo Light Met. Tech. Rep.* 22 (1981): p. 22.
12. D.M.M. Guyoncourt, S. Harper, BNF Metals Technology Centre Research Report, RRA1913, February 1977.
13. H.G. Appel, H. Urselmann, M. Tappen, W. Beckmann, M. Hecht, M. Jasner, "Suitability of CuNi 90/10 for Seawater Service," EUROCORR 2001 (Riva del Garda, Italy: European Federation of Corrosion, 2001).
14. J.P. Frick, L.R. Scharfstein, T.M. Parill, G. Haaland, *J. Met.* (March 1985): p. 28.
15. R. Francis, *The Selection of Materials for Seawater Cooling Systems: A Practical Guide for Engineers* (Houston, TX: NACE International, 2006).
16. W.W. Kirk, INCRA Report, PRJ 382, April 1986.
17. D.W. Townsend, British Non-Ferrous Metals Research Association Research Report, A1546, October 1965.
18. H.S. Campbell, D.W. Townsend, British Non-Ferrous Metals Research Association Research Report, A1547, October 1965.
19. D.W. Townsend, British Non-Ferrous Metals Research Association Research Report, A1671, December 1967.
20. D.B. Anderson, K.D. Efird, "The Influence of Chromium on the Corrosion Behaviour of Copper-Nickel Alloys in Seawater," Third International Congress on Marine Corrosion (NBS, 1972), p. 264.
21. N.W. Polan, M.A. Heine, J.M. Popplewell, C.J. Gaffoglio, CORROSION/82, paper no. 58 (Houston, TX: NACE International, 1982).
22. T. Nosetani, S. Sato, K. Kazama, Y. Yamaguchi, T. Yasui, *Sumitomo Light Met. Tech. Rep.* 22 (1971): p. 85.
23. H.T. Michels, R.M. Kain, CORROSION/05, paper no. 233 (Houston, TX: NACE International, 2005).
24. D.J. Mills, BNF Metals Technology Centre Research Report, RRA1899, September 1975.
25. H.S. Campbell, *Aluminium Bronze Alloys Corrosion Resistance Guide*, Publication 80 (Potters Bar, U.K.: Copper Development Association, 1981).
26. J.C. Rowlands, B. Angell, AMTE, Holton Heath, correspondence to author, 1978.
27. F.L. LaQue, *Marine Corrosion* (New York, NY: John Wiley, 1975).
28. H.S. Campbell, V.E. Carter, *J. Inst. Met.* 91 (1961–1962): p. 362.
29. W.K. Cheung, J.G.N. Thomas, "The Effect of Dissolved Copper on the Erosion-Corrosion of Copper Alloys in Flowing ASTM Seawater," ASTM STP 970 (1988): p. 190.
30. S. Sato, K. Nagata, S. Yamauchi, *Sumitomo Light Met. Tech. Rep.* 20 (1979): p. 32.
31. K.D. Efird, *Corrosion* 33 (1977): p. 3.

32. W.W. Kirk, INCRA Project Report, 396, July 1987.

33. R.J. Ferrara, J.P. Gudas, "Corrosion Behaviour of Copper Base Alloys with Respect to Seawater Velocity," Third International Congress on Marine Corrosion (NBS, 1972), p. 285.

34. R.J. Ferrara, J.P. Gudas, "Corrosion Behaviour of Copper Base Alloys with Respect to Seawater Velocity," Third International Congress on Marine Corrosion, J.B. Cotton, Discussion of Reference 32, ibid (NBS, 1972), p. 297.

35. R. Francis, *MP* 21 (1982): p. 44.

36. D.G. Melton, INCRA Project Report, 370, February 1986.

37. T.W. Bostwick, *Corrosion* 17 (1961): p. 12.

38. A.M. Lockhart, *Proc. Inst. Mech. Eng.* 179 (1964–1965): p. 459.

39. R. Francis, INCRA Project Report, 289, January 1979.

40. S. Sato, T. Nosetani, *Sumitomo Light Met. Tech. Rep.* 11 (1970): p. 271.

41. S. Sato, K. Nagata, "Factors Affecting Corrosion and Fouling of Condenser Tubes of Copper Alloys and Titanium," Conference on Corrosion in the Power Industry (Montreal, Canada: NACE International, 1977).

42. T. Nosetani, S. Sato, K. Onda, Y. Yamaguchi, S. Ando, *Sumitomo Light Met. Tech. Rep.* 15 (1974): p. 1.

43. S. Sato, M. Okawa, *Sumitomo Light Met. Tech. Rep.* 17 (1976): p. 17.

44. S. Sato, T. Nosetani, Y. Yamaguchi, K. Onda, *Sumitomo Light Met. Tech. Rep.* 16 (1975): p. 23.

45. J.P. Gudas, H.P. Hack, D.W. Taylor, CORROSION/78, paper no. 23 (Houston, TX: NACE International, 1978).

46. H.P. Hack, J.P. Gudas, D.W. Taylor, CORROSION/79, paper no. 234 (Houston, TX: NACE International, 1979).

47. E.B. Shone, G.C. Grimm, *Trans. Inst. Marine Eng.* 98 (1985): paper 11.

48. S. Sato, *Sumitomo Light Met. Tech. Rep.* 3 (1962): p. 276.

49. R. Francis, C.R. Maselkowski, BNF Metals Technology Centre Research Report, R506, May 1985.

50. R.S. Bem, P.T. Gilbert, British Non-Ferrous Metals Research Association Research Report, A1095, September 1955.

51. R.S. Bem, British Non-Ferrous Metals Research Association Research Report, A1132, December 1956.

52. J.P. Gudas, G.J. Danek, R.B. Niederberger, CORROSION/76, paper no. 76 (Houston, TX: NACE International, 1976).

53. J.P. Gudas, H.P. Hack, D.W. Taylor, CORROSION/78, paper no. 93 (Houston, TX: NACE International, 1978).

54. T.S. Lee, H.P. Hack, D.G. Tipton, "The Effect of Velocity on Sulphide-Induced Corrosion of Copper Base Condenser Alloys," Fifth International Congress on Marine Corrosion and Fouling (Barcelona, Spain: 1980).

55. J.C. Rowlands, *J. Appl. Chem.* 15 (1965): p. 57.

56. C. Christensen, "The Influence of Polluted Seawater on the Corrosion of Copper Alloys," Seventh Scandinavian Corrosion Congress (Trondheim, Norway, 1975).

57. M.W. Joseph, F.W. Hammond, INCRA Project, 349, Report A (1985), Report B (1986).

58. R.S. Bem, British Non-Ferrous Metals Research Association Research Report, A1233, March 1959.

59. T. Atsumi, A. Ogiso, K. Nagata, S. Sato, *Sumitomo Light Met. Tech. Rep.* 29 (1988): p. 23.

60. R. Francis, *Brit. Corros. J.* 20 (1985): p. 175.

61. A. Neville, T. Hodgekiess, *Surf. Eng.* 12 (1996): p. 303.

62. S. Sato, T. Nosetani, Y. Yamaguchi, K. Onda, *Sumitomo Light Met. Tech. Rep.* 16 (1975): p. 23.

63. J.G. Parker, J.A. Roscow, "The Service Performance of Condenser Tube Materials in Cooling Waters Containing Appreciable Solids," Proceedings of the Eighth International Congress on Metallic Corrosion (Mainz, Germany: 1981), p. 1285.

64. C.J. Gaffoglio, *Power Eng.* 86, 8 (1982): p. 60.

65. A. Neville, University of Leeds, correspondence to author, 2005.

66. J. Dallas, T.A. McConnell, "Design and Testing of a Protoype Slurry Pump for FGD Plants," Solids Pumping Conference (London: Inst. Mech. E., 1991).

67. J.P. Ault, G.A. Gehring, B.C. Syrett, *MP* 34 (1995): p. 63.

68. H.S. Campbell, "Erosion Corrosion in Copper Plumbing Systems," *Building Services Journal*, October 1984, p. 62.

69. BS MA18, "Salt Water Piping Systems in Ships" (London: BSI, 1973).

70. L. Caruso, "Cut Surface-Condenser Tube Problems by Paying Close Attention to Design," *Power*, August 1981.

71. G.G. Page, *N. Z. J. Sci.* 26 (1983): p. 415.

72. K. Nagata, S. Sato, *Sumitomo Light Met. Tech. Rep.* 27 (1986): p. 18.

73. T. Nosetani, Y. Hotta, S. Sato, *Sumitomo Light Met. Tech. Rep.* 28 (1987): p. 29.

74. B.C. Syrett, D.D. Macdonald, S.S. Wing, *Corrosion* 35 (1979): p. 409.

75. R. Cigna, K. De Ranter, O. Fumei, L. Giuliani, *Brit. Corros. J.* 23 (1988): p. 190.

CHAPTER 8

Cavitation

Cavitation is a corrosion phenomenon that can result in the loss of large amounts of metal over a relatively small area. It occurs when there is a sudden drop in pressure, usually at a high flow velocity. This creates a series of vapor cavities that may be up to 1 mm in diameter.[1] These cavities, or bubbles, migrate along the pressure gradient, until the external pressure is sufficient to cause their collapse. The energy release when they collapse can be sufficient to mechanically remove metal from the adjoining surface. This is termed *cavitation damage* and often presents a jagged, honeycomb appearance, as shown in Figure 8.1

Cavitation can occur on ships' propellers, where the bubbles form near the hub and migrate outward along the blades. Cavitation damage is usually seen about one-third of the way along the blade. This should not be confused with propeller wastage, which occurs at the tips of the propeller blades. The latter is erosion corrosion, and it should not be confused with cavitation damage, which has sometimes been termed *cavitation erosion.* Campbell discusses the differences in detail.[2-3]

Cavitation can also occur in pumps, particularly when they are being operated under conditions well removed from the design optimum. Again, cavitation damage occurs toward the eye of the impeller, while any damage at the outer edge of the impeller is usually erosion corrosion. Cavitation is also seen after partly throttled valves or orifice plates, where the pressure drop is large and the flow is often high. Vibration may also cause cavitation, and damage to the cylinder liners of large marine diesel engines has been attributed to cavitation.

Cavitation proceeds broadly in four stages, as shown in Figure 8.2.[4] The first stage is incubation, and the mass loss is small. However, pitting

FIGURE 8.1 Cavitation damage, showing typical jagged, honeycomb appearance.

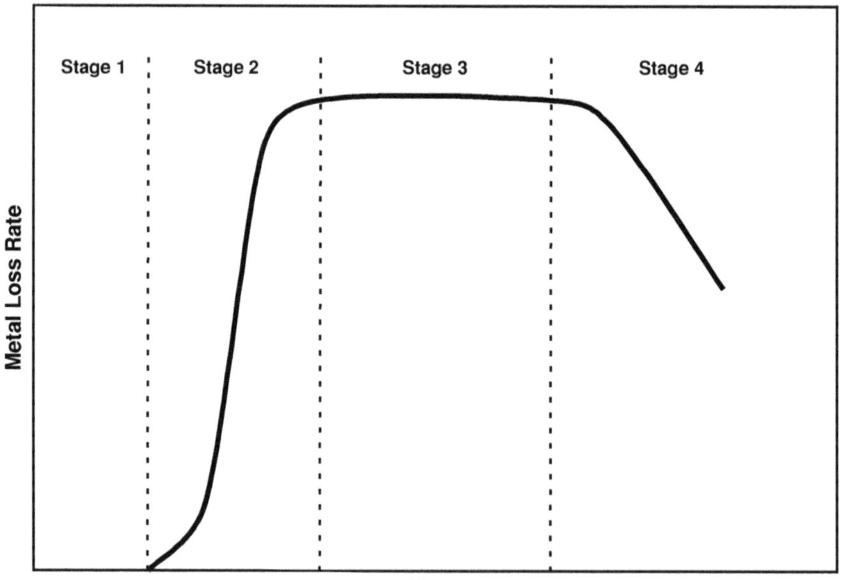

FIGURE 8.2 Schematic showing the four stages of metal loss during cavitation.[4]

will occur in ductile materials, whereas brittle materials will show cracking or pitting and cracking. In stage 2, the process accelerates, and significant metal loss occurs, which gradually spreads over the whole cavitation area. In the third stage, the metal loss rate is almost constant, corresponding to metal loss over the whole cavitation zone. In the fourth stage, the metal loss rate decreases, either because the loss of metal is now so great that the local flow conditions have changed and the cavitation intensity is reduced, or because the metal surface is now protected by a layer of stagnant water. Not all these four stages may show easily during any specific test for cavitation resistance.[4]

The resistance to cavitation shows a strong correlation with the hardness of the material. However, the direct correlation between the two shows a lot of scatter, and it is clear that material strength and ductility are also important.[1]

The foregoing description of cavitation suggests that it is purely mechanical and, as such, the application of cathodic protection would have no effect. However, in corrosive fluids, it has been observed that there is a synergistic effect between cavitation and corrosion, similar to that between erosion and corrosion, as described in Chapter 7. Hence, the total mass loss is given by

$$M_{\text{Total}} = M_{\text{Cav}} + M_{\text{Corr}} + M_{\text{Syn}} \tag{8.1}$$

where

M_{Total} = total mass loss
M_{Cav} = mass loss due to cavitation
M_{Corr} = mass loss due to corrosion
M_{Syn} = synergistic mass loss

Wood and Fry described cavitation tests of pure copper and 70/30 copper-nickel in seawater using a cavitation tunnel.[5] By separating each of the cavitation and corrosion effects, they demonstrated that there was a synergistic effect for both materials. Furthermore, the magnitude of the synergistic metal loss rate was approximately the same for both materials.

8.1 Cavitation Test Methods

There are a number of methods of studying cavitation, and it is important to understand the differences between the different techniques when comparing data derived by these methods. The simplest test method is based on a 20k Hz oscillator, in which the material under test is fastened to the end of the oscillating arm. This creates cavitation of high intensity, resulting in significant damage in a short time. It is useful for comparing materials, but the data cannot easily be related to hydrodynamic cavitation, say, in a pump. The test has been standardized and is described in ASTM G32.[6]

Another test, which is hydrodynamic, involves a disc rotating at high speed in the fluid of interest. A hole or pin is positioned near the rim of the disc, and this induces cavitation bubbles. In a variation of this test, a plate with a hole in it is placed in a high-velocity laminar fluid flow, again inducing cavitation bubbles.

In the cavitating jet test, a long orifice nozzle is submerged in liquid, and the liquid is pumped at high pressure through the nozzle, which causes cavitation bubbles to form. These impinge on a sample placed close to the jet. This test has been standardized and is described in ASTM G134.[7] The advantage of this test is that it is hydrodynamic, and also, it enables the cavitation number (σ) and the flow velocity to be varied, where

$$\sigma = (P_d - P_v)/0.5(\rho \times \mathrm{V}^2), \tag{8.2}$$

where

P_d = static pressure in the downstream chamber
P_v = vapor pressure
ρ = liquid density
V = jet velocity

It is relatively easy to calculate σ values for some service applications in which cavitation is occurring and so relate experimental data to service performance.

The final test method is a cavitation tunnel, in which the fluid is accelerated by a venturi and cavitation is induced by an obstruction, often a cylinder, at right angles to the flow. The specimen may be the obstruction, or it may be a panel that is flush mounted into the tunnel wall near the obstruction. This reproduces many cavitating service environments well, but testing is costly and time consuming.

Note that many reports on cavitation testing quote volume metal loss, rather than mass loss. This is because material densities can vary significantly, and volume loss gives a clearer idea of the scale of the damage.

8.2 Performance of Copper Alloys

It was stated previously that material hardness has a significant effect on cavitation resistance. Hence, pure copper will have relatively low

TABLE 8.1 Results of ultrasonic cavitation tests in fresh water[8]

Alloy	UNS no.	Cavitation rate (mm^3/h)
Manganese bronze	C86550	4.7
Admiralty gunmetal	C90600	4.9
Mn-Al-bronze	CMA1[(A)]	1.5
Al-bronze	C95200	0.8
NAB	C95800	0.6
Alloy K-500 (cold drawn)	N05500	2.8
Alloy K-500 (aged)	N05500	1.2
Type 316 stainless steel	S31600	1.7

[(A)] No UNS number, BS1400 designation quoted

resistance, while harder copper alloys, such as manganese bronze and aluminum bronze, will be significantly better.

Campbell describes the results of some ultrasonic cavitation tests (ASTM G32) in fresh water.[8] The results, given in Table 8.1, show that the resistances to cavitation of manganese bronze and admiralty gunmetal were much less than the resistances of the aluminum bronzes. The aluminum bronzes were superior not only to other copper alloys, but also to alloy K-500 and Type 316 stainless steel (SS) (UNS 531600). Some similar tests in 3% sodium chloride (NaCl) (Table 8.2) again showed nickel aluminum bronze (NAB) to be superior to manganese bronze and Type 321 SS (UNS S32100).[8]

Angell shows the results of tests on a range of copper and other alloys obtained in seawater in a cavitation tunnel.[1] The data were generated at a velocity of 40 m/s and a cavitation number (σ) of 0.08, which represents severe cavitation conditions. The results, given in Figure 8.3, show the range of metal loss rate for several alloys in some of the categories such as nickel alloys. It is clear that the ranking is somewhat different to the ranking obtained in the ultrasonic cavitation test. One reason is that Angell's data were obtained under real hydrodynamic cavitation conditions; a second is that Angell's tests were in natural seawater, where there would have been a significant contribution from corrosion and hence also the synergistic effect, as discussed previously.[5] The aluminum bronzes were significantly

TABLE 8.2 Depth of attack after ultrasonic cavitation tests in 3% sodium chloride solution[8]

Alloy	UNS no.	Depth of attack (mm)	Test time (h)
Manganese bronze	C86550	0.280	6
NAB	C95800	<0.025	7
Type 321 stainless steel	S32100	0.305	7

TABLE 8.3 Depth of cavitation after testing in seawater for 30 days

Alloy	Velocity (m/s)	Temperature (°C)	Depth (mm)
Manganese bronze	122	27	0.05
Mn-Al-Bronze	122	27	0.08
70/30 copper-nickel	121	6	0.56
90/10 copper-nickel	130	8	0.03
90/10 copper-nickel	128	27	0.03
Alloy 400	121	26	>1
Alloy K-500	132	6	0.23
Alloy K-500	121	27	0.03
Type 316 stainless steel	122	9	0.02
Type 316 stainless steel	122	12	0.25

more resistant to cavitation than the other copper alloys and were on a par with the nickel alloys, but they were not as resistant as the austenitic and duplex SS. Note that with a different cavitation number and flow velocity, the relative resistance of materials can change.

The LaQue Laboratory, in the United States, carried out unpublished cavitation tests on a wide range of alloys. The materials were tested as flat plates with a small hole 25 mm from the leading edge in a high-velocity laminar seawater flow for 30 days. The results, shown in Table 8.3, are difficult to compare with Angell's data because the latter were expressed as

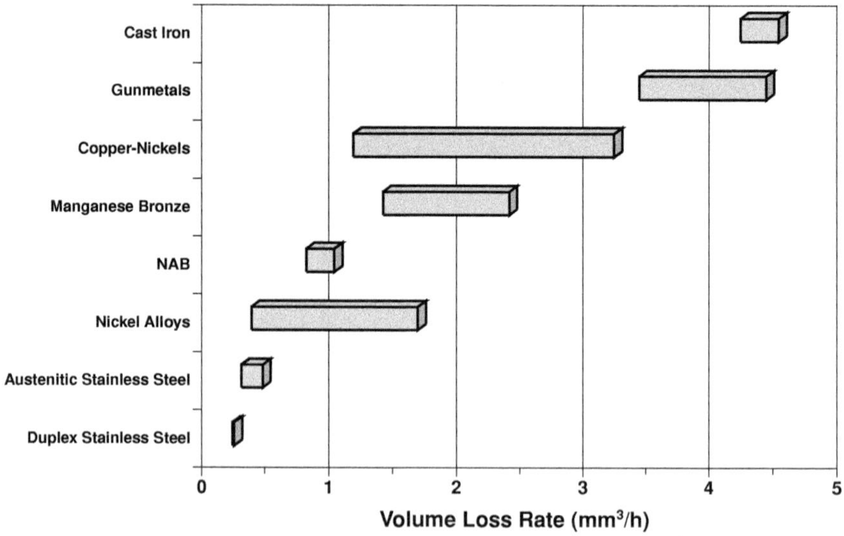

FIGURE 8.3 Cavitation data obtained in seawater with a cavitation tunnel at 40 m/s and a cavitation number of 0.08.[1]

volume loss rate, whereas the LaQue data are expressed as depth.[1] Hence, a shallow depth could represent a high metal loss if the attack was over a wide area. Similarly, deep attack might represent a low metal loss if it were very localized. Depending on the component in question, the depth of attack may be critical in some cases, whereas the extent of the attack may be important in others.

The LaQue data show Alloy 400 to have a low resistance to cavitation, whereas Alloy K-500, which is stronger and harder, has better resistance. The results also show 70/30 copper-nickel to be inferior to the 90/10 alloy and also much worse than manganese bronze and manganese aluminum bronze (CMA1). Unfortunately, aluminum bronze was not included in these tests. The results for Type 316 SS show the variability possible in this kind of test. It is not clear whether the results for Alloy K-500 show a real effect of temperature or whether they just represent typical test-to-test variation.

8.3 Avoidance of Cavitation

Where cavitation occurs in service, one option is to upgrade to a more resistant alloy. NAB is the most cavitation-resistant of the copper alloys, and it has given good service in pumps and valves around the world. Where the cavitation is intermittent, perhaps only under certain operating conditions, this is probably the simplest solution, although not necessarily the lowest-cost solution.

Where the cavitation is more or less constant, or it is intermittent but particularly severe, a design change is really the best solution. If the problem is with a valve or orifice plate, then decreasing the pressure more gradually will prevent cavitation. This might mean replacing a single orifice plate with a series of spaced plates, each producing a smaller pressure decrease than the original plate. Where the problem is in a pump, then the options are to change the operating conditions nearer to its design conditions or to replace the pump with one more suited to the changed operating conditions.

References

1. B. Angell, "Cavitation Damage," in *Corrosion*, 3rd ed., L.L. Shreir, R.A. Jarman, G.T. Burstein, eds. (Oxford: Butterworth Heinemann, 1994), p. 8:197.

2. H.S. Campbell, Report to the Inter Service Metallurgical Research Council, ISMET 2945, CED 526, August 1962.

3. H.S. Campbell, V.E. Carter, *J. Inst. Met.* 90 (1961–1962): p. 362.
4. P.A. Lush, *CME J.* 34 (1987): p. 31.
5. R.J.K. Wood, S.A. Fry, *J. Fluids Eng.* 111 (1989): p. 271.
6. ASTM G32, "Standard Test Method for Cavitation Erosion Using Vibratory Apparatus" (West Conshohocken, PA: ASTM).
7. ASTM G134, "Standard Test Method for Erosion of Solid Materials by a Cavitating Liquid Jet" (West Conshohocken, PA: ASTM).
8. H.S. Campbell, *Aluminium Bronze Alloys Corrosion Resistance Guide*, Publication 80 (Potters Bar, U.K.: Copper Development Association, 1981).

CHAPTER 9

Dealloying

Dealloying occurs when a component of an alloy is selectively dissolved. When one element is selectively removed from a metal, for example, zinc from brass, it is termed *dealloying*. Sometimes one phase in a multiphase alloy is corroded; this is known as *selective phase attack*. Since the two are closely related, they are treated together in this chapter.

9.1 Dezincification

Probably the most well-known form of dealloying is the selective removal of zinc from brass, which is known as *dezincification*. Lucey showed that both zinc and copper are dissolved together as the anodic reaction, and copper is redeposited at the advancing dezincified front as the cathodic reaction.[1] This leaves behind a spongy mass of copper with the same shape as the original material, but with no significant mechanical strength. Dezincification has been observed to form in layers over the whole surface, but it can also form plugs in localized regions. Lucey also showed that dezincification can be stimulated when crevice corrosion is occurring.[1]

Dezincification can lead to a number of problems. In certain soft waters, particularly hot waters, blockage of brass fittings can occur. They become filled with a voluminous, but hollow, white corrosion product, not unlike meringue (Figure 9.1). This is very different to hard water scale of calcium carbonate ($CaCO_3$), which is hard and compact.

If dezincification carries on for several months or years, then it can penetrate the wall of a brass fitting or a brazed joint. Usually, leakage is slow, unless the soft, residual copper plug is disturbed. Over a period of time, the mechanical strength of a component decreases significantly with

FIGURE 9.1 Meringue dezincification blocking a duplex brass fitting. (Photograph courtesy of Copper Development Association U.K.)

time as dezincification propagates. Figure 9.2A shows a brass wire from a toilet cistern in a soft water area. The fracture face has the typical coppery appearance associated with dezincification. The microsection of the wire (Figure 9.2B) shows that a large part of the cross section of the wire had dezincified.

Single-phase, alpha brasses containing up to about 15% zinc do not suffer from dezincification. This includes cast alloys, such as the gunmetals, which are all immune to dezincification. Single-phase brasses containing 15% to ~36% zinc will dezincify, but this can be prevented by the addition of small quantities of arsenic. This is why specifications for alloys like aluminum brass (C68700) require an addition of 0.02 to 0.06% arsenic.

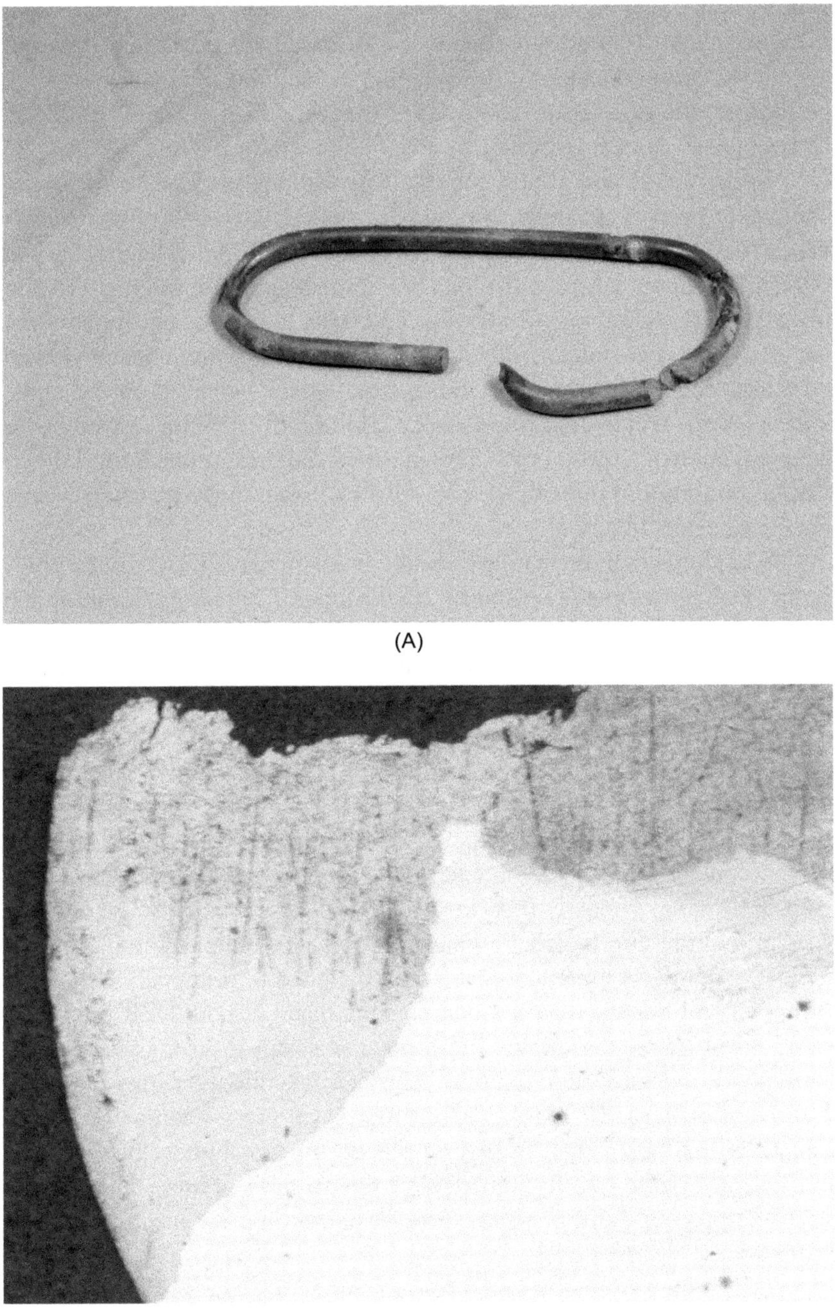

(A)

(B)

FIGURE 9.2 Dezincification of brass wire: (A) fractured wire; (B) microsection showing extent of dezincification.

Brasses containing more than ~36% zinc also contain beta phase such as muntz metal. It is not possible to make an alloying addition that will prevent the dezincification of the beta phase. Included in this category are the high-tensile brasses or manganese bronzes, which dezincify faster as the beta phase content increases.

In the 1950s and 1960s, much effort was made to find an alloying addition to prevent dezincification of brasses containing beta phase. The reason for this effort is that the presence of beta phase makes a brass easy to forge or hot stamp, whereas all alpha brasses may crack on forging. Weldon investigated a wide range of alloying additions to brasses but found none that totally prevented dealloying.[2] Reducing the beta phase content slowed the attack down, but after a period of time, attack occurred on the alpha phase as well. Weldon did find that the addition of ~1% tin slowed down dezincification of a 60/40 brass. This is why naval brass (containing 1% tin) is often preferred to muntz metal for heat exchanger tube plates in waters that cause dezincification.

The problem of preventing dezincification in brasses used for forgings or die-casting was finally solved by BNF Metals Technology Centre in the 1970s.[3] Because of the shape of the copper-zinc equilibrium diagram, a brass with ~36% zinc will be two-phase at 800°C (forging temperature) but all alpha phase at 500°C. A brass with this composition, with additions of lead for machinability, became dezincification-resistant brass, or DZR brass. DZR brass can be hot stamped at 800°C but is 100% alpha after a 2-hour heat treatment at 500°C. Because forging brass often contains impurities that can tie up arsenic, DZR brass requires a slightly higher arsenic content than brasses for heat exchanger tubes to ensure freedom from dealloying, typically 0.06 to 0.10%. DZR brass can also be die-cast if ~0.5% aluminum is added, which necessitates a slightly reduced zinc content. Both these DZR alloys are now included in European Standards. Campbell and Francis exposed components made of both DZR brass and rolled naval brass (the duplex brass most resistant to dezincification) in Portland Harbor seawater for one year.[4] On examination, there was no dealloying of the DZR brass, whereas the naval brass was dezincified to a depth of 0.10–0.15 mm. In several places, the alpha phase between the attacked beta phase was also dezincified. On the basis of these data, Lloyds Register approved DZR brass for through hull fittings on yachts and small craft.

Lucey showed that the effect of arsenic in a single-phase brass is to increase the potential above that at which copper can precipitate, as the cathodic reaction in dezincification.[5] Instead, the cuprous chloride (CuCl) hydrolyzes to form cuprous oxide (Cu_2O), which further acts to prevent dissolution. With beta brass containing arsenic, the potential remains below that at which copper can redeposit, thus explaining why arsenic does not prevent dezincification of beta brasses.

The rate of dezincification can be greatly increased if the brass is coupled to a large area of slightly more noble material such as copper. Hence, when brazing copper, it is usual to use alloys that are not susceptible to dezincification such as Cu-Ag-P. The acceleration occurs because the brazed joint has a small area compared to the large surrounding cathodic area of copper (see Chapter 18).

Dezincification can also occur externally on duplex brass fittings used underground, if the ground water is aggressive. Hence, it is necessary to select materials for fittings to be used underground with care where dezincification may occur.

Not all waters will cause dezincification. Turner examined a large number of waters to determine which ones would cause meringue dezincification.[6] The results in Figure 9.3 show that high-chloride, low-temporary hardness waters with a pH >8 caused meringue dezincification. The laboratory tests compared well with the known behavior of a number of waters from around the United Kingdom, as also shown in Figure 9.3. Note that these composition limits apply only to meringue dezincification. Dealloying can still occur with waters whose composition lies a little below the curve in Figure 9.3, but the corrosion products will not block the fittings. All waters with a high chloride content (such as seawater) cause dezincification, but the zinc corrosion products are soluble, and blockage does not occur. Hence, dezincification-resistant materials are required for seawater and similar brackish waters.

Dezincification is also more likely in hot waters than in cold. This is not just because the kinetics are greater at higher temperatures. Waters with a pH below 8, but which fall within the shaded area of Figure 9.3, are likely to cause dezincification when hot because heating usually causes a rise in pH.

When waters are base exchange softened, the chloride-bicarbonate ratio does not change; rather, only the sodium-calcium ratio changes. Hence, it would be expected that such waters would still cause dezincification. However, this is not the case because fully base exchange-softened waters rarely have a pH >8, that is, a pH sufficiently high to cause dezincification.

In central heating systems, duplex brass fittings are used without problems, despite the high temperatures. This is because the dissolved oxygen (DO) content falls to a very low level a few hours after first

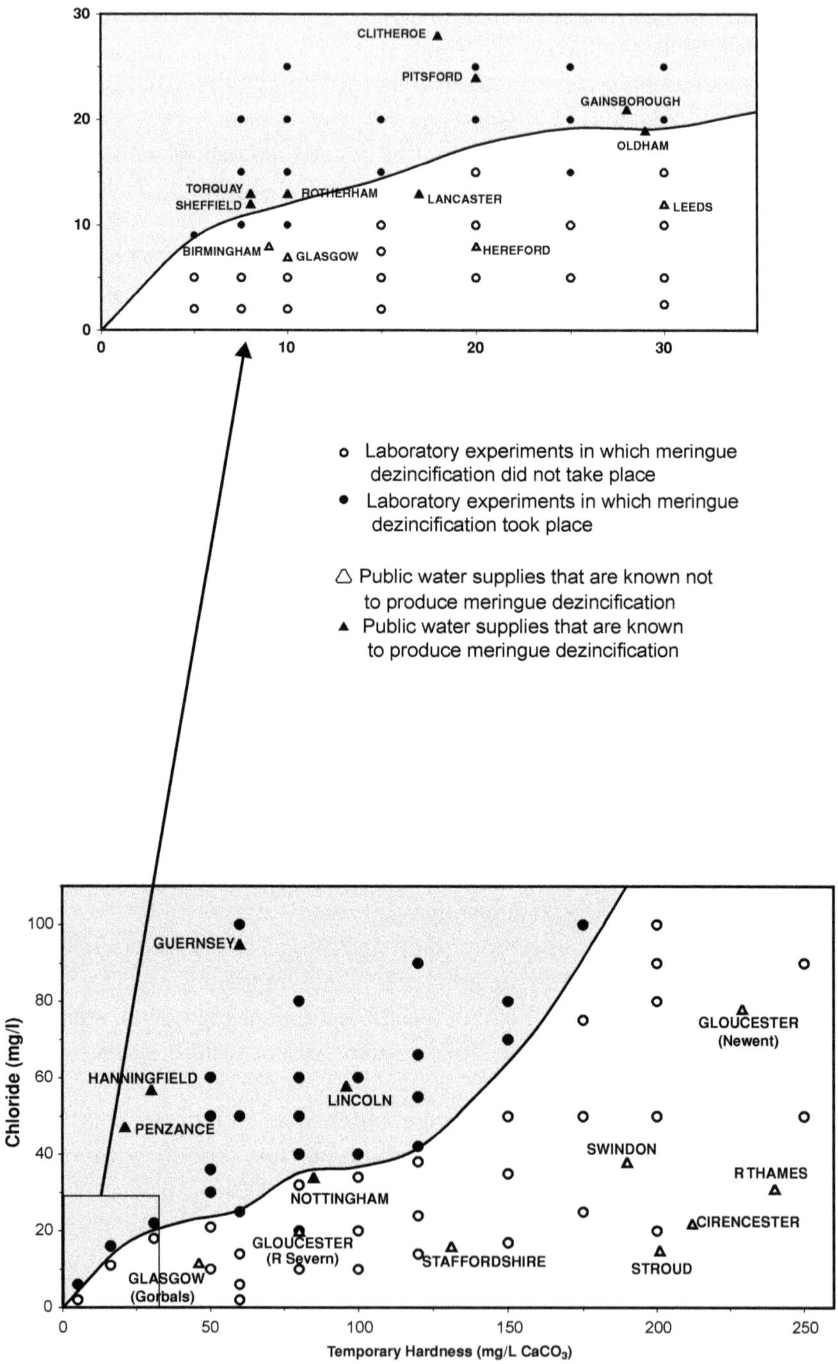

FIGURE 9.3 Influence of water composition on meringue dezincification.[6]

starting the system. Without DO in the water, for the primary cathodic reaction, dezincification does not occur, and duplex brasses can be used, even if the water composition is liable to cause dezincification in aerated waters.

9.2 Dealuminification

Aluminum bronzes, like brasses, can be two-phase. Dealloying of single-phase bronzes (up to ~8% aluminum) is unknown, but aluminum contents above 9.1% will cause formation of the gamma 2 phase in binary aluminum bronzes if they are cooled too slowly. Gamma 2 phase is not selectively attacked in free-flowing fresh water or seawater but dealloying can occur under crevice corrosion, under deposits or marine growths, or under the influence of galvanic corrosion. However, additions of 2% or more of iron suppress the formation of gamma 2 phase, resulting in an alpha-beta structure. Unlike brasses, the beta phase in aluminum bronzes is not detrimental, but rather improves the resistance to erosion corrosion and cavitation.[7]

Manganese stabilizes the beta phase, and in copper-manganese-aluminum alloys (containing typically 12% Mn and 8.5% Al), the beta phase has a different composition to that formed in alloys containing only ~1% manganese. In copper-manganese-aluminum alloys, the beta phase will dealloy in a similar manner to the beta phase in brass. The dealloying is more likely to occur under stagnant or shielded conditions. As the beta phase is continuous in these alloys, the attack can be severe, and the copper-manganese-aluminum alloys should be avoided in seawater when shielded areas or static conditions are likely.[7]

In the nickel aluminum bronzes, alpha and beta phases form at high temperatures, and the beta phase transforms to alpha and kappa on slow cooling.[8] If the alloy is cooled quickly, then residual beta phase can be retained, and this phase is highly anodic to the alpha phase. Welds are particularly sensitive to this as they usually cool very quickly, leaving some beta phase in the weld metal. It is not unknown for a seam-welded pipe that has not been postweld heat treated to unzip in seawater as the beta phase is corroded away (see Chapter 18). The transformation to alpha and kappa for castings can be achieved by a final heat treatment at 675°C for 6 hours, followed by air cooling. A higher temperature of 725°C is recommended for wrought products.[9]

Even when properly heat treated, nickel aluminum bronze (NAB) can suffer selective phase corrosion. This is sometimes on the alpha phase adjacent to the lamellar kappa III phase, but it is seen more frequently on the kappa III phase. Because the phase is present in long, thin stringers, penetration rates can exceed 1 mm/y.[10] The reason for this was explained by Rowlands, who showed that the alpha phase is anodic to the kappa III phase in normal seawater at pH 8.[10] However, at pH 3, there is a polarity reversal, and kappa III becomes anodic to the alpha phase. Hence, shielded areas, where the local pH can decrease to 3, are more likely to see preferential attack of the kappa III phase. Rowlands also demonstrated that if NAB is exposed immediately to flowing seawater, a protective film forms, and no selective phase attack occurs. However, if the NAB is exposed to very slow water flows (~0.1 m/s) or to alternating stagnant and flowing conditions (with stagnant first), then localized attack of the kappa III is likely. Heat treatment at 675°C did not prevent this happening, but it did delay the initiation of attack.

When the attack is on alpha phase adjacent to kappa III, this usually occurs in the heat-affected zone of a weld, and heat treatment does not seem to affect attack in this case.[11]

Maselkowski and Francis exposed in seawater plain and welded coupons of cast NAB that had been heat treated at 675°C and air cooled.[12] Some were in natural seawater with a nominal flow rate of 0.15 m/s, while others were in similar tanks that also contained 0.25 mg/L chlorine. The potentials were monitored with time over a period of 7 months, and the results are shown in Figure 9.4. Over the first 70 days or so, the potential moved electropositively from ~−220 mV to ~−150 mV Ag/AgCl (silver/silver chloride). The initiation of selective phase attack was marked by a sharp increase in potential from −150 mV to ~−50 mV. Welded material was more susceptible to corrosion than plain material. The addition of 0.25 mg/L chlorine retarded the initiation of selective phase attack, particularly for plain material, but after 150 days, all the samples in both tanks were suffering selective phase attack.

Michels and Kain investigated the effect of composition and heat treatment on the selective phase attack of NAB.[13] They tested an alloy with 9.4% aluminum to C95800 and a similar alloy with 8.1% aluminum. The second alloy actually falls outside the minimum aluminum content for C95800 (8.5%). These alloys were exposed in quiescent seawater for 6 months. The alloy with 8.1% aluminum was much more susceptible to selective phase attack, and the attack was also deeper. Material heat treated at 675°C for 6 hours generally suffered less attack, but was not immune.

It has been suggested that the resistance of NAB to selective phase attack is improved if the nickel content exceeds the iron content by 0.5% or more.[7] The nickel exceeded the iron content for both alloys tested by Michels and Kain, but selective phase attack still occurred.

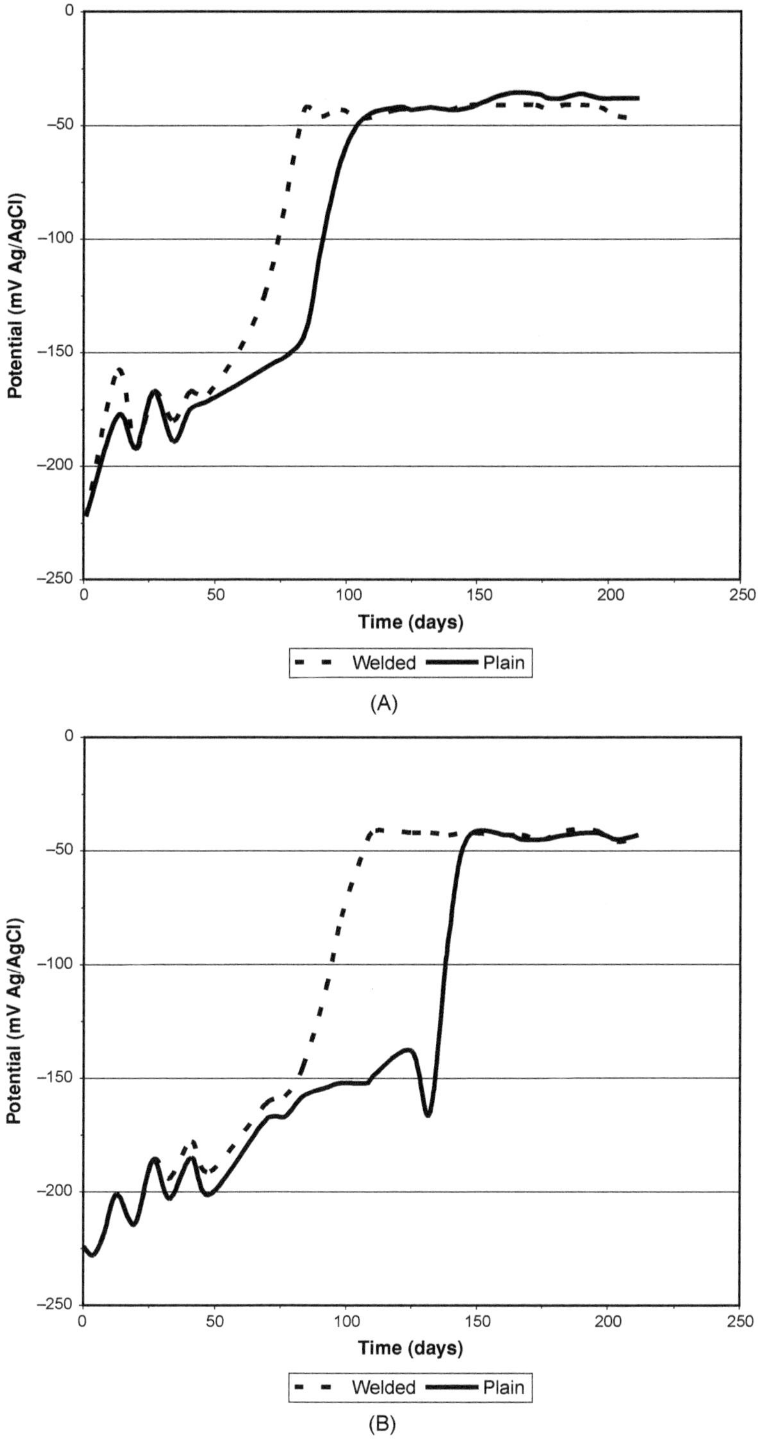

FIGURE 9.4 Potential-time curves for nickel aluminum bronze (NAB) in quiescent seawater at ambient temperature: (A) natural seawater; (B) chlorinated seawater.[12]

TABLE 9.1 Depth of attack of nickel aluminum bronze coupled to superduplex stainless steel after 60 days in seawater[14]

			Depth of attack (mm)		
Couple alloy	Area ratio	Chlorine (mg/L)	Crevice	Nut	Open surface
SD[(A)]	1:1	0	0.27	0.03	1.19
SD	1:1	0	0.35	0.01	0.29
SD	1:1	0	0.27	0	0.22
SD	1:1	0.7	0	0	0
SD	1:1	0.7	0	0	0
SD	1:1	0.7	0	0	0
NAB	1:1	0	0	0	0
NAB	1:1	0	0.01	0	0
NAB	1:1	0	0	0	0
NAB	1:1	0	0	0	0
NAB	1:1	0.7	0	0	0
NAB	1:1	0.7	0	0	0
NAB	1:1	0.7	0	0	0
NAB	1:1	0.7	0	0	0
SD	10:1	0	0.35	0	0.46
SD	10:1	0	0.36	0	0.47
SD	10:1	0	0.39	0	0.27
SD	10:1	0.7	0.20	0.01	0
SD	10:1	0.7	0.22	0	0
SD	10:1	0.7	0.49	0.01	0

[(A)] SD = superduplex.

When NAB is galvanically coupled to a more noble material, such as a high-alloy stainless steel (SS), then selective attack can be stimulated. The author tested cast NAB that had been heat treated, coupled to both itself and a superduplex SS,[14] with cathode-anode ratios of 1:1 and 10:1. The couples were exposed in slowly refreshed tanks of natural seawater or seawater with 0.7 mg/L chlorine at ~23°C for 60 days. Figure 9.5 shows the appearance of some of the samples on removal. The main form of corrosion was selective phase attack, and there was more of it on the samples exposed to natural seawater. The attack took the form of both crevice corrosion and pitting on open surfaces. Table 9.1 shows the maximum depth of attack of the NAB for each couple. The attack was all along the kappa III phase. Because the exposure was only for 60 days, no attack was seen on NAB coupled to itself as this takes some 70 or more days to initiate.[12] The results in Table 9.1 clearly show the severe attack on NAB in natural seawater, both on open surfaces and in crevices. With chlorine additions, open-surface

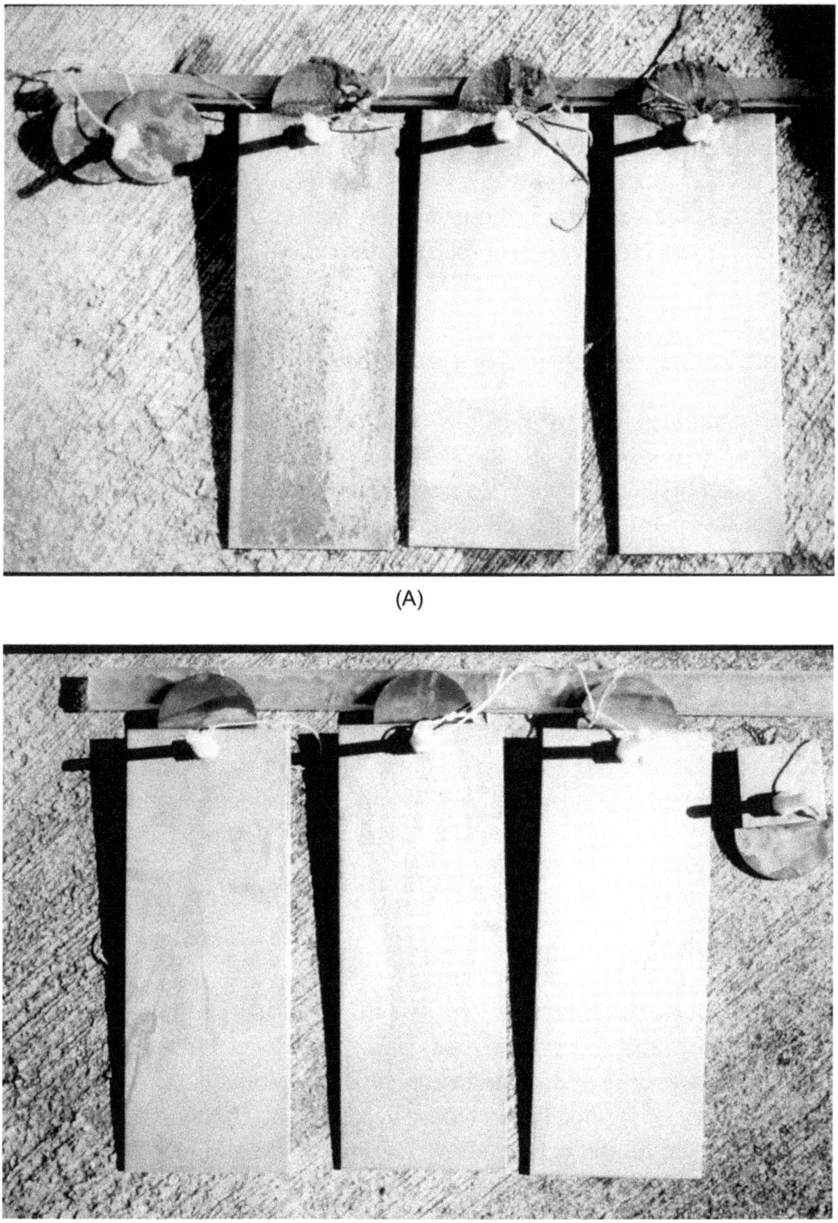

(A)

(B)

FIGURE 9.5 Appearance of NAB/superduplex stainless steel couples after 63 days' exposure: (A) natural seawater; (B) chlorinated seawater.

pitting was eliminated, and crevice attack only occurred at the higher area ratio. This is because the cathodic reaction is less efficient in the presence of chlorine compared with natural seawater.[14]

The laboratory results are borne out by service experience. NAB valves in a superduplex SS piping system on a North Sea oil platform showed severe selective phase attack. This platform was cooled by natural seawater. A similar platform, but using chlorinated seawater, has had few problems with its NAB valves in the superduplex SS pipe work.[15]

9.3 Dealloying of Other Copper Alloys

Denickelification is a term that has been used to describe the selective phase attack of copper-nickel alloys. This has mainly been seen with 70/30 copper-nickel in heat exchangers, where process-side temperatures are high and water flows are low.[16] Such conditions can occur in refineries, where process-side temperatures above 150°C are common. The problem is cured by increasing the water flow to remove local hot spots. This type of attack is rare, and because local high temperatures are required, it is discussed in more detail in Section 9.4.

Reports of dealloying of other copper alloys are even rarer, and all have occurred where local temperatures were high. There are a few reports of dealloying of tin in cast tin bronzes in hot brine or steam and dealloying of silicon from silicon bronze in high-temperature steam, containing acid species.[17]

9.4 Hot Spot Corrosion

Hot spot corrosion is characterized by pitting on the cooling-water side of copper alloy heat exchanger tubes at sites of local high temperature. Redeposited copper is often seen in the corrosion products.

A number of service failures were described by Breckon and Gilbert, who showed that all the copper alloys commonly used in heat exchangers could suffer from this type of attack.[18] Failures in some cases occurred in as little as a few months. BNF Metals Technology Centre examined a number of failed heat exchanger tubes from service where hot spot corrosion was believed to be the cause. Research was also undertaken to establish the conditions under which it could occur. The results led the author to believe that hot spot corrosion can be subdivided into three different types.[19]

9.4.1 Classical Hot Spot Corrosion

This type of corrosion occurs in essentially clean, unpolluted waters, where the temperature on the product side of the heat exchanger is >130°C. The corrosion is characterized by pitting on the cooling-water side of the tubes at sites of local high temperature, often with redeposited copper inside the pits. Bem and Campbell conducted a number of tests in clean seawater on the commonly used copper alloys and pure copper.[20] Only pure copper and 70/30 copper-nickel showed any appreciable attack at 130°C. Breckon and Gilbert carried out tests from 90 to 170°C, but they did not report detailed results, although they said that attack occurred in most cases.[18] However, initiation of this type of attack is a function of both flow velocity and temperature, as was demonstrated by Boden in his experiments on hot spot corrosion of pure copper in sodium chloride (NaCl) solutions.[21] Boden concluded that in aerated solutions, an oxygen concentration cell is established, and only under deaerated conditions would thermogalvanic effects become important in the initiation of hot spot corrosion.

9.4.2 Sulfide-Induced Hot Spot Corrosion

A number of the service failures reported by Breckon and Gilbert occurred in water polluted with hydrogen sulfide (H_2S), and the authors comment that the attack may be exacerbated in polluted waters.[18] The effect of sulfide was studied in more detail by Bem and Campbell, who showed that up to a temperature of 75°C, the rate of attack was very low, and that it greatly increased between 75 and 100°C.[20] Above 100°C, there was no significant change in the rate of attack with temperature. The critical temperature for the acceleration of attack between 75 and 100°C has not been determined.

More recently, work has been carried out by Lloyd et al. and Shone and Grimm on 66/30/2/2 Cu-Ni-Fe-Mn (C71640).[22-23] Some of the tankers operated by Shell had suffered perforations due to corrosion near the outlet end of both steam condensers and oil coolers. The corrosion appeared to be similar to classical hot spot corrosion, but the nominal tube wall temperature was only 60°C. Lloyd et al. tested tubes of the same alloy in a model heat exchanger cooled by an estuarine water containing 0.1 mg/L sulfide.[22] When the tubes were tested at steam temperatures of 40 to 60°C, superficial corrosion occurred at cooling-water velocities of 2.0 and 0.25 m/s. In a third test, the water velocity was reduced to 0.14 m/s, and the steam side of the condenser was partitioned so that half the tubes at the seawater inlet received steam at 50°C, while the other half received steam at

92°C. With such a low cooling-water velocity, air bubbles became trapped in the upper part of each tube. Corrosion similar in appearance to that seen in the service failures had occurred to a depth of 0.1 mm in the upper half of the tubes toward the outlet end. Lloyd et al. concluded that the presence of moving air pockets along the tubes would result in transient localized temperatures that would exceed those seen in areas in contact with cooling water. These results confirm the findings of Bem and Campbell that sulfide reduces the temperature at which hot spot corrosion can occur.[20]

The appearance of the pits is very similar to their appearance in classical hot spot corrosion in that the pits often contain redeposited copper, as shown in Figure 9.6A. To differentiate between classical and sulfide-induced hot spot corrosion, it is necessary to test the corrosion products for sulfides.

9.4.3 Ammonia-Induced Hot Spot Corrosion

The ability of ammonia (NH_3) to induce hot spot corrosion was suspected at BNF from an examination of service failures. The author examined the effects of ammonia on the corrosion of copper alloys in seawater and observed corrosion, similar to hot spot corrosion, occurring at very low flow rates and only when heat transfer was simultaneously taking place.[24] Pitting occurred on all the copper alloys tested when 2 mg/L ammonia was present in the water and there was a temperature difference of as little as 10°C between the outer tube wall and the cooling water. Because the corrosion is associated with low flows and very little heat transfer is required, the attack might also be described as ammonia-induced deposit attack. However, in the absence of a temperature differential, no localized corrosion occurred.[24]

The pits observed by Francis contained mostly copper oxide, with a variable amount of redeposited copper distributed throughout the oxide, as shown in Figure 9.6B.[24] However, tubes that have failed in service in the presence of ammonia often contain substantially more redeposited copper. This is not surprising as the quantity of redeposited copper is likely to depend on the temperature as well as the ammonia concentration and water flow rate. From the preceding observations, it seems likely that there is no lower temperature limit for hot spot corrosion in the presence of ammonia, provided that there is some heat transfer and an as yet undetermined minimum concentration of ammonia present. Both the laboratory tests and the service failures indicate that ammonia-induced hot spot corrosion is only likely to occur when there is little or no cooling water flow, for example, in so-called dead areas in oil coolers.

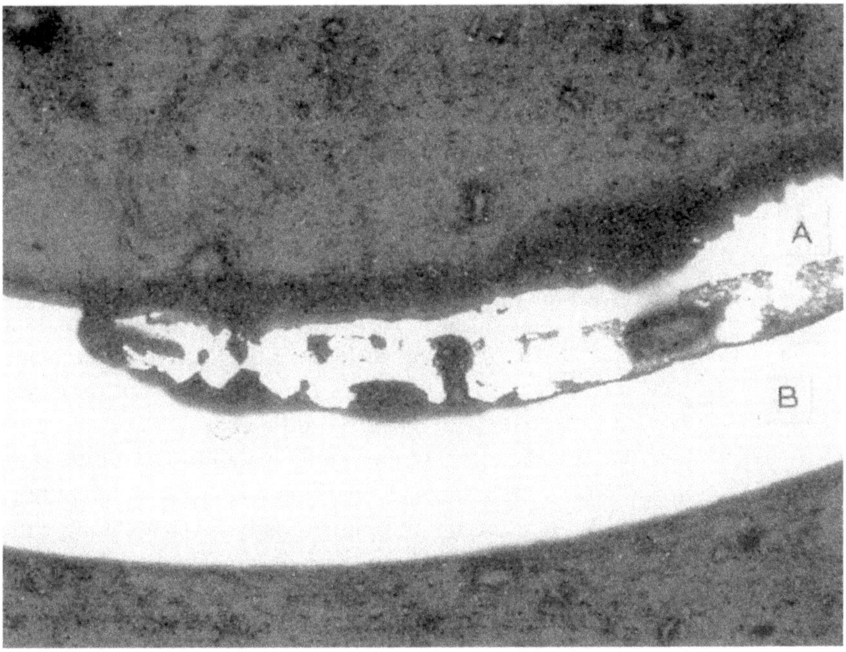

(A) (A = copper; B = parent metal)

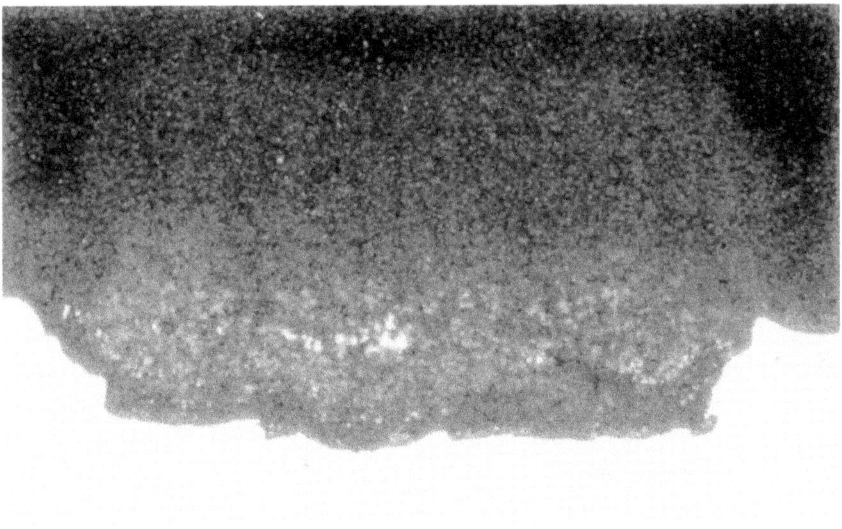

(B)

FIGURE 9.6 Microsections showing redeposited copper during hot spot corrosion: (A) sulfide-induced; (B) ammonia-induced.[19]

A low product-side temperature usually indicates that the presence of ammonia has aided the initiation of hot spot corrosion. Ammonia can sometimes be detected in the corrosion products within the pits, but because of its volatile nature, its retention within the pit depends very much on the tube history between removal and examination in the laboratory.

The effect of mixtures of ammonia and sulfide on the initiation of hot spot corrosion has not been studied in detail, but Francis has reported results obtained with 1 mg/L ammonia and 0.1 mg/L sulfide in aerated seawater.[25] All the copper alloys tested suffered pitting in the same area of the test samples and to the same extent, as was observed with 2 mg/L ammonia alone.[24] In both cases, the temperature difference between the tube outer wall and the seawater was ~10°C. Hot spot corrosion was not observed with either pollutant added singly at that concentration.

Thus, it appears that the effect of ammonia in addition to sulfide is to reduce the concentration of pollutants and the temperature at which pitting can occur. Ammonia is often present in polluted harbor waters, along with hydrogen sulfide, and service failures have been observed in the oil coolers from ferries, where oil temperatures can vary from 50 to 80°C and water flow rates are often low.

9.5 Avoidance of Dealloying

The avoidance of dezincification lies in the selection of dezincification-resistant alloys for use in waters, or underground, where dealloying may occur. The following guidelines should be followed:

0% to 15% zinc	will not dezincify
15% to 36% zinc	will resist dezincification if arsenic is added
>36% zinc	will dezincify, although 1% tin will slow it down

Note that some elements have a high zinc equivalence such as silicon and aluminum. Silicon has a zinc equivalence of 10; that is, 1% silicon is equivalent to 10% zinc. Hence, it is important that these elements be controlled when a single-phase, alpha structure is required that can be inhibited with arsenic.

The author investigated the failure of some die-cast brass water meter bodies that had dezincified in an aggressive water in the United Kingdom. The bodies were supposed to be in DZR brass to BS EN 1982 (1988),[26] but the silicon content was 0.23%, well in excess of the 0.02% maximum permitted by the standard. The aluminum content was 0.3%, well within the

FIGURE 9.7 Half section of die-cast brass water meter body with fracture around base due to dezincification.

permitted limit, but the silicon content was enough to make the brass two-phase, with significant beta content. The plastic filter in the meter created a crevice, just in a region where the meter body was thinnest. The dezincification in this area weakened the body so much that fracture occurred (Figure 9.7).

It is advisable, when procuring dezincification-resistant alloys, to specify testing to BS EN ISO 6509 (1995).[27] This will demonstrate resistance to dezincification and can also show if DZR brass has not been given a final heat treatment to remove any residual beta phase.

With aluminum bronzes, there is no risk of dealuminification with single-phase alloys such as C61400. The formation of gamma 2 phase can be avoided in two-phase aluminum bronzes by ensuring that they contain a minimum of 2% iron. With NAB, it is advisable to specify that the nickel content must exceed the iron content by 0.5% or more. This is written in some standards, but not all. This measure does not prevent selective phase attack, but it does reduce the risk. As a final operation, that is, after hot forming or welding, it is recommended that components be heat treated for 6 hours, followed by air cooling. The temperature should be 675 $\pm$ 25°C for

TABLE 9.2 Relative susceptibility to hot spot corrosion of some common copper alloys

ALLOY	UNS No.	SUSCEPTIBILITY TO HOT SPOT ATTACK		
		Classical	Sulphide	Ammonia
Admiralty Brass	C44300	Erratic results	Erratic results	NT
Aluminium Brass	C68700	Most resistant	Most resistant	Intermediate
90/10 Cu-Ni	C70600	Intermediate	Intermediate	Most susceptible
70/30 Cu-Ni	C71500	Most susceptible	Most susceptible	Intermediate
66/30/2/2 Cu-Ni-Fe-Mn	C71640	Most susceptible	NT	Most susceptible
Alloy 722	C72200	NT	NT	Most resistant

NT = Not Tested

castings and 725 ± 25°C for wrought products. This ensures freedom from residual beta phase and also improves the mechanical properties slightly.

To prevent selective phase attack of the kappa III phase in NAB, it is advisable to put the component into flowing water soon after first immersion so that a protective film forms quickly. This seems to confer a high resistance to subsequent dealloying.[10] It is interesting to note that industry, which usually puts materials into service as quickly as possible, reports few problems with selective phase corrosion of NAB, while the Royal Navy, whose vessels spend a long time with intermittent flow in fitting-out basins, has had numerous problems.

Dealloying in other alloys appears to be associated with high temperatures and low water flows. Hence, dealloying can be avoided by keeping water flows high, ≥1.5 m/s.

The relative susceptibility of the copper condenser alloys to hot spot corrosion is shown in Table 9.2. Generally, aluminum brass has the best resistance, particularly when ferrous sulfate ($FeSO_4{\cdot}7H_2O$) dosing is being carried out. The best way of preventing hot spot corrosion of all types is to increase the water flow. A minimum of 1.5 m/s should ensure that hot spot corrosion cannot initiate. The problems of hot spot corrosion in tankers, described by Lloyd et al., were also remedied by increasing the cooling-water flow rate.[22] In large tankers, this is achieved by switching from scoop feed to pump feed when traveling at slow speed. Where increasing the flow is not possible, sometimes a modification is possible to eliminate dead or slow-flow areas. Francis quotes two cases in which ammonia-induced hot spot corrosion occurred in heat exchangers tubed with 70/30 copper-nickel.[24] In both cases, remedies were effected by small changes in the heat exchanger design and mode of operation, to eliminate slow-flow or dead areas. This should be equally effective in combating classical or sulfide-induced hot spot corrosion.

Francis showed that tubes that were dosed with iron did not suffer hot spot corrosion, even with 2 mg/L ammonia in the water.[24] The author used ferrous sulfate dosing for 2 months at the start of a multistage flash (MSF) desalination plant to form more protective films on 90/10 copper-nickel heat exchanger tubes. The water was intermittently polluted with ammonia, which had caused hot spot corrosion in the heat rejection section when the first unit was started. This was retubed, and ferrous sulfate dosing was started from new at a level of 1 mg/L Fe^{2+} for 1 hour per day. This was continued for 60 days, until protective films were judged to have formed. The plant has now been in operation since 1994 with no further failures. At another MSF plant in the Middle East, the ammonia pollution was almost continuous because the water intake was near the outfall of a fertilizer plant. Daily dosing of ferrous sulfate throughout its life was recommended for this plant, as it was tubed with aluminum brass. There had been failures of 90/10 copper-nickel tubes in an earlier plant on the same site. The MSF plant has now been in operation since 1998 with no problems.

Where sulfide or ammonia and sulfide are present, ferrous sulfate dosing may not be successful. An alternative could be to use low doses of chlorine or sodium hypochlorite (NaOCl) at the cooling-water inlet to increase the redox potential. This technique was used by Cigna et al. to solve corrosion problems due to sulfide in a Belgian power station.[28] The aluminum brass tubes had suffered corrosion during the first year of operation due to sulfides. The tubes were acid cleaned to remove the harmful films, and the redox control system was instituted. There have been no further corrosion problems. Some initial work would be required to identify the optimum redox potential for each specific application.

References

1. V.F. Lucey, *Brit. Corros. J.* 1 (1965): p. 9.

2. B.A. Weldon, BNF Metals Research Association Research Report, A1148, April 1957.

3. J.E. Bowers, P.W.R. Oseland, G.C. Davies, *Brit. Corros. J.* 13 (1978): p. 177.

4. H.S. Campbell, R. Francis, "Dezincification Resistant Brass, CZ132, in Seawater," *The Metallurgist*, April 1983.

5. V.F. Lucey, *Brit. Corros. J.* 1 (1965): p. 53.

6. M.E.D. Turner, *Proc. Soc. Water Treat. Exam.* 10 (1961): p. 162.

7. H.S. Campbell, *Aluminium Bronze Alloys Corrosion Resistance Guide*, Publication 80 (Potters Bar, U.K.: Copper Development Association, 1981).

8. G.W. Lorimer, F. Hasan, J. Iqbal, N. Ridley, *Brit. Corros. J.* 21 (1986): p. 244.

9. R. Francis, *The Selection of Materials for Seawater Cooling Systems—A Practical Guide for Engineers* (Houston, TX: NACE International, 2006).

10. J. Rowlands, "Studies of the Preferential Phase Corrosion of Cast Nickel Aluminium Bronze in Seawater," Eighth International Congress on Marine Corrosion and Fouling (Mainz, Germany, 1981).

11. E.A. Calpan, G. Rose, *Brit. Corros. J.* 14 (1979): p. 160.

12. C.R. Maselkowski, R. Francis, BNF Metals Technology Centre Research Report, R506 (Supplementary Report), July 1985.

13. H.T. Michels, R.M. Kain, CORROSION 2003, paper no. 262 (Houston, TX: NACE International, 2003).

14. R. Francis, *Brit. Corros. J.* 34 (1999): p. 139.

15. R. Francis, *Galvanic Corrosion—A Practical Guide for Engineers* (Houston, TX: NACE International, 2001).

16. *A Guide to the Selection of Marine Materials*, 2nd ed. (INCO, 1973).

17. R. Steigerwald, *Metallurgically Influenced Corrosion*, ASM Handbook, vol. 13 (Materials Park, OH: ASM International, 1987), p. 130.

18. C. Breckon, P.T. Gilbert, "Corrosion of Condenser Tubes Under Conditions of Local High Temperatures," *Proceedings of the First International Congress on Metallic Corrosion* (London, U.K.: Butterworths, 1961), p. 625.

19. R. Francis, *Brit. Corros. J.* 22 (1987): p. 199.

20. R.S. Bem, H.S. Campbell, "Laboratory Hot Spot Corrosion Tests of Condenser Tube Alloys," Proceedings of the First International Congress on Metallic Corrosion (London: Butterworths, 1961), p. 630.

21. P.J. Boden, "Experimental Investigation and Evaluation of the Hot Spot Corrosion of Copper" Proceedings of the Second International Congress on Metallic Corrosion (Houston, TX: NACE International, 1963), p. 771.

22. K.A. Lloyd, R.A. Shaw, E.B. Shone, "Effects of Operating Temperatures and Seawater Flow Rates on the Corrosion of 70/30 Copper-Nickel Heat Exchanger Tubes" Proceedings of the Symposium "Corrosion in Heat Transfer Conditions" (London, U.K.: SCI, 1982), p. 65.

23. E.B. Shone, G.C. Grimm, "25 Years Experience with Seawater Cooled Heat Transfer Equipment in the Shell Fleets," *Trans. Inst. Marine Eng.* 98 (1985): paper 11.

24. R. Francis, *Brit. Corros. J.* 20 (1985): p. 167.

25. R. Francis, *Brit. Corros. J.* 20 (1985): p. 175.

26. BS EN 1982, "Copper and Copper Alloys: Ingots and Castings" (London, U.K.: BSI, 1988).

27. BS EN ISO 6509, "Corrosion of Metals and Alloys: Determination of Dezincification Resistance of Brass" (London, U.K.: BSI, 1995).

28. R. Cigna, K. De Ranter, O. Fumei, L. Giuliani, *Brit. Corros. J.* 23 (1988): p. 190.

CHAPTER 10

Stress Corrosion Cracking

Stress corrosion cracking (SCC), or *environmentally assisted cracking*, requires a specific corrosive environment and a simultaneous stress for it to occur. The environments that can cause SCC are often specific to a metal or alloy, for example, the cracking of titanium in methanol. There are several environments that can cause SCC of copper and its alloys, of which the best known is ammonia (NH_3). This and others of commercial significance are reviewed in this chapter.

The stress to cause cracking can be an externally applied load due to service conditions or an internal stress due, for instance, to welding or cold working. SCC is not a new phenomenon; so-called season cracking of brass cartridge cases and other brass items has been known for over 150 years. The environment that caused the cracking was airborne ammonia, and the stress was the residual stress in the deep-drawn cartridge cases.

There are a number of reviews of SCC of copper alloys that are useful, particularly where they evaluate apparently conflicting results. Whittaker reviewed SCC data up to 1965 for the International Copper Research Association (INCRA) in the United Kingdom (now the International Copper Association, New York).[1] Campbell updated this review in 1973, while Ledheiser did an update for INCRA in 1985.[2-3] McEvily produced an atlas of stress corrosion curves in 1990, but there is a paucity of data for some alloys and some environments.[4] Whittaker and Ledheiser both commented on the shortage of data in many areas and, particularly, for many cast copper alloys, and also some common wrought ones. In the 20+ years since Ledheiser's review, there has been little work done to fill this gap with engineering data.

There are some general comments about SCC that are worth making before discussing specific environments. The first is that there are a number of environments that can cause SCC of copper alloys, and the alloy ranking order with regard to SCC resistance can change with the environment. When SCC occurs under natural atmospheric conditions, it is common for ammonia, sulfur dioxide (SO_2), and chloride all to be present, and the effects of each one are not easily separated. In the atmosphere, a high humidity and a low temperature favor cracking because condensation is more likely. Rates of reaction increase with temperature, and this can increase the risk of SCC. However, this is more important under immersed conditions, rather than atmospheric exposure, in which higher temperatures increase the rate of evaporation and decrease the time of wetness. Cracking may also correlate with rainfall because a thin film of moisture is necessary for corrosion. However, high corrosion rates mean that SCC is unlikely because it usually occurs where a protective film suffers local breakdown.[1]

The nature of the films that form is also very important. Under normal atmospheric conditions, the more hygroscopic a film is, the more water it will retain, and the more likely it is that SCC can initiate.[1] The orientation of the samples can also affect the susceptibility to SCC. Samples cut transverse to the rolling direction can perform differently from those cut longitudinally. This is because the metallurgical properties (for example, strength) vary with the rolling direction.

Although there is a large body of data published on the SCC resistance of copper and its alloys, much of it has been generated under a wide range of test conditions, stressing, and so on. This makes comparison of data from different laboratories very difficult.

10.1 Ammonia

Copper and its alloys are susceptible to SCC in moist ammonia vapor and also in the presence of certain ammonia compounds, although there are no known cases of SCC of the single-phase copper-nickel alloys.[5] Figure 10.1 shows a typical intergranular SCC crack in dezincification-resistant (DZR) brass after 20 hours in Mattsson's solution. (Tests in Mattsson's solution are discussed in Section 10.7.) In addition, amines can also cause SCC, either by direct attack of the metal or by breakdown of the amines to release ammonia.[1] All of these are common in service environments. Amines can be found as additives to boiler feed water or as additives to concrete. Ammonia is present in some fertilizers and cleaners and can also be produced by the decomposition of some organisms, both in the soil and in the sea.

Pure copper is immune to SCC by ammonia, but commercially pure grades of copper, such as phosphorus-deoxidized (PDO) copper, with

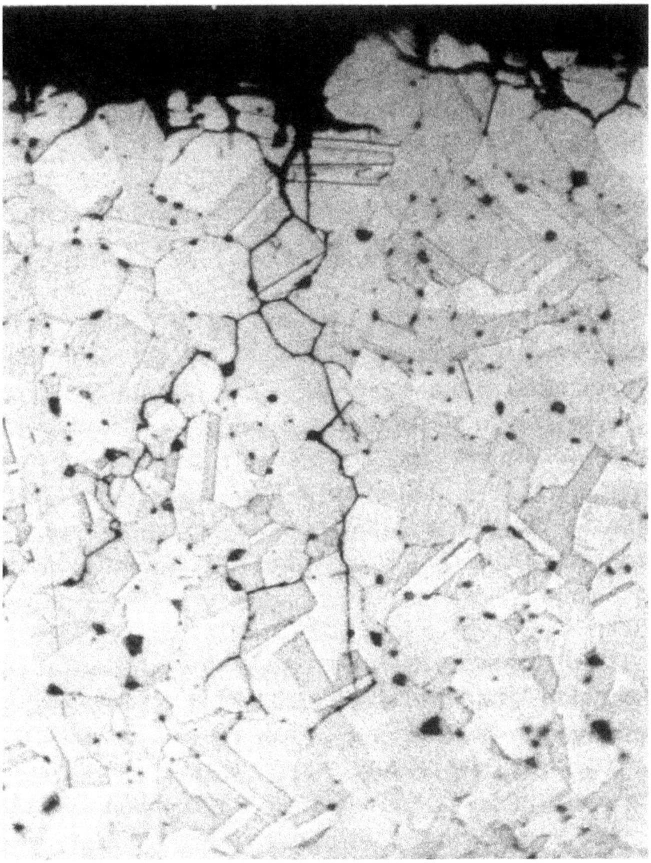

FIGURE 10.1 Typical stress corrosion crack in DZR brass exposed to Mattsson's solution for 20 hours.

maybe 0.04% phosphorus, and arsenical copper (0.3% As), are susceptible.[2,6]

PDO copper hot water pipes in a district heating scheme failed by SCC where they were laid in foamed concrete with an organic foaming agent, which released ammoniacal compounds. This agent is no longer used, but SCC failures still occur occasionally with copper pipes laid in concrete floors. Ammonia is always detectable on the surface of the tubes. The source could be latex cements used for flooring or the decomposition of amines that were used in curing the concrete.[2]

The effects of alloying elements on SCC resistance were studied by Thompson and Tracey using a season cracker.[7] This is a sealed chamber at constant temperature (35°C) with a controlled atmosphere of 80% air, 16%

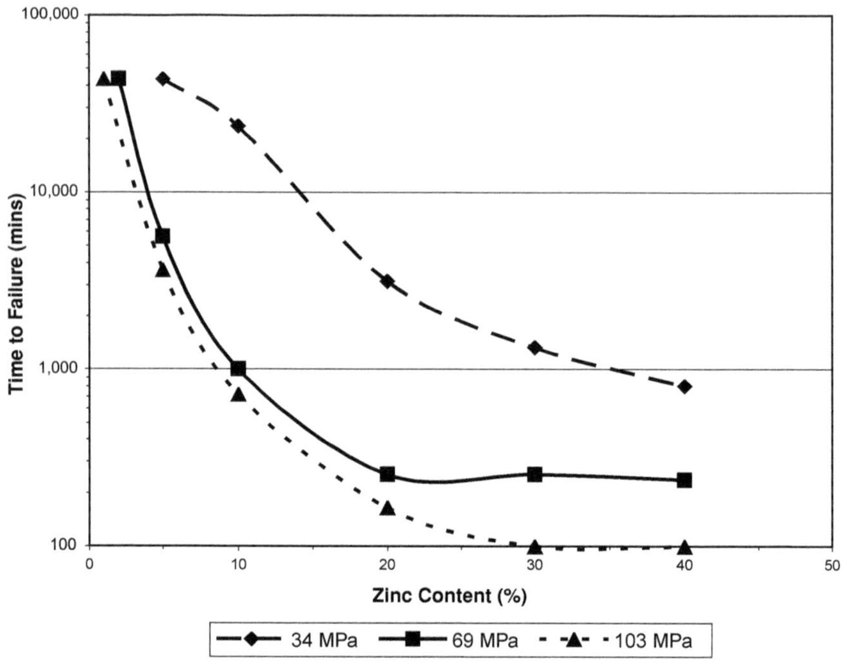

FIGURE 10.2 Time to SCC failure versus zinc content.[7]

ammonia, and 4% water vapor. They alloyed copper with various concentrations of zinc, phosphorus, arsenic, antimony, silicon, nickel, aluminum, and tin. Figures 10.2 and 10.3 show the time to failure vs. alloying content for zinc and tin. Increasing zinc additions clearly reduce the SCC resistance, particularly above 10 to 15% zinc. Shreir et al. commented that failures of brass with up to 10 or 15% zinc are not common.[5] However, low-zinc brasses can suffer SCC in ammonia if the stresses are high enough. Additions of tin increase the SCC resistance, which supports Campbell's comment that he had never seen an ammonia SCC failure of a gunmetal (5% to 10% tin).[2] Note that gunmetals are used as castings, often with thick sections, so that stresses will normally be low.

Thompson and Tracey's data for the other elements generally show a decrease in the SCC resistance with a small alloying addition, followed by a restoration of SCC resistance at higher concentrations.[7] For details, the reader is referred to Whittaker's review because his graphs contain data not included in Thompson and Tracey's original paper.[1]

Brasses are commonly used alloys of copper, and Whittaker commented that the addition of most third elements has no great beneficial

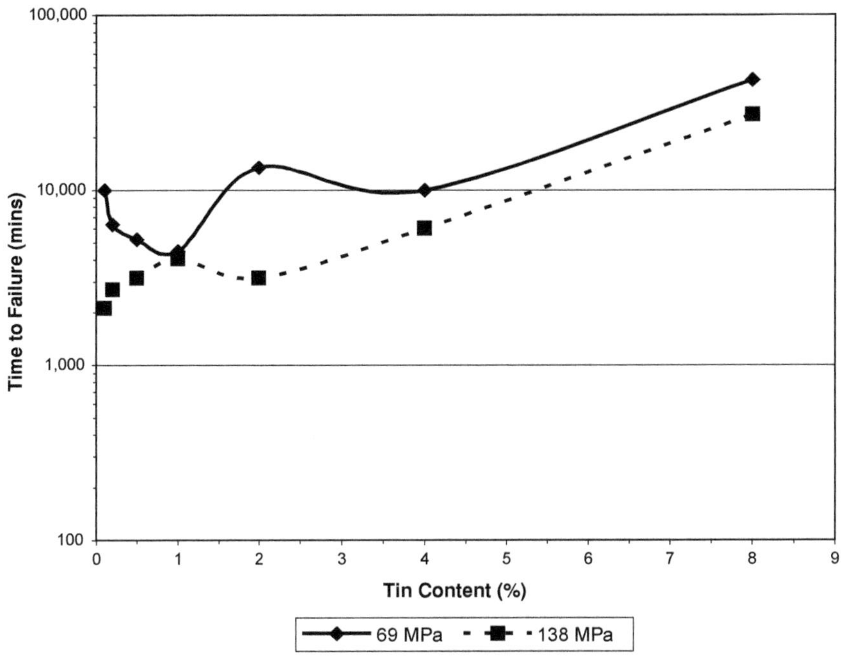

FIGURE 10.3 Time to SCC failure versus tin content.[7]

effect.[1] The results from various authors vary as to whether the effect is beneficial or detrimental, and to what degree. The two exceptions are silicon and nickel, both of which have been observed to increase the SCC resistance of single-phase brasses.

More recently, the effects of arsenic and tin additions to 70/30 brass were investigated in Mattsson's solution at pH 7.3 (for details of Mattsson's solution, see Section 10.7).[8] Both elements increased the SCC resistance, and this was ascribed to the effects of the arsenic and tin on the nature of the protective film.

In some unpublished work at BNF Metals Technology Centre, the SCC resistance of DZR brass was compared with that of a two-phase leaded hot stamping brass in ammoniacal environments.[9] The purpose was to assess the suitability of DZR brass for pipe fittings for use underground. The test samples were forged, three-quarter-inch female couplings, into which taper plugs were screwed to a torque of 38 Nm. Tests were conducted in a simulated ground water containing from 1,000 to 10,000 mg/L chloride and 500 mg/L ammonium (NH_4^+) ions. Small, shallow cracks were seen in the DZR brass after 28 days with 10,000 mg/L chloride, but not at lower chloride

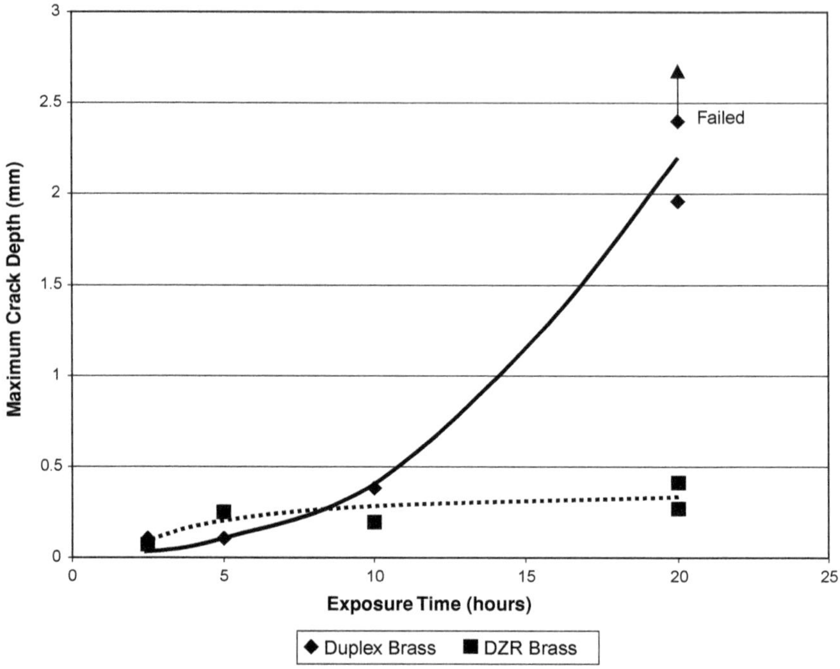

FIGURE 10.4 Crack depth vs. exposure time in Mattson's solution.[9]

concentrations. With 780 mg/L chloride and 1,000 mg/L ammonium ions, there was also some shallow cracking of the DZR brass. No cracking of the duplex brass occurred in any environment, but dezincification occurred in all of them.

Tests were also conducted in Mattsson's solution at pH 7.5 for various times. Figure 10.4 shows the crack depth versus exposure time. The results show that cracking occurs quickly in DZR brass, but once the immediate stress is relieved, cracking stops. In contrast, cracking was a little slower to initiate in the duplex brass, but once started, the rate of propagation increased dramatically. This was presumably due to the higher internal stresses in the duplex alloy. These results suggest that the DZR brass would be satisfactory for plumbing fittings for use underground.

All-beta brass is extremely prone to SCC in the presence of ammonia, the cracking usually being transcrystalline. Two-phase alpha/beta brasses are comparatively less susceptible to SCC by ammonia, and when they do crack, it is usually transgranular through the beta phase.[2] This is why high-tensile brasses should not be used for critical applications where ammonia might be present. Although these alloys are intended to be two-phase, it is possible to finish with an all-beta microstructure that cracks readily.[2]

TABLE 10.1 Time to failure for some aluminium bronzes in ammonium sulfate and copper sulfate at various pH values[3]

	Time to failure (hours)	
pH	Cu/4Al	Cu/8Al
4.0	NF[(A)]	NF
5.8	NF	NF
6.8	NF	NF
7.3	NF	NF
8.3	37.5	21
10.2	37.5	21
11.2	15.5	3.8
12.1	2.2	1.6

[(A)] NF = no failure.

Campbell commented that tin bronzes (7% to 10% tin) occasionally crack in the presence of ammonia, but gunmetals do not.[2] This probably has to do with stresses. Cast gunmetals generally experience low stresses in service due to their thickness, whereas wrought tin bronzes are sometimes highly stressed.

The aluminum bronzes are less susceptible to SCC by ammonia than brasses.[2] This was also shown by the tests on binary Cu-Al alloys in the season cracker, where 5% aluminum alloys gave the same resistance to SCC as copper.[7] Ledheiser presented data on two aluminum bronzes that were constantly strained (0.33 per second) in a solution of ammonium sulfate ($[NH_4]_2SO_4$) and copper sulfate ($CuSO_4$) over a range of pH values.[3] The results (Table 10.1) show that cracking only occurred in solutions with an alkaline pH and that cracking occurred more readily as the pH increased. In addition, the Cu/4Al alloy was slightly more resistant to SCC than the Cu/8Al alloy. However, this final finding was contradicted by other work, in which Cu/1.5Al, Cu/4Al, and Cu/7Al alloys were tested and the Cu/4Al alloy was the most susceptible to SCC.[10]

Atmospheric exposure tests at two sites in the United States showed the relative resistance to SCC of some common copper alloys and the high resistance of nickel aluminum bronze (NAB; Table 10.2).[11] All the alloys were cold rolled and exposed as U-bend specimens, so the stresses were high.

Silicon bronze (Cu-3%Si-1%Mn) has been used to replace high-tensile brass because of its high corrosion resistance and virtual immunity to SCC by ammonia.[2] The other copper alloys that are essentially immune to SCC by ammonia are the single-phase copper-nickel alloys (for example, C70600 and C71500). Note that if the iron content exceeds 1% and it is

TABLE 10.2 Time to failure of some common copper alloys in atmospheric exposure tests[11]

Alloy	Temper (% cold rolled)	Time to failure	
		New Haven	Brooklyn
70/30 brass	50	35–47 days	0–23 days
Leaded duplex brass	50	51–136 days	70–104 days
Admiralty brass	40	51–95 days	41–70 days
Aluminium brass	40	221–495 days	311–362 days
Aluminium bronze (Cu/9.7Al/3.7Fe)	40	>8.5 years	>8.5 years

not all in solution, then 90/10 copper-nickel can be susceptible to SCC by ammonia.[12] If aluminum is added to increase the strength of these alloys, it makes them liable to intergranular SCC in ammoniacal environments.[2] The high-strength copper-nickel alloys are discussed in more detail in Section 10.6.

During the 1960s, BNF conducted a series of tests on heat exchanger alloys in seawater containing ammonia (added as ammonium chloride [NH_4Cl]).[13] Samples of both aluminum brass and 90/10 copper-nickel tubing were held in tension at various stresses, while seawater containing up to 0.75% ammonium ions was circulated around them. The results (Table 10.3) show, not surprisingly, that aluminum brass is susceptible to SCC under these conditions. The threshold stress with 0.75% ammonium ions was ~50% of the 0.1% proof stress, but this increased to >75% of the 0.1% proof stress when the ammonium ion content was reduced to 0.1%. As expected, 90/10 copper-nickel did not suffer any SCC, even at 95% of the 0.1% proof stress. All of the cracking originated in a shielded area beneath a PVC seal. This is not surprising as both Campbell and the author have reported the concentration of ammonia in areas of localized corrosion on copper alloys.[2,14]

The nature of the cracking is affected by several parameters. Cracking is mostly intergranular in straight alpha brasses, although it is transgranular in aluminum brass and mixed mode in admiralty brass. Besides composition effects, the nature of the cracking can also be affected by the stacking fault energy, as changed by cold work, as well as the environment, in terms of the composition and pH.[5] It is well documented that increasing the pH of Mattsson's solution can change the cracking from intergranular to transgranular for many copper alloys.[1,2] The grain size can also be important. Shreir et al. reported increased susceptibility to SCC of large-grained alpha brasses.[5]

TABLE 10.3 Results of SCC tests in seawater plus ammonia[13]

Alloy	NH_4^+(%)	Stress (% of 0.1% PS)	Life (weeks)
Aluminium brass	0.75	95	4
		95	4
		85	8
		85	>34
		75	7
		75	8
		50	38
		50	>95
		30	>52
		30	>95
	0.1	95	32
		95	>40
		75	>40
		75	>40
	0	95	>80
		95	>80
90/10 copper-nickel	0.75	95	>52
		95	>52
	0.1	95	>65
		95	>65
		75	>40
		75	>40

Clearly the higher the stress, the greater is the susceptibility to SCC. The effect of cold work is somewhat varied. With some alloys, it can increase susceptibility, whereas in others, it can reduce it.[1] This is because cold work can produce more than one effect on the microstructure, depending on the alloy. Whittaker described tests on a Cu-Ni-Sn age-hardening alloy.[1] Cold working before aging the alloy significantly increased the SCC resistance to ammonia vapor, as shown in Figure 10.5. Cold working after aging produced only a small increase in SCC resistance.

The Chase Brass Company produced a large number of stress vs. time to failure curves for many alloys using data from a season cracker. Most of the alloys were tested in several different heat treatment conditions. Whittaker reproduced much of these data in his review, but unfortunately, many of the alloys tested are no longer in common use, and there are no data for some recent copper alloys.[1] The data for some common alloys that were included are summarized in Table 10.4; Figure 10.6 shows the data for silicon bronze, which shows the increase in the threshold stress for SCC obtained in the cold-worked condition.

TABLE 10.4 Some results from exposure tests by Chase Brass in a season cracker[1]

Alloy	Nominal composition (wt%)	Temper	Loss in UTS (%)	Endurance stress (MPa)
Commercial brass	Cu/10Zn	Hard	41	276
		Annealed	80	55
Cartridge brass	Cu/30Zn	Hard	99	14
		Annealed	96	21
Muntz metal	Cu/40Zn	Hard	98	14
		Annealed	95	21
Admiralty brass	Cu/30Zn/1Sn	Hard	98	14
		Annealed	96	34
Copper-nickel	Cu/30Ni	Hard	NC[A]	–
		Annealed	NC	–
Silicon bronze	Cu/3Si	Hard	49	345
		Annealed	76	172

[A] NC = no cracking.

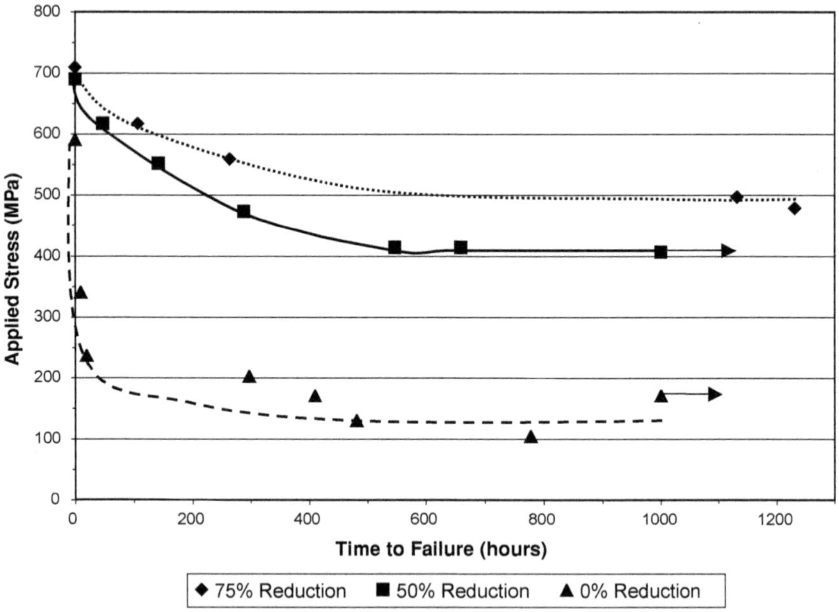

FIGURE 10.5 Applied stress vs. time to failure for Cu-Ni-Si alloy as a function of cold working prior to age hardening.[1]

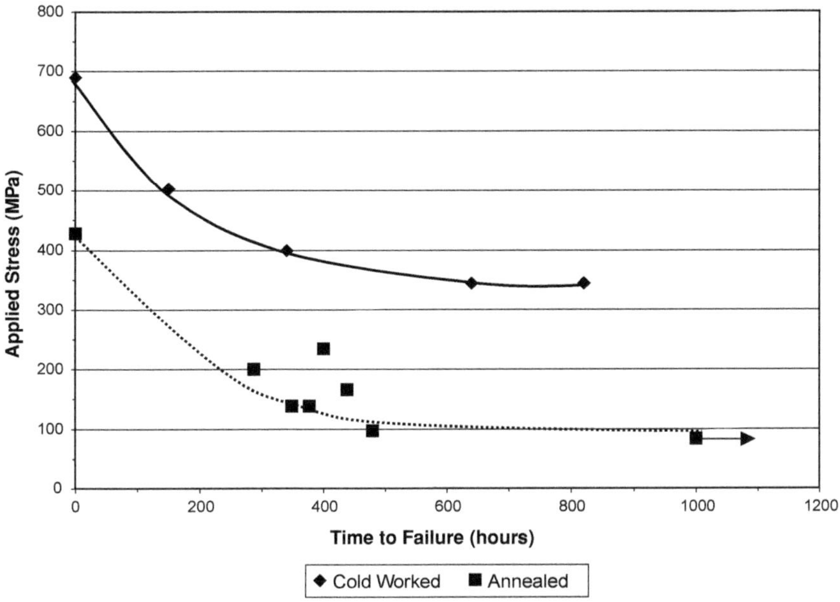

FIGURE 10.6 Stress vs. time to failure in a season cracker for 3% silicon bronze.[1]

It is well known that copper and ammonia readily form complexes. In tests on the SCC resistance of brass, it was found that 3–6 g/L of copper in ammonia greatly increased the corrosion rate and shortened the time to failure.[15] Clearly copper will frequently be present in service due to dissolution of the parent metal, unless it has already formed a very protective film.

The mechanism of SCC in ammoniacal environments has been linked to the films that form. A film formation and rupture mechanism has often been suggested for the propagation of SCC in copper alloys. Hoar and Rothwell refined this model when they recalculated the potential pH diagram for copper, ammonia, and water, taking the oxidation of ammonia to nitrite into account.[16] They showed that at pH 7.3 and 11.3, there can either be a film of cuprous oxide (Cu_2O) or a soluble cuproammonium complex, depending on the potential. They suggested that active attack of the yielding metal at the crack tip produces the soluble complex, whereas a protective cuprous oxide film forms on the static metal at the crack sides. This mechanism seems more plausible than a simple brittle film rupture theory.

One final feature of SCC in ammoniacal environments is the role of localized pitting. Campbell observed SCC from the base of pits in aluminum brass tubes carrying drinking water.[2] Ammonia was readily detected in the

TABLE 10.5 Copper alloys tested by Goodman et al. in mercurous nitrate[17]

Alloy	Description	UNS equivalent
CZ114	Manganese bronze	C67600 (lower Mn)
DGS 1043	NAB	C63200
CZ106	70/30 brass	C26000
CZ132	DZR brass	No equivalent
CZ121	Leaded duplex brass (free cutting)	C38500
CZ122	Leaded duplex brass (forging)	C37700

corrosion product in the pits, whereas only very small amounts could be detected in the bulk water. The author has seen a similar failure of admiralty brass heat exchanger tubes, handling a brackish water. In this case, the pitting was caused by carbon films (see Chapter 5), but the leakage was due to SCC from the base of the pits. As before, ammonia was detected in the corrosion products in the pits, but not in the products on the open surfaces. Clearly ammonia can concentrate in pits, presumably by a mechanism involving the formation of cuproammonium complexes.

10.2 Mercury

Copper and its alloys are all susceptible to SCC by mercury and its compounds. The attack is liquid metal embrittlement, but the threshold stress for cracking, compared with the 0.2% proof stress, varies markedly from alloy to alloy. Mercury is used as the aggressive species in the mercurous nitrate test, which is discussed in more detail in Section 10.7. In this test, samples are exposed in a solution containing 10 g/L mercurous nitrate and 10 mL/L nitric acid (HNO_3) at room temperature.

Mercury and its compounds are the only species that will cause SCC of single-phase copper-nickel alloys when they are in the correct heat treatment condition.

Goodman et al. investigated alternatives to the mercurous nitrate test for the detection of excessive residual stresses.[17] As part of this work, they looked at the applied stress vs. time to failure for some common copper alloys in the mercurous nitrate test. Table 10.5 lists the alloys tested and their approximate UNS equivalents, if they exist. The results, plotted in Figure 10.7, show that all of the alloys have a fairly clear threshold stress, which can be related to the mechanical properties of the alloy. Figures 10.8 and 10.9 show the stress ratio vs. the 0.2% proof stress and the tensile strength, respectively. In these figures, the threshold stress was taken as the

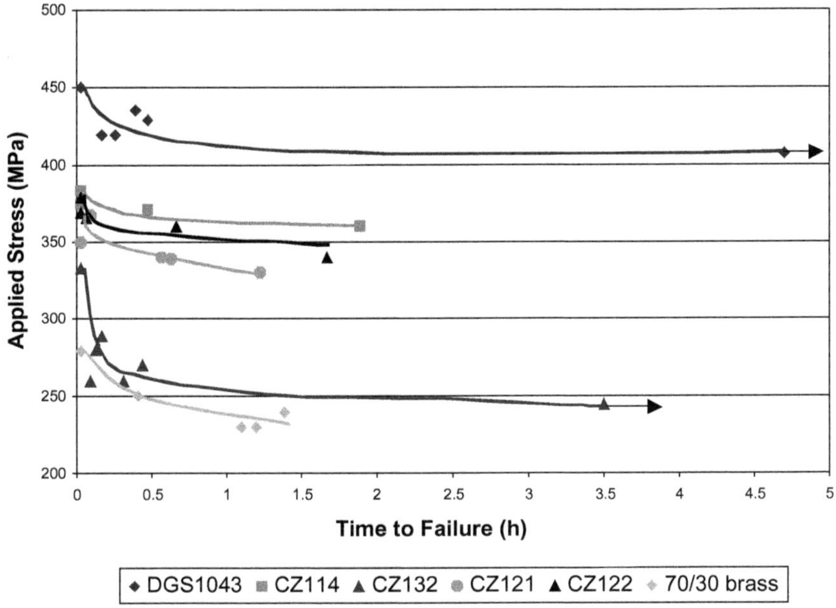

FIGURE 10.7 Applied stress vs. time to failure in a mercurous nitrate test for some stress-relieved copper alloys.

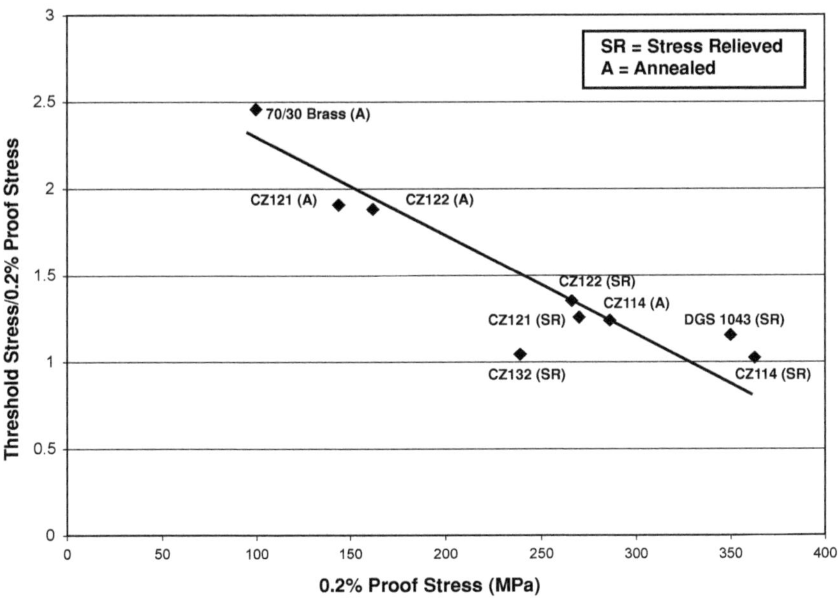

FIGURE 10.8 Stress ratio vs. 0.2% proof stress for some copper alloys in the mercurous nitrate test.[17]

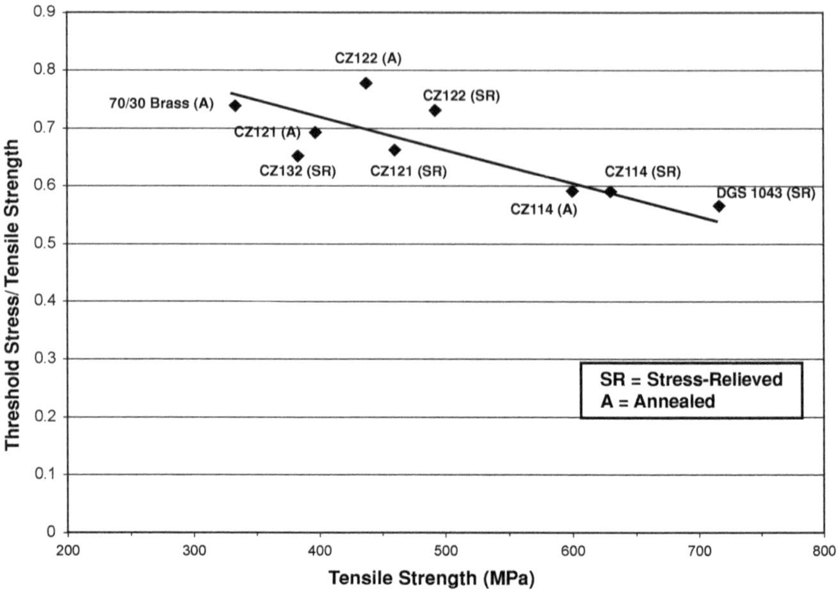

FIGURE 10.9 Stress ratio vs. tensile strength for some copper alloys in the mercurous nitrate test.[17]

stress at which cracking occurred after 0.5 hours because this is the standard exposure time in the mercurous nitrate test for many copper alloys. Once cracking occurs in a tensile sample in mercurous nitrate, failure follows shortly after because the crack propagation rate is high. Figures 10.8 and 10.9 show a strong correlation between the mechanical properties and the threshold stress ratio, except for CZ132 (DZR brass), which consistently fails to follow the trend. DZR brass is a leaded, single-phase brass, and it is not clear why it does not follow the trend when 70/30 brass (also single-phase) does. The threshold stress is around the 0.2% proof stress for the strongest alloys, whereas it is twice that or more for the very ductile alloys.

Goodman et al. also investigated the effect of lowering the mercury concentration to determine whether similar results to the standard test could be obtained.[17] The results, given in Table 10.6, show that similar threshold stresses could be achieved for some alloys, but with significantly longer exposure times. The 70/30 brass showed unusual results in that the alloy was more susceptible to cracking with 10 mg/L mercurous nitrate than with 100 mg/L. Goodman et al. were unable to explain these anomalous results. However, they show that even small quantities of mercury compounds can cause cracking of brasses in a relatively short time.

TABLE 10.6 Effect of reducing the mercury concentration on the exposure time for cracking of some brasses[16]

	Exposure time (hours)			Threshold stress (MPa)		
	$HgNO_3$ (mg/L)			$HgNO_3$ (mg/L)		
Alloy	0.01	0.1	10.0	0.01	0.1	10.0
CZ 121 (stress relieved)	5	–	0.5	360	–	340
70/30 brass (annealed)	2.5	4	0.5	246	240	246
CZ114 (annealed)	–	22	0.5	–	355	355

10.3 Nitrites

Despite the fact that SCC of copper alloys by nitrites has been known for well over 25 years, it is rarely mentioned in most corrosion textbooks. This is possibly because the effect of nitrites was first discovered in the laboratory, and it was thought to have little practical relevance. However, this is far from the case.

Failures of copper heating pipes by SCC led to the DIN specification for coatings and insulation forbidding the use of all materials containing nitrites. Slusser et al. reported the failure of gas cylinder rupture discs in 90/10 brass (C22000) by SCC from the outside.[18] They showed that nitrites readily caused SCC of this brass and several other copper alloys that were being considered as replacements for C22000. Their work is described in detail later.

A failure examined by the author a few years ago also serves as a useful example of the role of nitrites. It concerns an electronics package in the North Sea that was encased in coated carbon steel (CS) and was attached to the flotation buoy by several nickel-aluminum bronze (NAB) bolts. To avoid galvanic corrosion of the CS at defects in the coating, the operators elected to coat the NAB bolts adjacent to the steel. This is in accord with best practice to avoid galvanic corrosion. However, the bolts all failed within a few months, the package sank, and retrieval was an expensive operation. The bolts failed by SCC because the coating method chosen was heat shrink tubing. It is common practice with heat shrink tubing to add nitrite inhibitors to limit corrosion when it is used with CS. However, some of the nitrite reacted with the copper in the NAB to form ammonia, and the combination of nitrite and ammonia caused SCC of the NAB at relatively low stresses. The solution was to use nitrite-free heat shrink tubing.

Whittaker reported some Russian work on simple binary alloys that produced cracking in 10–30 min in a copper nitrite solution.[1] However, it is the later work of Parkins and his coworkers that really defined the susceptibility of copper and brasses to SCC in nitrite solutions.[19-20]

Ammonia does not readily crack copper, but nitrites do; this suggests that nitrites are more potent in terms of SCC.[18] Parkins and colleagues used the slow strain rate technique and defined the potential ranges at which SCC could occur and also the minimum concentration of nitrite to cause cracking.[19-20] Rebak et al. looked at the mechanism of SCC by nitrites for a 37% zinc alpha brass.[21] They investigated a range of solution concentrations and pH values and concluded that SCC only occurred where passivity breakdown had occurred.

When copper or brass is abraded into 1M sodium nitrite solutions, ammonia can be produced by the reduction of nitrite.[22] When Slusser et al. examined failed brass rupture discs, they found ammonia present after exposure to sodium nitrite.[18] Copper cracked under these conditions, whereas it did not crack in the presence of ammonia alone. Hence, the ammonia was not the cause of the SCC. This does suggest that in service failures where ammonia is detected, nitrite may also be involved. As nitrite is readily oxidized to nitrate in the atmosphere, it is useful to check for all three species when examining service failures.

Wu et al. examined cyclic loading of 70/30 brass in sodium nitrite solutions. If the film cleavage mechanism is correct, then the crack growth rate should be dependent on the frequency of cyclic loading.[23] They found no such correlation and concluded that the film cleavage model did not apply to 70/30 brass in nitrite solutions. It seems more likely that the mechanism of SCC involves the formation of cupronitrite complexes at the crack tip, similar to the mechanism proposed for SCC of copper alloys by ammonia.[16]

Slusser et al. tested gas cylinder rupture discs at a pressure of 2,000 psi (138 bar).[18] Every day, the disc was depressurized, sprayed with test solution, and then repressurized. Most tests were conducted for 25 cycles, or until failure occurred. Their results are summarized in Table 10.7, which shows that nitrites readily caused cracking of 90/10 brass and copper, but the susceptibility decreased as the temperature increased. They selected either beryllium copper or nickel 201 as their first choices for replacement of 90/10 brass for the rupture discs.

Goodman et al. investigated alternatives to the mercurous nitrate test.[17] They tested a solution of 1M sodium nitrite in 0.2% nitric acid and the same solution with an addition of 5 g/L copper nitrate. This addition was to form the cupronitrite complex more readily and thus promote SCC more rapidly, as was observed with copper and ammonia, described in Section 10.1. The addition of copper nitrate shortened the time to failure

TABLE 10.7 Results of SCC tests on gas cylinder rupture discs at 2,000 psi (138 bar)[18]

	Result (temperature °C)			
Test solution	90/10 brass (C22000)	Copper (C10100)	Cu-Ni-Si (C72500)	Cu-Be (C17200)
1% $NaNO_2$	F4C[(A)]; F10C; NF24C[(B)]	F6C; F8C; F11C	NF22C	NF22C
1% $NaNO_2$ (pH 8–9)	F4C; F8C; F10C; F11C; F9C; F18C	F4C; F8C; NF15C	NF22C	NF22C
1% $NaNO_2$ +1% Na_2PO_4	NF34C	–	–	–
1% $NaNO_2$ + 1% $NaSiO_3$	NF39C	–	–	–
1% $NaNO_2$ + 1% $NaSO_4$	F16C; F8C; NF15C	–	–	–
1% $NaNO_2$ + 1% NaCl	NF98C	–	–	–
Detergent + $NaNO_3$	NF39C	–	–	–
Spray disinfectant (contains $NaNO_2$)	NF39C	–	–	–

[(A)]F = failure.
[(B)]NF = no failure.

under constant load compared with the copper-free solution for alloys CZ114, CZ121, and 70/30 brass. The results for CZ122 are shown in Figure 10.10, where it can be seen that the addition of copper nitrate totally changed the shape of the stress vs. time to failure curve. However, even with a copper nitrate addition, no cracking was observed with DGS1043 (NAB) after 139 hours at a stress of 450 MPa (1.4 × 0.2% proof stress). This is surprising in view of the failure of NAB bolts described previously. Despite the fact that CZ122 readily cracked in nitrite solutions, CZ121 only cracked after 54.7 hours at a stress of 340 MPa (1.3 × 0.2% proof stress). However, the CZ122 was exposed at much higher stresses before rapid cracking occurred (~1.8 × 0.2% proof stress). This study used commercially produced alloys and did not include any effects of grain size, phase distribution, impurities, and so on, on SCC.

10.4 Sulfur Dioxide

Whittaker reported that the Russians were the first to demonstrate that some copper alloys can suffer SCC in moist atmospheres containing sulfur

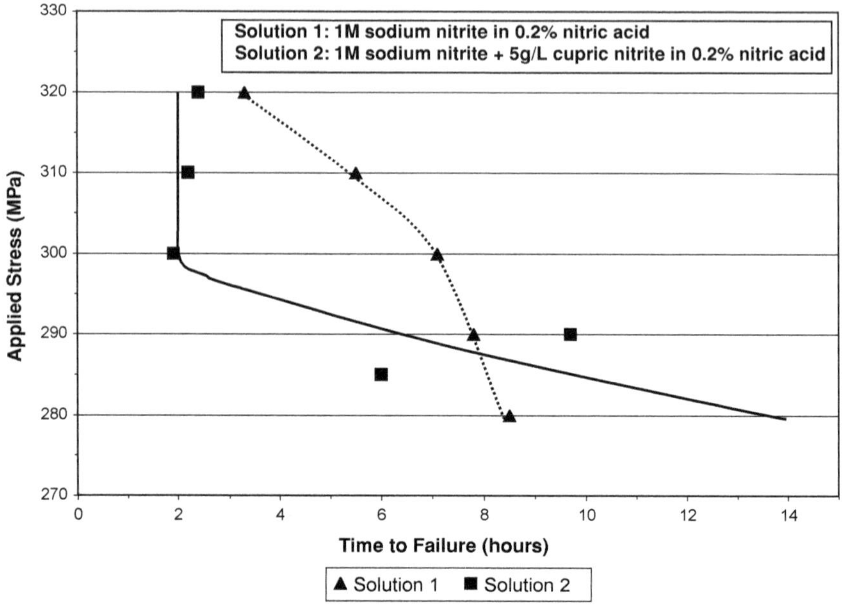

FIGURE 10.10 Applied stress vs. time to failure of annealed CZ122 in nitrate solutions.[17]

dioxide.[1] The attack is much slower than that in the presence of ammonia, and there is no hard quantifiable data on threshold stresses. Whittaker presented previously unpublished data on U-bends of cold-rolled brass strip.[1] The results showed that cracking only occurred when the SO_2 content in the test atmosphere was between 0.05% and 0.5%. At lower SO_2 concentrations, there was no attack, and at higher SO_2 concentrations, the attack became general corrosion everywhere.

Campbell described the work of Bobylev and Manzienko, who tested some copper-zinc-silicon alloys in moist sulfur dioxide.[2,24] They obtained similar results in the laboratory to atmospheric exposure tests, but the laboratory tests were of long duration. Unfortunately, their results do not properly show the effect of silicon because their control alloys, without silicon, had a very different copper content.

Slusser et al. tested a number of alloys as gas cylinder rupture discs (the tests were described in the previous section).[18] One of the environments that was sprayed onto the discs was 1% sodium sulfite (Na_2SO_3). This produced cracking of 90/10 brass (C22000) at 1°C, but not at 22°C or 38°C. No cracking occurred on copper (C10100), beryllium-copper (C17200), or copper-nickel-tin (C72500), although these alloys were only tested at 25°C.

The production of SO_2 atmospheres in the laboratory with a known SO_2 concentration is difficult because it is not possible to calculate this. Cotton, in a private communication to BNF, reported that his laboratory had determined a method by trial and error.[25] They found that a concentration of 0.5% was most useful for determining susceptibility to SCC, but it was possible to produce a range of SO_2 concentrations in an 8-in desiccator (volume of 6 L), as shown:

SO_2 (%)	Na_2SO_3 (g)	+	2%v/v H_2SO_4 (mL)
0.1	0.07	+	10
0.5	0.35	+	10
1.0	0.7	+	10

Cotton also pointed out that the chemicals should be mixed after the desiccator is sealed.

10.5 Other Environments

There have been many studies of SCC of copper alloys, mostly brasses, in a range of environments. One of the most common is chloride, and copper alloys are generally resistant to SCC in chloride solutions. The exception is all-beta brass, which Campbell described as being very susceptible to chloride SCC.[2] Whittaker reported the results of an extended U.S. Navy test of copper alloys in both full immersion seawater and a marine atmosphere.[1] The only alloys to suffer SCC were a cast beta brass and a nickel silver (Cu-27Zn-18Ni).

Attempts to modify the composition of manganese bronze to increase its resistance to chloride SCC were reviewed by Whittaker.[1] The chloride SCC resistance of Cu-33Zn-4Al was greatly improved by additions of zirconium or titanium. A cast manganese bronze (BS1400-HTB3) resisted chloride SCC in seawater when 1.0% to 2.5% nickel was added to the alloy. Without the nickel addition, both samples failed in under an hour.

Sato and Nagata reported the failure of aluminum brass tubes in pure, high-temperature steam.[26] They showed that at steam temperatures from 120 to 210°C, aluminum brass suffered intergranular corrosion that developed into cracking under an applied stress. They found that cracking was worse with a commercial aluminum brass compared with a simple laboratory Cu-Zn-Al ternary alloy. They ascribed the difference to the presence of arsenic in the commercial alloy (added to prevent dezincification). They also found that the coarser the grain size, the worse was the cracking. An

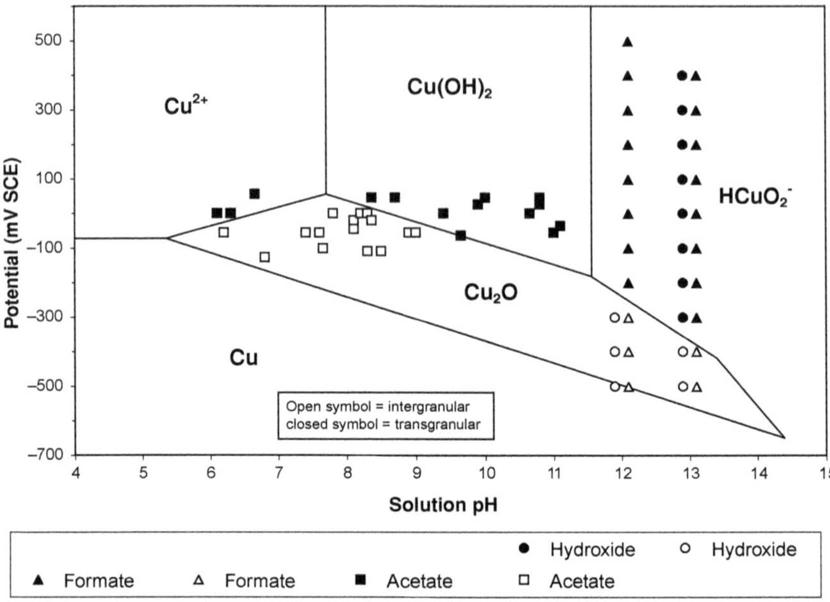

FIGURE 10.11 Potential pH diagram showing the domain of SCC failure for 70/30 brass in various chemical solutions.[27]

alloy of 70/30 brass with 0.03% phosphorus was totally resistant to cracking in this environment.

Cracking of single-phase aluminum bronze (C61400) has also occurred in high-pressure steam service.[11] The addition of 0.25% tin to this alloy completely eliminated the susceptibility to SCC in this environment. The modified alloy is designated C61300. Campbell commented that C61400 is also susceptible to SCC in hot brines and recommended the use of the tin-modified alloy for this service.[11]

Campbell reported tests on 70/30 brass in both citrate and tartrate solutions, both of which produced cracking.[2] Citrate produced the more rapid cracking. Parkins and Holroyd investigated 70/30 brass at various potentials in acetate, formate, and tartrate solutions.[27] They demonstrated that intergranular cracking could occur in all three solutions when the potential and pH were such that cuprous oxide (Cu_2O) was the stable phase. At more positive potentials, the cracking was transgranular, with mixed-mode cracking at the cuprous oxide domain boundary. The results are summarized in Figure 10.11. At very positive potentials, the corrosion was either pitting or general attack. Parkins and Holroyd used electrochemical techniques to determine the potentials where cracking was likely to occur.[27] These potentials

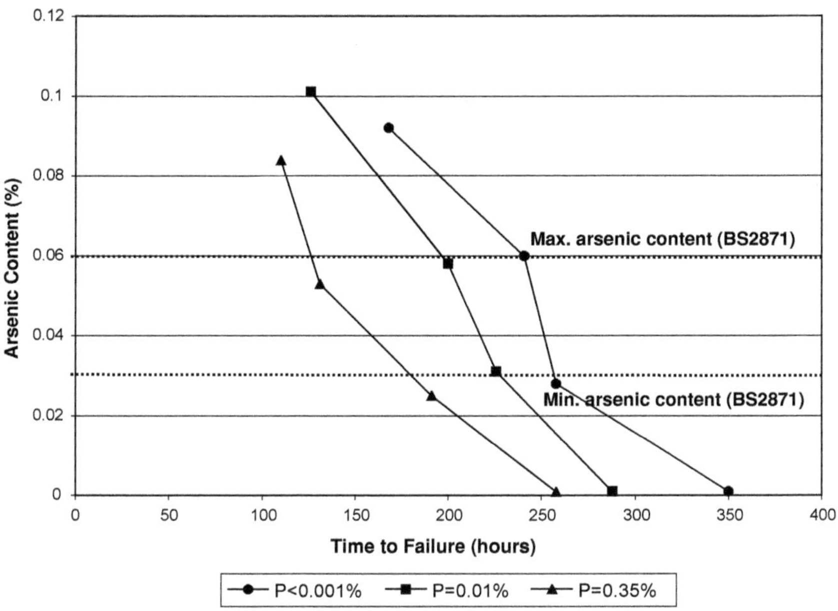

FIGURE 10.12 Arsenic content vs. time to failure of aluminum brass in a seawater-citrate solution.[29]

were at the boundary of active and passive regions, where the scan speed of a potentiodynamic sweep caused a change in the threshold potential for the change from active to passive. It is presumed that the mechanism in all three of these solutions involves the formation of complexes with copper, in a similar manner to that of copper and ammonia, as described in Section 10.1.

In the early 1980s, concern was expressed about the differing ranges of arsenic permitted in arsenical brasses by various standards and its possible effect on susceptibility to SCC. Most national standards permit 0.02% to 0.06% arsenic, but the ASTM standard B111[28] permits up to 0.1%, while the German DIN standard 17660[29] restricts the maximum arsenic to 0.035%. Campbell reviewed the data on arsenic and SCC resistance and concluded that there was no significant difference between the SCC susceptibility with 0.03% arsenic compared with 0.06% at typical phosphorous levels of 0.01%.[30] Figure 10.12 shows the data of Torchio for aluminum brass, obtained in a solution of sodium chloride (NaCl) and sodium citrate ($Ca_6H_5Na_3O_72H_2O$), with cuprous chloride (CuCl) additions.[31]

Campbell also reviewed the effect of arsenic content on intergranular corrosion.[30] The data show that phosphorus has five times the effect that arsenic does on intergranular corrosion, and it is much more important to keep phosphorus levels low, rather than arsenic levels. In the arsenic range of commercial interest, 0.03% to 0.1%, no significant intergranular corrosion was observed on stressed samples exposed to ammoniacal solutions containing copper.

Veld reported the failure of 90/10 copper-nickel overhead condenser tubes in a steam cracker.[32] The environment contained both hydrogen sulfide (H_2S) (1–2 wt%) and ammonia (1–2 wt%) at 120°C maximum and had a pH of 9–9.5. Failure of the cold-worked tubes was by intergranular corrosion after just a few months of operation. The authors argued that solution-annealed tubes would have been more resistant to H_2S under these conditions. An adjacent condenser had been in service without problems for 20 years, but this one was tubed with aluminum brass.

In alkaline solutions, from 10^{-1} to 10^{-5}N sodium hydroxide (NaOH), Whittaker reported cracking of a beta brass at all concentrations and cracking of a duplex brass at 10^{-3}N and 10^{-5}N concentrations of sodium hydroxide.[1] No cracking was seen with an all-alpha brass after 50 days at all concentrations. Parkins and Holroyd carried out SCC tests of a 70/30 brass in caustic solutions at pH 12 and 13 using the slow strain rate test.[27] The results were the same as those found for acetate solutions, reported previously. Where cuprous oxide was the stable phase, the cracking was intergranular, and at more positive potentials, the cracking was transgranular. The results are summarized in Figure 10.11.

Whittaker also presented some results showing that a range of brasses can crack when exposed over nitric acid.[1] Alloys with 70% copper or less were susceptible to SCC.

10.6 High-Strength Copper-Nickel Alloys

The strength of single-phase copper-nickel alloys is relatively low, and there are two main methods of alloying to increase the strength. The first is to alloy with tin, which increases the mechanical properties by a spinodal mechanism. The alternative is to add aluminum, which increases the strength by the formation of a gamma prime (Ni_3Al) precipitate. Both types of alloy have been used where high strength and corrosion resistance are required, for example, in marine fasteners.

Campbell commented that high-strength copper-nickel alloys containing aluminum have been observed to suffer SCC in ammonia environments, but he gave no details.[2] In the early 1990s, there were a series of failures of a copper-nickel-aluminum alloy, mostly as fasteners, on a number of offshore

TABLE 10.8 Summary of the test environments and results from SSRT tests for C72420, high-strength copper-nickel alloy[32]

Environment	Result
Amine paints and lubricants	Cracking
Mercury	Cracking
Ammonia	Cracking
Natural seawater (no CP), $T \geq 50°C$	Cracking
MoS_2 – containing lubricants (except those with a heavy paraffin base)	Cracking
Seawater with CP (all temperatures)	No cracking
Natural seawater (no CP), $T < 50°C$	No cracking

platforms. These were all in the same alloy, C72420. Eight of these were described by Andersen et al., who noted that SCC had been caused by a variety of species.[33] In some cases, the aggressive element was not identified. Those that were included mercury (from a cap seal), hydrogen sulfide (from the decomposition of a molybdenum disulfide-containing lubricant), and amines (from paints and greases). The various operators were concerned with the wide variety of chemicals that could cause SCC of this alloy, all of which are commonly found on offshore platforms.

A series of slow strain rate tests were undertaken in a wide range of chemicals found on offshore platforms.[33] The environments and the results are summarized in Table 10.8. One surprising result was that some of the greases containing bentone or lithium compounds were the most aggressive chemicals causing SCC of C72420. As the production of bentone involves a quaternary amine compound, which can decompose to form ammonia, this is not surprising. Cracking in the presence of lithium compounds has not previously been reported. This paper highlights the need to check carefully all the chemicals that will come into contact with critical components to ensure that none of the ingredients can cause SCC.

Andersen et al. suggested that it might be the presence of grain boundary precipitates that was making the C72420 alloy so susceptible to SCC.[33] This was investigated further by Tuck, who studied the effect of various environments on a copper-nickel-tin alloy (C72900), a copper-nickel-aluminum alloy (Werk. no. 2.1504), and a copper-nickel-aluminum-manganese-niobium alloy (C72420).[34]

In hydrogen charging tests at −1.0 V SCE (saturated calomel electrode), at room temperature, none of these alloys showed any signs of embrittlement. However, in NACE TM0177[35] tests (for sulfide SCC), the

TABLE 10.9 Details of some common tests used to assess the SCC resistance of copper and its alloys

Test	Standard	Solution	Type[(A)]	Duration (hours)
Mercurous nitrate	ASTM B154	10 g/L $HgNO_3$ 10 ml/L nitric acid	I	0.5[(B)]
Mattsson	ASTM G37	55.5 g/L $(NH_4)_2SO_4$ 12.6 g/L $CuSO_4.5H_2O$ 2N NaOH[(C)]	I	24
Eichorn	–	12.5 wt% ammonia solution	A	48
Aebi	ISO 6957	2M NH_4Cl NaOH[(D)]	A	24

(A) A = atmospheric; I = immersion.
(B) Twenty-four hours for Cu-Ni alloys.
(C) Sufficient to adjust to pH 7.5.
(D) Sufficient to adjust to pH 10.

copper-nickel-tin alloy was brittle, and the copper-nickel-aluminum alloy showed secondary cracking. Only Alloy C72420, showed no signs of SCC in this environment. Similar results for all three alloys were found in 10% ammonium chloride at 82°C.

In a series of tests on Alloy C72420, Tuck found that overaging, which increases the amount of gamma prime precipitate, also increases the susceptibility to SCC.[34] Heat treatments that increased the grain size also increased the amount of gamma prime on grain boundaries and increased susceptibility to SCC. Finally, the iron content was also critical and, where it was not all in solution, the susceptibility to SCC increased. Tuck suggested heat treatment to keep the grain size small and the alloy underaged. He also suggested limiting the maximum iron content to 0.85% to increase the SCC resistance (the UNS composition permits up to 1.2% iron).

10.7 Test Methods and Testing

A number of methods are commonly used in the laboratory to assess the susceptibility of copper alloys to SCC. The main ones are summarized in Table 10.9, where it can be seen that apart from the mercurous nitrate test, the tests are based on ammonia. However, these tests do not all have the same purpose, and it is important to be aware of their limitations before doing any testing.

The mercurous nitrate test (ASTM B154[36]) is designed to determine whether there are excessive residual stresses in a copper alloy. As such, it is a production test, with the main requirements that it be simple and rapid. For all copper alloys, except the copper-nickel alloys, a 30-min exposure is sufficient, and cracking can be seen with the naked eye. The copper-nickel alloys are not so susceptible to SCC by mercury, and a 24-hour exposure is required. This is the one big advantage of the mercurous nitrate test; it is the only medium that will produce SCC of copper-nickel alloys. The disadvantages of the test are that the chemicals are hazardous and disposal of spent solution is difficult and expensive. However, it remains the only test that can be used with copper-nickel alloys.

Because of its purpose, the threshold stress for cracking is high for all alloys.[17] Goodman et al.'s attempts to find a replacement for the mercurous nitrate test using one based on nitrites (see Section 10.3) were not successful because there was no solution that would work equally well on all, or even most, copper alloys.[17] The exposure time required varied with the alloy, and a low-power microscope was needed to see the cracking.

Whittaker commented on the high stresses needed for an alloy to fail the mercurous nitrate test.[1] While an alloy that fails the test clearly should not go into service, there is no guarantee that an alloy that passes the test will not fail by SCC in service, particularly if ammonia is present. The same comment is made by Shreir et al.[6]

Mattsson et al. compared the results of tests in mercurous nitrate, Mattsson's solution, and the Eichorn test with long-term atmospheric exposure tests.[37] The samples were deep-drawn duplex brass cups, and some were stress relieved at 250°C for 2 hours. These cups passed the mercurous nitrate test, but many failed within a year during atmospheric exposure.

The Eichorn test cracked all the samples, irrespective of stress relief heat treatment, although cups stress relieved at 275°C or greater did not crack in the atmospheric exposure test. Clearly the Eichorn test is exceedingly aggressive. The results for the tests in Mattsson's solution, at pH 7.5, showed a better correlation with the atmospheric exposure tests. However, some of the samples stress relieved at 275°C and higher temperatures cracked in Mattsson's solution but did not crack in the 1-year atmospheric exposure test. Clearly Mattsson's solution is somewhat more aggressive than atmospheric exposure.

The Aebi test was designed to be less aggressive than the Eichorn test and is controllable by adjusting the pH with sodium hydroxide. Mattsson et al. evaluated the Aebi test as the basis for a standard test to assess the SCC resistance of copper alloys.[38] They compared the results with those from 1-year atmospheric exposure tests at different locations in Sweden. The test worked well with brasses, but more work would be required to demonstrate its suitability with other copper alloys, for example, bronzes. At pH 9.5, there was a good correlation between most alloys in the Aebi test and the atmospheric exposure tests. The test was more aggressive at pH 10.5, although it was not as severe as the Eichorn test. Premoistening the samples increased the severity of the test to close to that of the Eichorn test. Testing at pH 10 gave similar results to the mercurous nitrate test when applied to brasses.

Mattsson not only showed the importance of pH in this test, but also showed the need for good temperature control (±1°C). Although some samples cracked very quickly in the Aebi test, 24 hours was felt to be the optimum exposure time to obtain results that were reproducible. The test has since been adopted as an International Organization for Standardization method as ISO 6957.[39]

Thompson and Tracey's season cracker was described in Section 10.1, but this test requires complex and expensive equipment, and the testing time is much longer than in the Aebi test. However, the results from the season cracker do reflect severe atmospheric exposure fairly well.

It has been pointed out by several workers that the addition of copper ions to a test solution that can form complexes with copper (for example, ammonia) makes the test more aggressive.[1,15] This can obviously shorten testing times when trying to reproduce a service failure.

The method for obtaining a sulfur dioxide atmosphere to test for SCC was described in Section 10.4.

Parkins reviewed the important factors to be taken into account when trying to simulate, in the laboratory, service environments that have caused, or may cause, SCC.[40] Not only is the solution composition important, but the potential and the pH are also important. He also pointed out the usefulness of Pourbaix diagrams and electrochemical testing for identifying regions where SCC may occur. The choice of material for testing is very important because small changes in composition can strongly affect SCC susceptibility in some environments. The surface finish can also be important in some circumstances, as can the sample orientation with respect to the rolling direction.[1]

Whittaker pointed out the need to include unstressed samples as controls. Some alloys suffer intergranular corrosion without undergoing SCC.[1] He also said that, in atmospheric exposure tests over a solution (for example, Aebi), the height of the samples above the liquid affects the aggressivity. The closer the samples are to the liquid surface, the more aggressive is the vapor and the greater the risk of SCC. Hence, it is important to maintain a constant sample height over the solution in laboratory tests to get good reproducibility.

TABLE 10.10 A guide to the susceptibility of copper alloys to SCC[(A)]

	Environment				
Alloy	Ammonia	Chloride	Mercury	Nitrite	SO_2
Copper	R	R	S	S	
90/10 brass	M	R	S	S	S
70/30 brass	S	R	S	S	S
60/40 brass	S	R	S	S	
High-tensile brass	S	M	S		
Beta brass	S	S	S		
Phosphor bronze	R	R	S		
Silicon bronze	R	R	S		
Aluminium bronze	M	R	S	M	
High-strength Cu-Ni	M	R	S		
90/10 Cu-Ni	R	R	S	R	
70/30 Cu-Ni	R	R	S	R	

[(A)] S = high risk of SCC; R = high resistance to SCC; M = may suffer SCC. Empty cells are unknown.

10.8 Prevention of Stress Corrosion Cracking

Where it is known in advance that an environment may cause SCC, it is advisable to select a material that is resistant in that environment. To this end, it is useful to know the relative resistance of copper alloys to SCC in various environments. Table 10.10 gives some guidelines on the relative resistance of some common alloys in some common environments that can cause SCC of copper alloys. Remember that it is not only composition that affects the susceptibility to SCC. The applied stress (as a function of the tensile strength) and the degree of cold work also affect the susceptibility to SCC. Generally, the higher the applied stress, the greater is the susceptibility to SCC. As mentioned previously, cold work can increase SSC susceptibility for some alloys and can decrease it for others. Note that there are significant blanks in Table 10.10, where data are not available. Do not assume that a blank means that an alloy is suitable.

An example of changing alloy is the use of manganese bronze (high-tensile brass) for wall ties, particularly for concrete panels in large buildings. The alloy was selected to replace CS, which had a relatively short life due to corrosion. Unfortunately, there were numerous failures of manganese bronze ties due to SCC, either by ammonia or chlorides.[41] As mentioned in a previous section, manganese bronze is very susceptible to SCC when the beta phase content is high. This can be caused by poor alloying control and/or heat treatment. The solution was to use silicon bronze or aluminum bronze. Both of these have given excellent service.

However, there are sometimes reasons why changing material is not possible, and then methods of minimizing risk must be adopted. One way is to reduce the stress, and Mattsson et al. showed that annealing at 300°C for 2 hours greatly reduces the internal stresses.[37] The atmospheric exposure test results showed much better performance for stress-relieved samples than for as-drawn samples.

Cathodic protection, either impressed current or by means of a sacrificial anode, can depress the potential sufficiently that SCC cannot occur. Separating the metal from the environment is an alternative preventative method that may be attempted in a number of ways.

One way is to use inhibitors, but it is important to test these to ensure that they are satisfactory for any specific environment. Whittaker presented results from a brass exposed to a 0.5% moist sulfur dioxide atmosphere.[1] The untreated material cracked, but material treated with benzotriazole remained shiny and unattacked. However, benzotriazole had no effect on samples exposed to moist ammonia, in which both treated and untreated samples cracked. Whittaker suggested that this was because the inhibitor was soluble in ammonia.

The other way of separating the metal and the environment is to use a coating. Not only must the coating be free of defects, where the environment can penetrate to the metal, but the coating must also resist mechanical cracking on stressing of the base metal.

Laub investigated the use of electroplated coatings on a Cu/35Zn brass.[42] His testing included both stressing of the substrate and the effect of heat treatment. The best coatings were nickel and nickel plus silver (6 μm of each), heat treated for 1 hour at 350°C after coating.

Laub also investigated the use of absorbents as a method of preventing SCC.[43] Silica gel was found to be very useful in the presence of ammonia, and it was recommended as a packing agent for use with brass. Coatings of metallic sulfides, water-repellent layers, and air-drying paints showed little success in preventing SCC.[44] Campbell reported that coating with an air-drying acrylic-ester lacquer containing phenylthiourea prevented SCC of brass in both ammonia and sulfur dioxide atmospheres.[2]

References

1. J.A. Whittaker, "A Survey on the Stress Corrosion of Copper Based Alloys," International Copper Research Association Report, 1966.

2. H.S. Campbell, *J. Inst. Met.* 101 (1973): p. 232.

3. H. Ledheiser, "Stress Corrosion Cracking of Copper Alloys," International Copper Research Association Report, 1985.

4. A.J. McEvily Jr., *Atlas of Stress Corrosion and Corrosion Fatigue Curves* (Materials Park, OH: ASM International, 1990).

5. L.L. Shreir, R.A. Jarman, G.T. Burstein, eds., *Corrosion*, 3rd ed. (Oxford, U.K.: Butterworth Heinemann, 1994), p. 1:50.

6. L.L. Shreir, R.A. Jarman, G.T. Burstein, eds., *Corrosion*, 3rd ed. (Oxford, U.K.: Butterworth Heinemann, 1994), p. 4:63.

7. D.H. Thompson, A.W. Tracey, *Trans. Am. Inst. Min. Met. Eng.* 185 (1949): p. 100.

8. B.C. Syrett, R.N. Parkins, *Corros. Sci.* 10 (1970): p. 197.

9. R. Francis, BNF Metals Technology Centre Research Report, A1969, June 1980.

10. A.W. Blackwood, N.S. Stoloff, *Trans. ASM* 62 (1969): p. 677.

11. H.S. Campbell, *Aluminium Bronze Alloys Corrosion Resistance Guide*, Publication 80 (Potters Bar, U.K.: Copper Development Association, 1981).

12. J.M. Popplewell, *Corros. Sci.* 13 (1973): p. 593.

13. H.S. Campbell, BNF Metals Research Association Report, A1577, February 1966.

14. R. Francis, *Brit. Corros. J.* 20 (1985): p. 167.

15. E.N. Pugh, W.G. Montague, A.R.C. Westwood, *Trans. ASM* 58 (1965): p. 665.

16. T.P. Hoar, G.P. Rothwell, *Electrochem. Acta* 15 (1970): p. 1037.

17. P.D. Goodman, V.F. Lucey, J.J.B. Ward, BNF Metals Technology Centre Research Report, R503/4, August 1985.

18. J.W. Slusser, S.W. Dean, D.M. Drummer, CORROSION/86, paper no. 330 (Houston, TX: NACE International, 1986).

19. J. Yu, J.H. Holroyd, R.N. Parkins, "Application of Slow Strain Rate Tests to Defining the Stress Corrosion Crack Initiation in 70/30 Brass," ASTM STP 821 (1984), p. 288.

20. L.A. Benjamin, D. Hardie, R.N. Parkins, *Brit. Corros. J.* 23 (1988): p. 89.

21. R.B. Rebak, R.M. Carronza, J.R. Galvele, *Corros. Sci.* 28 (1988): p. 1089.

22. R.C. Newman, G.T. Burstein, *J. Electrochem. Soc.* 127 (1980): p. 2527.

23. D. Wu, H.S. Ahluwalia, H. Cai, J.T. Evans, R.N. Parkins, *Corros. Sci.* 32 (1991): p. 769.

24. A.V. Bobylev, Y.P. Manzienko, *Sov. J. Non-Ferrous Met.* 7 (1966): p. 97.

25. J.B. Cotton, IMI, correspondence to BNF Metals Research Association, July 1967.

26. S. Sato, K. Nagata, *Sumitomo Light Met. Tech. Rep.* 15 (1974): p. 73.

27. R.N. Parkins, N.J.H. Holroyd, *Corrosion* 38 (1982): p. 245.

28. ASTM B111, "Standard Specification for Copper and Copper-Alloy Seamless Condenser Tubes and Ferrule Stock" (West Conshohocken, PA: ASTM).

29. DIN 17660 (Brussels, Belgium: DIN).
30. H.S. Campbell, BNF Metals Technology Centre Miscellaneous Publication, MP600, December 1982.
31. S. Torchio, *Corros. Sci.* 21 (1981): pp. 59, 425.
32. A. Veld, *MP* 45, 4 (2006): p. 52.
33. O. Andersen, M.W. Joosten, J. Murali, D.E. Milliams, CORROSION/96, paper no. 78 (Houston, TX: NACE International, 1998).
34. C. Tuck, CORROSION 2005, paper no. 462 (Houston, TX: NACE International, 2005).
35. NACE TM0177, "Laboratory Testing of Metals for Resistance to Sulfide Stress Cracking in Hydrogen Sulfide (H_2S) Environments" (Houston, TX: NACE International).
36. ASTM B154, "Standard Test Method for Mercurous Nitrate Test for Copper Alloys" (West Conshohocken, PA: ASTM).
37. E. Mattsson, S. Lindgren, S. Rask, G. Wennstrom, "Stress Corrosion in Brass," in *Current Corrosion Research in Scandinavia* (Helsinki, Finland: Scandinavian Corrosion Congress, 1964).
38. E. Mattsson, R. Holm, L. Hassel, "Ammonia Test for Stress Corrosion Resistance of Copper Alloys," ASTM STP 970 (1988): p. 152.
39. ISO 6957, "Ammonia Test for Stress Corrosion Resistance" (Geneva, Switzerland: ISO).
40. R.N. Parkins, "Synthetic Solutions and Environment Sensitive Fracture," ASTM STP 970 (1988): p. 132.
41. *Guidance Note on the Security of the Outer Leaf of Large Concrete Panels of Sandwich Construction* (London: Institution of Structural Engineers, 1989).
42. H. Laub, *Metalloberflache* 20 (1966): pp. 413, 453.
43. H. Laub, *Metalloberflache* 21 (1967): p. 587.
44. H. Laub, *Metalloberflache* 20 (1966): p. 493.

CHAPTER 11

Corrosion Fatigue

Fatigue occurs when a material is subject to a repetitive cyclic stress, usually at stresses below that at which failure would occur under constant load. As the peak applied stress is reduced, the number of cycles to failure increases until a threshold stress is reached at which failure does not occur, even after a large number of cycles ($>10^8$), as shown schematically in Figure 11.1. When the material is also exposed to a corrosive environment, failure often occurs after fewer cycles than in air, and this is termed *corrosion fatigue*. With many materials, there is no clear threshold stress for corrosion fatigue (Figure 11.1).

Corrosion fatigue is not just a function of the peak stress; it is also affected by the stress range, which is often defined by the R value, where

$$R = \frac{\text{minimum stress}}{\text{maximum stress}} \tag{11.1}$$

Corrosion fatigue is also influenced by the frequency of the stress cycle and the amount of strain induced under the stress.

Corrosion fatigue has several phases, the first of which is initiation, followed by propagation. Propagation may proceed differently as the crack develops. The stress/number of cycles to failure curves, or S/N curves (for example, Figure 11.1), show the total life to failure, usually for a plain sample. Other tests take precracked samples and look at the effect of various factors on crack propagation. Unfortunately, there is no simple relationship between S/N data and crack propagation data for many practical applications.[1]

Corrosion fatigue cracks propagate by the action of local strains deforming or damaging the protective film, followed by corrosion of the underlying metal. Hence, the environment does not need to be particularly

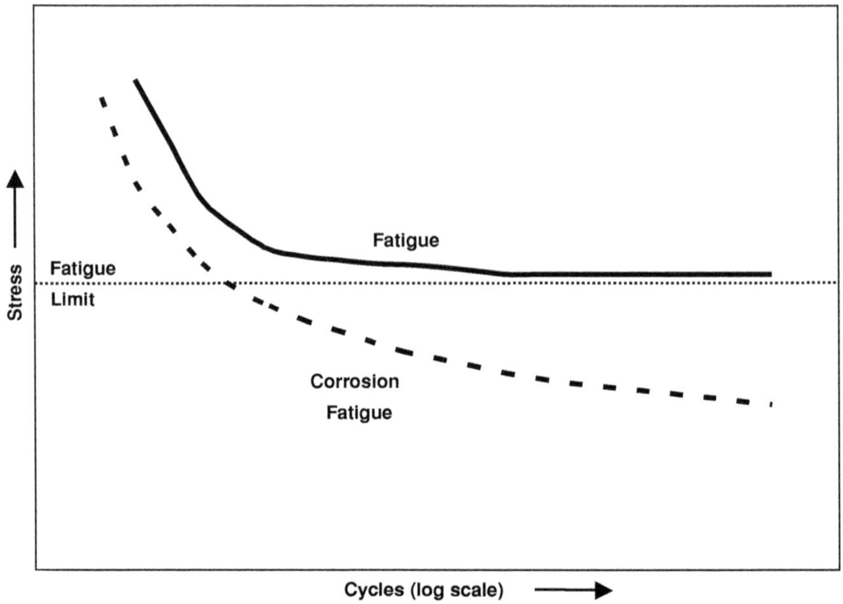

FIGURE 11.1 Schematic comparing the typical stress vs. number of cycles for fatigue and corrosion fatigue.

corrosive for corrosion fatigue to occur.[1] The mechanisms of corrosion fatigue are discussed in detail by Scott, but corrosion fatigue cracks are often blunt ended compared with stress corrosion cracks.[1] Propagation is by a series of crack growth and arrest steps, which produce a classic series of curves on the fracture face, sometimes called *beach marks*. Figure 11.2 shows a fatigue failure of a pump shaft with the curves radiating from the left-hand corner of the groove in the half on the left-hand side, where the crack initiated.

Scott commented that surprisingly little has been published on corrosion fatigue of copper alloys.[1] The following sections group the available data by alloy type.

11.1 Brasses

Jaske et al. reported the results of corrosion fatigue tests of manganese bronze in seawater.[2] The environment had little effect on fatigue crack propagation, although the frequencies were quite high.

Campbell carried out corrosion fatigue tests on aluminum brass in both seawater and seawater containing 0.75% ammonia (NH_3) (added as

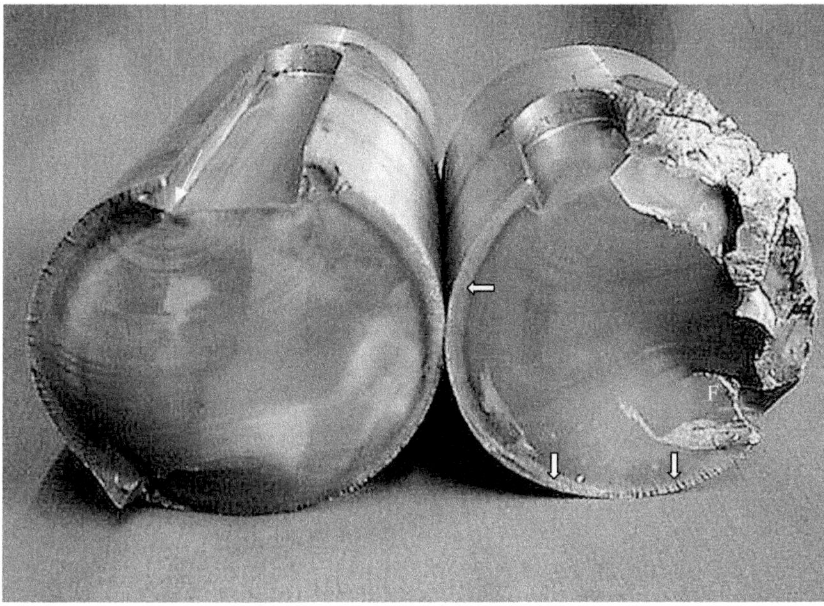

FIGURE 11.2 Typical appearance of a fatigue fracture, including beach marks.

ammonium chloride [NH_4Cl]), using reverse bend tests at low frequency (<1Hz).[3] The results, in Figures 11.3 and 11.4, show the life as a function of the mean stress and the alternating stress. The corrosion fatigue limit in seawater was ±85 MPa for zero mean stress, falling to ±55 MPa with a mean stress of about 100 MPa. In seawater with 0.75% ammonia, the corrosion fatigue limit decreased, being ±70 MPa with zero mean load. This decreased sharply to about ±30 MPa for mean stresses >40 MPa, but this, then, remained constant up to the 0.1% proof stress.

Work described in the previous chapter showed that aluminum brass is susceptible to stress corrosion cracking (SCC) in seawater containing ammonia, so the reduction in corrosion fatigue strength in the presence of ammonia is not surprising. There appears to be a fatigue limit in the presence of ammonia of about 30 MPa, irrespective of mean stress.

11.2 Copper-Nickels

Spiedel suggested that in seawater and aerated salt solutions, the copper-nickel alloys have a corrosion fatigue strength in the range 75–125 MPa.[4]

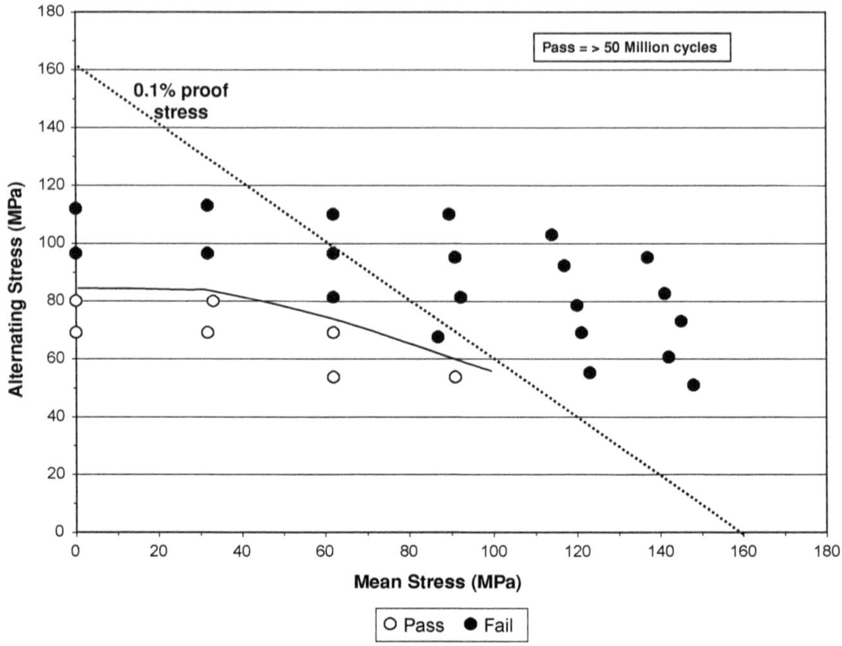

FIGURE 11.3 Corrosion fatigue of aluminum brass in seawater.[3]

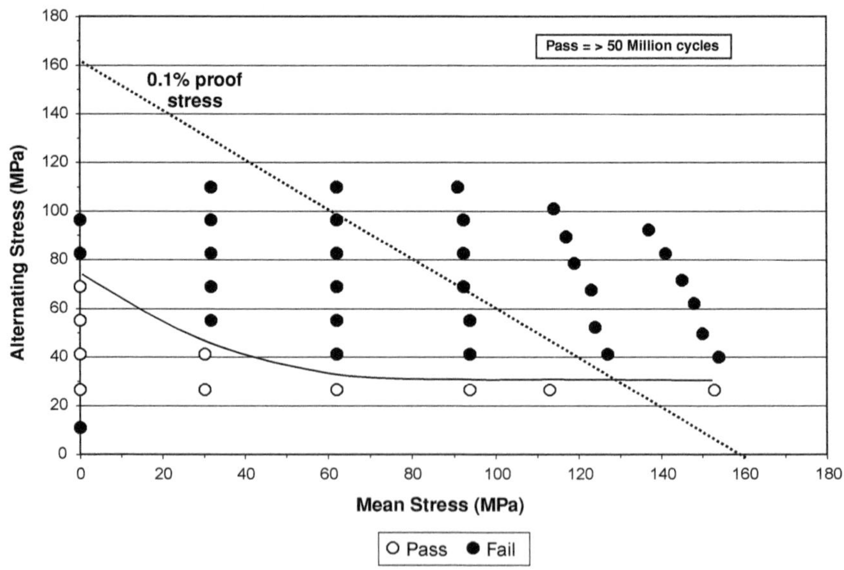

FIGURE 11.4 Corrosion fatigue of aluminum brass in seawater plus 0.75% ammonia.[3]

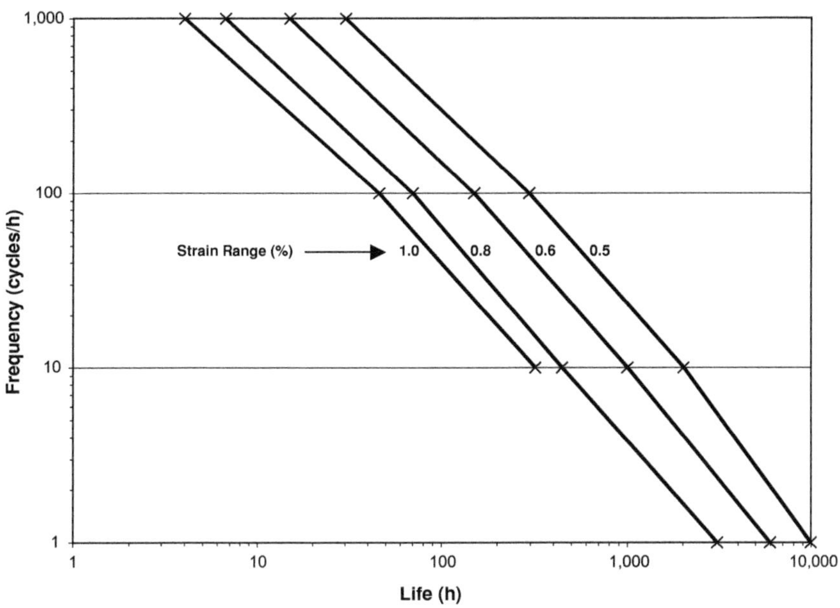

FIGURE 11.5 Life vs. strain frequency for 70/30 copper-nickel in seawater at 10–15°C.[5]

Bowers reported the results of high-strain, reverse bend tests on 70/30 copper-nickel (C71500) over a range of frequencies.[5] The results, given in Figure 11.5, show that the life increases as the frequency decreases, but not in a linear manner. At very low frequencies, the curves show a marked departure from linearity, with an increase in life, but not as much as would be expected from the data at higher frequencies. The introduction of a notch, with a theoretical stress concentration factor of 2.35 or a weld with a matching filler, gave similar reductions in fatigue life compared with plain samples, as shown in Figure 11.6 for a frequency of 10 cycles/h.

Townsend conducted Vibraphore tests on 70/30 copper-nickel alloys at frequencies in the range 50–100 Hz.[6] These results in air would not be expected to be very different in seawater because of the high corrosion resistance of these alloys and the fact that corrosion has a lesser effect on corrosion fatigue at higher frequencies. Figure 11.7 shows the data for 70/30 copper-nickel (C71500) and a high-strength version containing 0.54% silicon and 1.21% niobium. This alloy was discussed in Chapter 7 in terms of its erosion corrosion resistance. Figure 11.5 shows that although the alloy with silicon and niobium had a higher strength, its fatigue limit was only a little better than the conventional C71500 alloy.

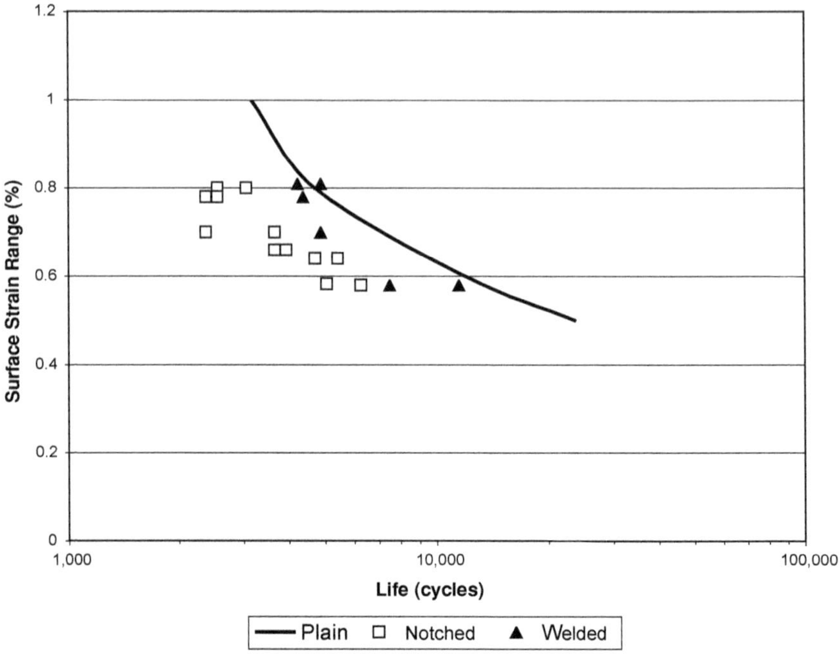

FIGURE 11.6 High strain fatigue curves for 70/30 copper-nickel in seawater at 10–15°C at 10 cycles/h.[5]

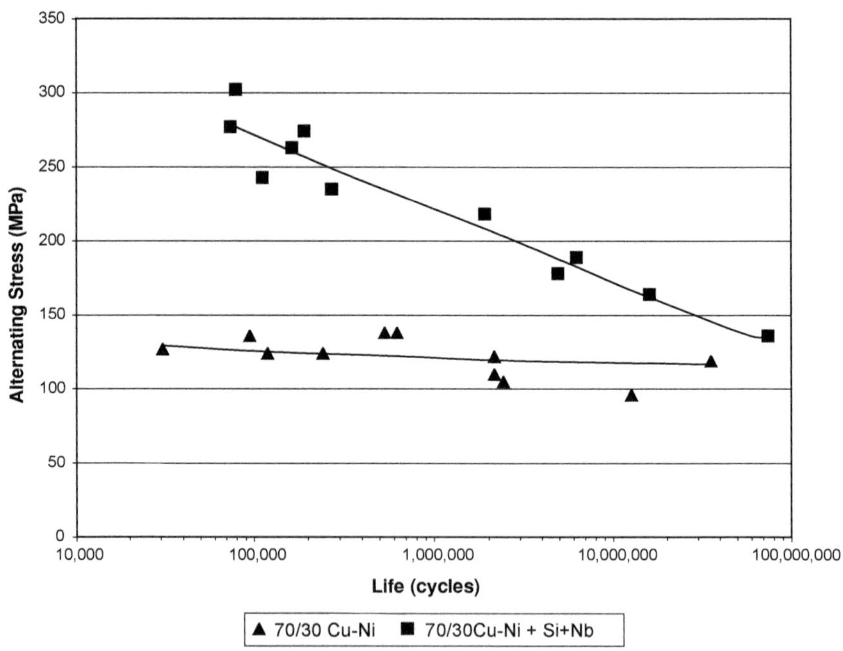

FIGURE 11.7 Fatigue of 70/30 copper-nickel alloys in air at 50–100 Hz.[6]

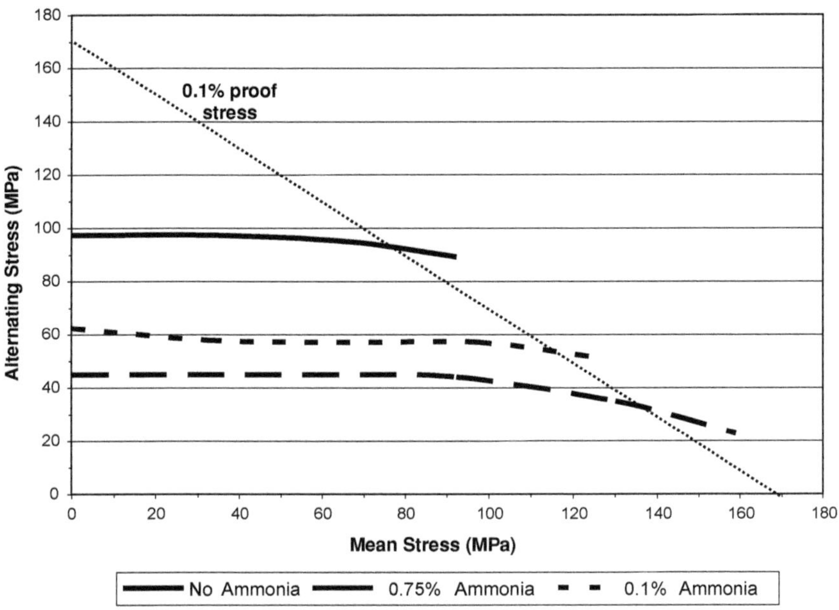

FIGURE 11.8 Corrosion fatigue of 90/10 copper-nickel in seawater with ammonia.[3]

In the previous section, the results of Campbell on the corrosion fatigue resistance of aluminum brass in seawater and seawater containing ammonia were described.[3] Campbell also carried out tests on 90/10 and 70/30 copper-nickel, the results of which are shown in Figures 11.8 and 11.9. All the test conditions below the lines suffered no corrosion fatigue failure after 50 million cycles. The data for 90/10 copper-nickel show that 0.75% ammonia produced a big reduction in the corrosion fatigue strength. Although the loss in corrosion fatigue strength was less with 0.1% ammonia, it was only with 0.01% ammonia that no effect on corrosion fatigue strength compared with clean seawater was seen. This is surprising because in the previous chapter, data were presented showing that 90/10 copper-nickel did not increase its susceptibility to SCC in seawater containing 0.75% ammonia. This suggests that the protective film on 90/10 copper-nickel may be more easily disrupted by a cyclic stress than by a constant one.

70/30 copper-nickel proved to be even more susceptible to corrosion fatigue in seawater containing ammonia than 90/10 copper-nickel. There was a large decrease in the corrosion fatigue strength with 0.1% ammonia, and there was still a significant decrease with 0.01% ammonia (Figure 11.9). 70/30 copper-nickel also had very good resistance to SCC in

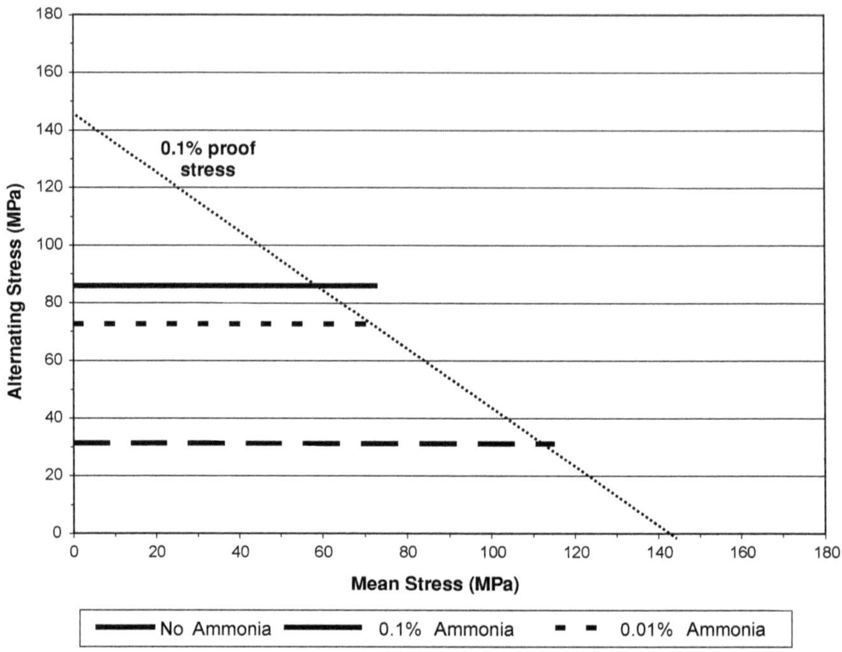

FIGURE 11.9 Corrosion fatigue of 70/30 copper-nickel in seawater with ammonia.[3]

seawater containing ammonia, and its effect on the corrosion fatigue strength suggests that the protective film is more easily disrupted by cyclic stresses than by constant ones.

In the 1980s, when 90/10 copper-nickel was first being considered for cladding for the legs of offshore platforms, concern was expressed about corrosion fatigue of the cladding. Harvey et al. reported tests on 90/10 copper-nickel in 3.5% sodium chloride (NaCl) at 15 Hz.[7] They used an *R* value of 0.4, with axially loaded compact tension specimens and applied anodic currents from 0 to 1,800 $\mu A/cm^2$. However, normally, the copper-nickel would be welded to the carbon steel (CS) legs, and Harvey et al. also conducted fatigue tests on 90/10 copper-nickel coupled to an equal area of CS. As the anodic current increased, the number of cycles to failure decreased, with the biggest decrease in the range 0 to 300 $\mu A/cm^2$ (Figure 11.10). However, when coupled to CS, the number of cycles to failure increased from ~2.5 million to ~4 million. The cathodic protection conferred by the CS significantly reduced the effect of corrosion on fatigue life. These results demonstrate that the use of 90/10 copper-nickel for cladding offshore platform legs poses no corrosion fatigue risk.

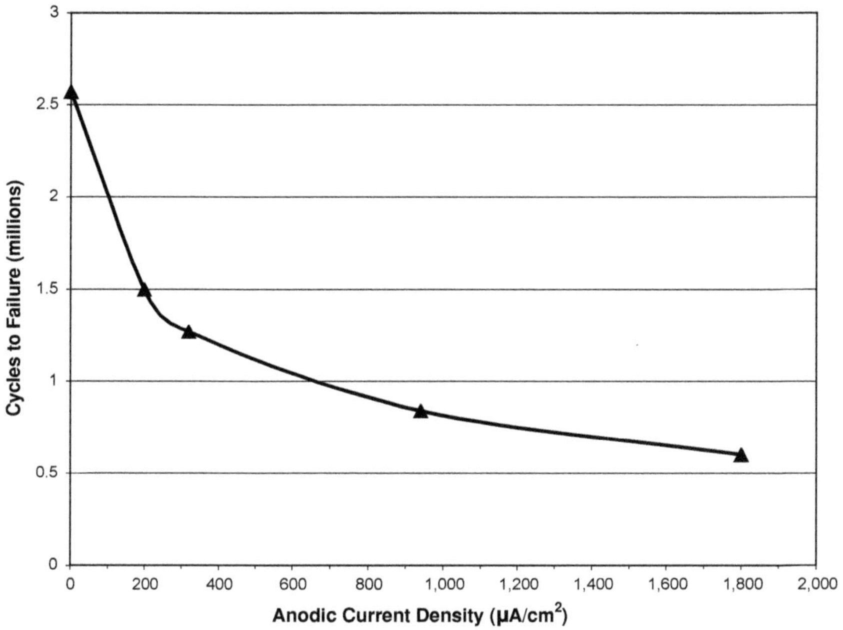

FIGURE 11.10 Cycles to failure for 90/10 copper-nickel as a function of anodic current density.[7]

11.3 Nickel Aluminum Bronze

Because of its high strength, nickel aluminum bronze (NAB) is often chosen for applications in which the alloy may be subject to high strains at low-frequency cyclic stressing. Campbell summarized the data generated at BNF Metals Research Association that was originally presented by Bowers.[5,8] The test method was the same as that used for copper-nickel alloys, as described in the previous section. The data are summarized in Figure 11.11, where it can be seen that the life at low frequencies is much greater for cast NAB than for wrought 70/30 copper-nickel (Figure 11.5). The excellent resistance of NAB to corrosion fatigue at high frequencies is due mainly to its high strength, whereas its resistance at low frequencies is due to a combination of high strength and good resistance to corrosion.[8] Bowers pointed out that at frequencies below 10 cycles/h, there is little increase in fatigue life, and a 10-fold reduction in the frequency only doubles the corrosion fatigue life.[5]

Bentley and Duquette examined the corrosion fatigue of cast NAB (C95800) in air and 0.5 M sodium chloride at 20 Hz.[9] Under freely corroding

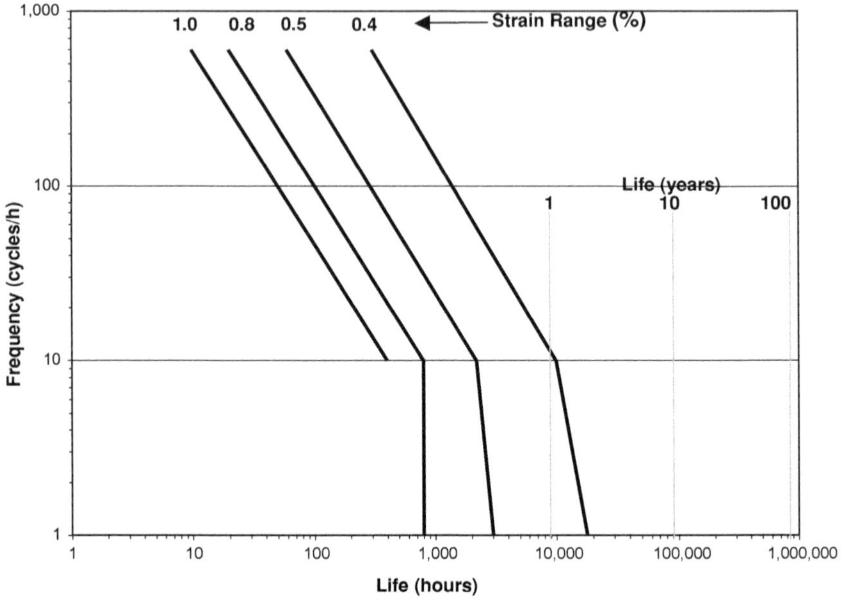

FIGURE 11.11 High-strain, low-cycle fatigue data for cast NAB in seawater at 32°C.[5,8]

conditions, there was a reduction in the threshold fatigue stress compared with air, but only of the order of 15 MPa. Under applied currents, the corrosion fatigue behavior was variable. With small cathodic currents, the corrosion fatigue life increased beyond the corrosion fatigue life seen in air. If the cathodic current was increased to 1.0 mA/cm^2, there was an increase in corrosion fatigue life of 38% above the corrosion fatigue life seen in air. The authors postulated that this was due to a thin alumina film on the NAB, but alumina films have never been detected on NAB in seawater (see Chapter 3).

Anodic currents <0.2 mA/cm^2 increased the fatigue life compared to anodic currents under freely corroding conditions. In the anodic current range 0.2–1.0 mA/cm^2, the corrosion fatigue life was decreased because of the creation of etching notches by anodic dissolution. However, at even higher anodic current densities, the fatigue life was restored because the notches were blunted by anodic dissolution and cracks did not form.

11.4 Prevention of Corrosion Fatigue

Where corrosion fatigue may occur, it is important to use good design practice at the outset, using known corrosion fatigue properties of the alloy being considered and adhering to the requirements of the design code that is in force. It is also important to avoid designs that increase the risk of localized corrosion, such as crevices, which can act as stress raisers.

However, there are occasions when the stresses are higher than anticipated and cracks are detected on inspection. If the component must be replaced, then a change to a more fatigue-resistant alloy will be needed, or perhaps a change in design that reduces the stresses will have to be made. The aluminum bronzes, and NAB in particular, offer a significant increase in corrosion fatigue resistance compared with most other copper alloys.

Where the corrosion is more severe than anticipated, a corrosion prevention method may be an alternative to a change of alloy. The beneficial effects of even small amounts of cathodic protection on 90/10 copper-nickel and NAB were described previously, where corrosion fatigue lives in excess of those in air were possible. Where this is impractical, an organic coating may be adequate. Where this is to be applied in situ, it is important that the surface preparation requirements are adhered to. A poorly applied coating may induce localized corrosion more severe than that which led to the initial problem. In some cases, a metallic coating may be preferred, applied either by electroplating or thermal spray. Such coatings usually contain minor defects, and it is important that the metal chosen for the coating be anodic to the copper alloy substrate to prevent localized corrosion at defects.

In closed-loop systems, corrosion inhibitors can be used to prevent corrosion, remembering that the alloy chosen should still have an adequate resistance to fatigue in air to give a satisfactory life.

References

1. P.M. Scott, "Corrosion Fatigue," in *Corrosion*, 3rd ed., L.L. Shreir, R.A. Jarman, G.T. Burstein, eds. (Oxford, U.K.: Butterworth Heinemann, 1994), p. 8:144.

2. C.E. Jaske, J.H. Payer, V.S. Balint, *Corrosion Fatigue of Metals in Marine Environments* (New York, NY: Springer/Batelle, 1981).

3. H.S. Campbell, BNF Metals Research Association Research Report, A1577, February 1966.

4. M.O. Spiedel, "Influence of Environment on Fracture," in *Proceedings of the Fifth International Conference on Fracture*, vol. 6 (New York, NY: Pergamon, 1982), p. 2685.

5. J.E. Bowers, "The High Strain Fatigue Properties in Seawater of 70/30 Cupro-nickels and Cast Aluminium Bronze," BNF Metals Research Association Research Report, May 1974.

6. D.W. Townsend, BNF Metals Research Association Research Report, A1546, October 1965.

7. D.P. Harvey, T.S. Sudershan, M.R. Louthan, R.E. Swanson, "Corrosion Fatigue Behaviour of Copper-Nickel Cladding for Marine Structures," Coatings and Bimetallics for Aggressive Environments (Hilton Head, SC: ASM International, 1985).

8. H.S. Campbell, *Aluminium Bronze Alloys Corrosion Resistance Guide*, Publication 80 (Potters Bar, U.K.: Copper Development Association, 1981).

9. R.M. Bentley, D.J. Duquette, International Copper Research Association Project Report, 241, November 1979.

CHAPTER 12

Hydrogen Embrittlement

When carbon steel (CS) is used underground or subsea, it is common to apply cathodic protection (CP) to prevent the steel from corroding. The potential is usually in the range −850 to −1,040 mV SCE (saturated calomel electrode), and any other metals connected to the CS will also receive this level of CP. At these potentials, hydrogen evolution is the main cathodic reaction, and if the hydrogen diffuses into the metal, it can become embrittled. Materials well known to suffer hydrogen embrittlement (HE) include high-strength steels, ferritic and duplex stainless steels, and some nickel alloys. The following sections examine the resistance of copper alloys to HE.

12.1 Copper and Its Alloys

When HE occurs, hydrogen atoms are created as part of the cathodic reaction. Instead of combining to form hydrogen molecules and evolving as gas, the atomic hydrogen diffuses into the metal. Where it accumulates or becomes trapped, embrittlement occurs. Copper and its alloys have a very low hydrogen adsorption coefficient, making hydrogen entry into copper very difficult. Hence, copper and its alloys are usually regarded as being immune to HE.

High-strength copper alloys (for example, UNS C72420) have been used for subsea fasteners on cathodically protected structures. In the 1980s, BNF Metals Technology Centre tested two of these alloys following failures of Alloy K-500 bolts (see the next section).[1-2] One alloy was Hiduron 191 (C72420), and the other was Marinel (also C72420, but with a niobium addition to further increase strength).

The test method, which was described in detail by Campbell and Francis, involved constant load tests of tensile samples exposed to natural seawater containing 5 mg/L sulfide (as hydrogen sulfide [H_2S]) to poison the hydrogen recombination reaction.[3] The samples were polarized to −1,040 mV SCE for 60 days, after which they were removed, tensile tested, and examined for indications of embrittlement.

Samples of Hiduron 191 were tested at 90% of the 0.2% proof stress and showed no signs of embrittlement after 60 days' exposure.[1] Marinel was tested at 75% of the 0.2% proof stress and showed no susceptibility to embrittlement in either the standard aged condition or in material that was aged and cold reduced 10%.[2]

Marinel has also been tested using the slow strain rate test, and no embrittlement was seen under CP in seawater at 20 and 50°C.[4] Tuck also reported no embrittlement of C72420 under cathodic charging in slow strain rate tests.[5]

12.2 Nickel-Copper Alloys

The two most common nickel-copper alloys are Alloy 400 (UNS N04400) and K-500 (UNS N05500). The latter is a high-strength version of Alloy 400 and was widely used for corrosion-resistant fasteners subsea in the North Sea during the 1970s. However, in the early 1980s, there were failures of several K-500 bolts subsea in a major North Sea oil field. The failures were identified by BNF Metals Technology Centre as being due to HE. The failures were reproduced in the laboratory, and a program of research was undertaken to investigate the effects of various factors on the embrittlement of K-500 and also to test some alternative alloys.[3]

Efird reported K-500 bolt failures by HE from another North Sea oil field, and he suggested that the problem could be overcome by keeping the hardness of the K-500 below HRC35.[6] However, Francis and Campbell tested K-500 with hardnesses of 336 HV and 296 HV, the latter well below HRC35, and found that both samples embrittled equally.[1] Stress was also not important, as unstressed samples also embrittled. Figure 12.1 shows a typical fracture face of K-500 after 60 days' exposure and tensile testing. The embrittled area around the perimeter of the sample is clearly visible, as are the secondary cracks along the rest of the gauge length. Embrittlement was observed to begin at potentials of ~−750 mV SCE, and it became more severe as the potential became more negative.

Tests on samples that were aged and cold worked showed a similar depth of embrittlement to normally aged material. Campbell and Francis also investigated the effect of time on the properties of cathodically charged

FIGURE 12.1 Scanning electron microscopy image of fracture face of Alloy K-500 after 60 days' hydrogen charging in seawater and tensile testing.[3]

K-500.[3] Figure 12.2 shows the elongation and reduction of area for exposures up to 200 days. The results show that the elongation decreased fairly steadily over the period, but the reduction in area decreased rapidly over the first 60 days and was almost constant thereafter. Figure 12.3 shows the depth of embrittlement for both stressed and unstressed samples, and it can be seen that embrittlement increased almost linearly with time over the 200 days. Francis and Campbell used these data to estimate the time to failure of a 1.25-in (31.75-mm) diameter K-500 bolt under CP.[2] With a stress around the 0.2% proof stress, failure was estimated to occur in 2 to 4 years, while at 50% of the 0.2% proof stress, failure would occur in 10 to 11 years. For subsea structures, the design life is usually at least 25 years, and this means that K-500 is not a suitable material for the bolting.

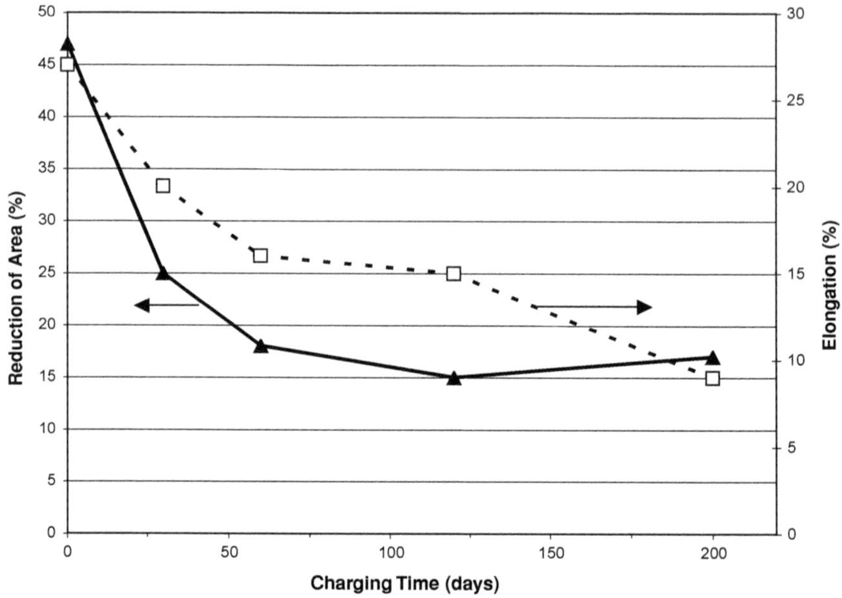

FIGURE 12.2 Reduction of area and elongation of K-500 after hydrogen charging.[3]

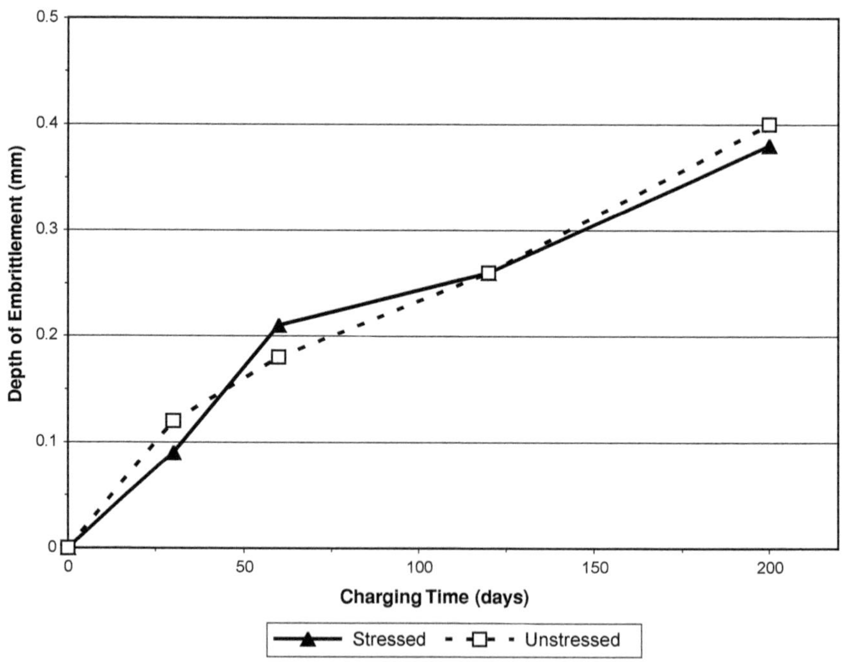

FIGURE 12.3 Depth of embrittlement of K-500 vs. charging time.[3]

Erlings et al. observed HE of K-500 drill collars on CS downhole tubulars in a magnesium chloride packer fluid.[7] The hydrogen evolution occurred on the K-500 as the cathodic reaction in the galvanic couple, with the CS under downhole conditions.

Alloy 400 was tested by Price and colleagues and was also shown to be very susceptible to HE—more than other nickel alloys tested.[8,9]

12.3 Avoidance of Hydrogen Embrittlement

All the common copper alloys appear to be immune to HE and can be used freely in equipment to which CP is applied. It is clear from the data in the previous section that Alloys 400 and K-500 should not be used in situations where they are under stress and exposed to hydrogen evolution from CP or galvanic coupling, for example subsea marine bolting.

References

1. R. Francis, H.S. Campbell, BNF Metals Technology Centre Research Report, R538/1, June 1986.
2. R. Francis, H.S. Campbell, BNF Metals Technology Centre Research Report, R611/14, March 1990.
3. H.S. Campbell, R. Francis, *Brit. Corros. J.* 30 (1995): p. 154.
4. O. Andersen, M.W. Joosten, J. Murali, D.E. Milliams, CORROSION/96, paper no. 78 (Houston, TX: NACE International, 1996).
5. C.D.S. Tuck, CORROSION 2005, paper no. 462 (Houston, TX: NACE International, 2005).
6. K.D. Efird, *MP* 24, 4 (1985): p. 37.
7. J.G. Erlings, H.W. de Groot, J.M.M. van Roy, *MP* 25, 10 (1986): p. 28.
8. C.E. Price, R.K. King, "The Embrittlement of Monel 400 by Hydrogen and Mercury as a Function of Temperature," International Conference and Expo on Fatigue Corrosion Cracking, Fracture Mechanics, and Failure Analysis (Salt Lake City, UT: ASM International/ASTM, 1986): p. 81.
9. C.E. Price, J.A. Morris, "The Comparative Embrittlement of Nickel Base Alloys by Hydrogen and Mercury," Corrosion of Nickel Base Alloys (Columbus, OH: ASM International/INCO, 1985): p. 69.

CHAPTER 13

Fouling and Microbially Influenced Corrosion

This chapter is concerned with the effect of biological attachments onto copper and copper alloy surfaces, principally in waters. The effects may be mechanical and/or corrosion related and may occur on both the macro and micro scale. Not all biological effects are detrimental, and some beneficial effects are described in the following sections.

13.1 Macrofouling

When almost any solid material is immersed in seawater, marine growths will gradually attach themselves over a period of weeks or months. These can take the form of shellfish, for example, mussels, or plants, such as seaweeds; they are termed *fouling*. More correctly, such attachment of marine growths is termed *macrofouling*, whereas *microfouling* is the colonization of metal surfaces by bacteria and will be discussed in the next section. Note that some macrofouling only occurs at certain times of the year (for example, mussels), whereas other types can occur all year round, (for example, limnoria and tunicates). Macrofouling can also occur in fresh water, but it is usually much less of a problem.

Macrofouling can cause a number of problems, one of which is the increase of weight on offshore structures and also the increase in the stresses on the structure under wave action. Fouling in heat exchangers can block the tubes, dramatically reducing heat transfer, and can cause severe erosion corrosion at partial blockages. Severe fouling can also significantly reduce the water flow in piping systems. Finally, the environment beneath the fouling can be very different to the one in bulk seawater, causing rapid or localized corrosion.

Copper is toxic to many organisms, and it is well able to resist fouling in seawater—so much so that the Royal Navy began cladding the immersed parts of wooden warships in copper in the 18th century. However, it was not until the second half of the 20th century that the fouling of copper and its alloys was studied in detail.

Efird carried out exposure trials on flat panels at the LaQue Laboratory, North Carolina, over a period of 5 years.[1] The materials were carbon steel (CS), copper (C11000), and 90/10 and 70/30 copper-nickels. The appearances of each material after 3, 9, 18, 36, 48, and 60 months' exposure are shown in Figures 13.1–13.4. CS showed increasingly heavy fouling with time (Figure 13.1), but this sloughed off from time to time, when the fouling was heavy and the corrosion products beneath were not adherent enough to the metal surface. After 60 months, part of one panel had completely corroded away. With copper (Figure 13.2), there was no significant fouling after 9 months; rather, there was just a green layer of copper hydroxychloride ($Cu_2Cl[OH]_3{\cdot}4H_2O$) over a thin layer of copper oxide. After 36 months, there was some attachment of fouling, and 60% of the surface was covered after 60 months. The appearance of the 90/10 copper-nickel panels (Figure 13.3) was somewhat similar to copper, with no macrofouling after 9 months and 70% coverage after 60 months. The 70/30 copper-nickel panels were also similar in appearance (Figure 13.4). There was no significant difference in the fouling behavior of pure copper and the copper-nickel alloys. A close examination of Figures 13.2–13.4 reveals that after 36 months, the fouling was attached to the Alloy 400 mounting bolts. After 48 months, it was still attached to the bolts but was beginning to encroach onto the panels. Only after 5 years was the fouling well established on the panels.

The corrosion rate of all three copper alloys was in the range 2.5–7.5 μm/y. The copper metal release from the alloys over this range of corrosion rates is too low to cause the fouling resistance observed. In fact, the corrosion rate, and hence, the copper release rate, for 70/30 copper-nickel is about one-third that of pure copper, but the fouling resistance is about the same. It is commonly stated that a minimum corrosion rate of 18 μm/y is required to release enough copper to suppress fouling.[1]

Over many years, copper and its alloys lose and regain their fouling resistance, as shown by the periodic attachment of fouling and its sloughing off. This was observed over 14 years at the LaQue Laboratory.[2] Efird suggested that in addition to the release of copper, the copper oxide layer was also toxic.[1] After it became covered with a layer of copper hydroxychloride, the fouling could attach, but as this layer is poorly adherent, the fouling

FIGURE 13.1 The 5-year sequence of fouling on CS in seawater (exposure periods from left to right): (top) 3 months, 9 months, 18 months; (bottom) 36 months, 48 months, 60 months.

FIGURE 13.2 The 5-year sequence of fouling on copper in seawater (exposure periods from left to right): (top) 3 months, 9 months, 18 months; (bottom) 36 months, 48 months, 60 months.

FIGURE 13.3 The 5-year sequence of fouling on 90/10 Cu-Ni in seawater (exposure periods from left to right): (top) 3 months, 9 months, 18 months; (bottom) 36 months, 48 months, 60 months.

FIGURE 13.4 The 5-year sequence of fouling on 70/30 Cu-Ni in seawater (exposure periods from left to right): (top) 3 months, 9 months, 18 months; (bottom) 36 months, 48 months, 60 months.

would slough off when it became too heavy. The copper oxide would then be exposed, and the surface would resist fouling until the copper hydroxychloride layer reformed.

The mechanism was also supported by the results from exposures of half-coated panels of 90/10 copper-nickel, where fouling was seen up to the boundary of the coating, but not on the metal. If fouling resistance was conferred by copper release alone, then some resistance to fouling on the coating adjacent to the metal should have been seen.

Copper-containing antifouling paint was also evaluated, and though it had an effect, it was not as effective at resisting fouling as solid metal.[1] Copper and copper-nickels were also very resistant to fouling at the water line compared with plastics and other inert materials.[1] Even when copper, or its alloys, did foul, the corrosion rate was no different to that on material exposed in flowing seawater, where fouling was much less; that is, the fouling was having no significant effect on the corrosion rate.[1]

The antifouling properties of 90/10 copper-nickel mean that it has applications where resistance to fouling is essential. In many cases, solid copper-nickel would be uneconomical, and a number of methods of applying copper-nickel coating onto steel have been examined. The most common are hot-rolled, clad-steel sheets, adhesive-backed foil, and granules embedded in rubber. The latter are cylinders approximately 1 mm diameter and 1 mm long, with a density such that approximately 30% of the exposed surface is metal and the total coating thickness is about 3 mm. Campbell et al. exposed these three product forms as well as expanded 90/10 copper-nickel mesh bonded onto a rubber base (EPDM) in Langstone Harbor, United Kingdom, for up to 8 years.[3] All four products showed excellent resistance to fouling. The performance of the granules in rubber is surprising as only 30% of the surface is metal. However, the particles are small and close together, so perhaps this small area of rubber is prevented from fouling by copper release from the metal.

One of the first modern uses of copper-nickel to prevent fouling was for the hulls of small ships. The earliest of these was the *Asperida II*, launched in 1968, which is still afloat and has excellent resistance to fouling. Between 1968 and 1991, 10 ships were built with solid hulls, mostly of 90/10 copper-nickel, and these have resisted fouling well. Not only does the copper-nickel reduce downtime for the removal of fouling, but also, a cleaner hull means that the boats are more economical on fuel. The average fuel savings were calculated at 25% in 1977 and are probably greater now with higher fuel costs.[4] Corrosion on all these vessels has been minimal.

Although the 70/30 copper-nickel welds on the 90/10 copper-nickel hulls are slightly more noble than the base 90/10 metal, they were no more prone to fouling. Even when fouling does attach on copper-nickel, it is very easy to remove, compared with fouling on CS and inert materials.[3,4]

Some seven boats have been built with copper-nickel-clad steel hulls, using 2 mm of 90/10 copper-nickel. Most of these have performed well, but some had poor fouling resistance. In some cases, this was thought to be due to the boats being tied up for long periods in stagnant and/or polluted water, where the toxic copper oxide film would not form properly.

Another method of obtaining antifouling is to weld copper-nickel sheets onto steel hulls. Twelve large panels were welded onto the hull of the 16-knot oil tanker *Arco Texas* and examined after 2 years. The tanker had made seven trips through the Panama Canal, and the panels were intact and performing well, even though several showed mechanical damage from forceful impacts on the sides during passage through the canal.[4]

Adhesive 90/10 copper-nickel foil comprises sheets that are 0.15–0.3 mm thick and can be molded to curved surfaces. This technique has mostly been used on fiberglass-hulled vessels using an overlapping wall-papering technique, which also confers excellent resistance to fouling.[3-4] Unfortunately, this product was no longer available at the time of writing. A more detailed evaluation of the use of copper-nickel for ships' hulls and its performance up to 1997 is contained in the work of Powell and its appendixes.[4]

If copper-nickel is used to protect steel structures offshore, the areas that are of chief concern are the splash zone, the water line, and the fouling zone, which usually extends several meters below the water line. If the copper-nickel is welded onto the steel, it will be galvanically coupled to it and corrosion will be reduced, which would be expected to reduce fouling resistance. Powell and Michels reported the results of a 10-year trial in which the fouling resistance of bare steel piles was compared with that of piles with directly welded copper-nickel cladding, piles with copper-nickel cladding insulated from the steel with concrete, and piles with rubber-insulated cladding.[5] The results (Table 13.1) show no significant difference between using concrete and rubber as an insulator. Both of these resisted fouling well. Although directly welded copper-nickel cladding lost some of its fouling resistance, it had only half the fouling seen on bare steel.

Copper-nickel cladding was first used offshore on platform legs on the Morecambe Bay Gas Platform in the United Kingdom in the 1980s. After 20 years, the 4-mm-thick cladding was in good condition, and divers reported that removal of fouling from the directly welded cladding was much easier

TABLE 13.1 Biofouling mass on copper-nickel-sheathed steel pilings after 10 years' exposure[5]

Piling	Exposure (years)	Biofouling (kg/m^2)	Coverage (%)
Bare steel	5	18.0	100.0
	10	12.0	100.0
Concrete insulated	5	0.36	1.9
	10	0.40	1.2
Directly welded	5	7.95	44.3
	10	4.43	36.8
Rubber insulated	5	0.26	1.4
	10	0.55	4.5
	10	0.35	2.8
	10	0.62	5.3

than from bare steel. Since that time, the rubber coating embedded with copper-nickel granules has also become available, and several offshore platforms have used this for protection of structural legs, cross-bracing, and piping.[3-4]

90/10 copper-nickel has been used for seawater piping for many years, and its resistance to fouling means that other antifouling measures may not be required, or only intermittently. Albaugh reported experience from the Gulf of Mexico, where 8-in (203-mm) CS piping was severely fouled in just 22 months of operation, while 90/10 copper-nickel piping showed no fouling.[6] It was clear that 6-in (152-mm) 90/10 copper-nickel piping would be adequate for the water flow because of the alloy's resistance to fouling. Albaugh carried out a 20-year life cycle cost analysis and showed that 8-in copper-nickel piping was 73% of the cost of CS, while the cost of 6-in copper-nickel piping would be only 65% that of steel.[6]

13.2 Microfouling and Microbially Influenced Corrosion

For microbially influenced corrosion (MIC) to occur, it is necessary for bacteria to colonize the metal surface. Despite the toxicity of copper, all organisms require it to sustain them, usually in small quantities. Some bacteria are more copper-tolerant than others, and it would be expected that they would colonize the metal surface first.

Costerton et al. reviewed the formation of biofilms on metal surfaces.[7] Once the first bacteria colonize a surface, they produce extracellular material that typically consists of polysaccharide polymers. As the biofilm grows

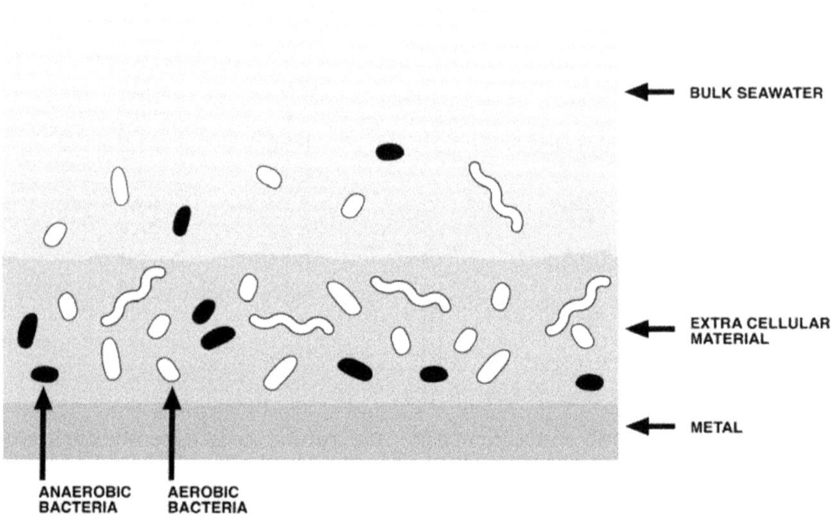

FIGURE 13.5 Schematic of biofilm on an immersed metal surface.

and thickens, the aerobic bacteria consume the oxygen near the metal-biofilm interface, and anaerobic bacteria become active. In the outer layer of the biofilm, oxygen can still penetrate, and aerobic bacteria are active. This is shown schematically in Figure 13.5. The secondary colonization of the biofilm can promote or inhibit the activity of the primary bacteria, depending on the type.[7] Corrosion only occurs once an effective microcolony is established and a potential difference is established between different areas on the metal surface. The growth and development of biofilms are affected by the temperature, pH, water velocity, and surface roughness.[8] Most bacteria thrive best within a limited range of temperature and pH, although these can vary significantly from species to species. Bacteria do not adhere well to metal surfaces if the water flow is high or the surface is very smooth.

Chamberlain and Garner followed the development of a biofilm on 90/10 copper-nickel in natural seawater and found that after 4 weeks, the surface was colonized by bacteria and diatoms.[9] After 30 weeks, *Ectocarpus*, a copper-tolerant species, was present, and after this, other species were able to form secondary colonies. After 9–10 months, sloughing of the biofilm occurred, exposing fresh metal. Chamberlain and Garner concluded that the high copper concentration at the metal surface initially prevented colonization and enabled the protective oxide film to grow. The subsequent colonization by *Ectocarpus* and other species prevented sulfate-reducing bacteria (SRB) from colonizing the metal surface and accelerating corrosion.

Bacteria cause corrosion not by directly reacting with the metal, but because some of their metabolic by-products are aggressive to the metal. There are three common types of reaction that lead to corrosion: one of these is iron-oxidizing bacteria, which mainly affect steel and cast iron; a second type is sulfur-oxidizing bacteria, which produce sulfuric acid (H_2SO_4) as a by-product, and any metal that is attacked by sulfuric acid can suffer MIC by these bacteria; the third type is SRB, and this has received the most study. These reduce sulfate in the environment to sulfide as part of their metabolic processes. Unlike the first two types, SRB are only active under anaerobic conditions.

Not all bacteria are detrimental to metal surfaces. In Chapter 3, the development of protective hydrotalcite ($Mg_6Al_2CO_3[OH]_{16}.4H_2O$) films on aluminum brass in seawater was described. These films are rich in aluminum and magnesium, minor constituents of the alloy and seawater, respectively, and they can only form if there are bacteria present to chelate the copper and prevent it from forming a copper oxide film, as happens in 3% sodium chloride (NaCl) solution.[10] The hydrotalcite is able to buffer the pH at the metal surface, as described in Chapter 3. Not only does the hydrotalcite form on the metal surface, but it also forms within the biofilm, growing as the biofilm grows.[10]

Although the role of biofilms in corrosion has been discussed almost as long as corrosion has been studied, work on copper alloys was slow to take off, presumably because of the known toxicity of copper ions. In the 1930s and 1940s, a series of research programs were conducted at BNF Metals Research Association that included elements of MIC. Many of these elements were summarized by Rogers in a 1948 paper that is seldom referenced.[11] Following are some of the findings from this review.

Bacteria isolated from various estuarine waters showed a range of copper tolerance. Some could only tolerate 10 mg/L or so, but others could tolerate in excess of 100 mg/L copper, and one species was found to tolerate 2,000 mg/L copper. Bacteria with a high copper tolerance were always found in waters where MIC of copper alloys was occurring.

Cultures isolated from an estuarine water greatly increased the corrosion of 70/30 brass in stagnant solution, more so than occurred under inert deposits under sterile conditions. The addition of organic acids that are found in seaweeds and can act as bacterial food increased pit depths by a factor of 2–10. The addition of peptone to the water strengthened the protective film so that some bacteria could not break it down, although the most aggressive still did.

Rogers postulated that the acceleration of attack was caused by an increase of hydrogen acceptors and the establishment of alternative oxygen and reduction systems during bacterial metabolism, that is, depolarization of the cathodic reaction.[11] Samples of seawater that caused rapid pitting of copper alloys contained an organic sulfur compound that could be reduced to a mercaptan and oxidized back to a disulfide very rapidly. Hence, this compound could act as an oxygen carrier, depolarizing the cathodic reaction, the reduction of dissolved oxygen (DO), and hence accelerating the rate of pitting.

It was also found that pigments could cause accelerated corrosion of copper alloys in seawater. Melanin was one such pigment, which is produced by the action of the enzyme tyrosinase on tyrosine. It was shown that bacteria producing tyrosinase could accelerate corrosion in direct proportion to the amount of tyrosine present.

Tests were also conducted using organic sulfur compounds that are known to be metabolic by-products or breakdown products. Additions of cystine or glutathione greatly increased the depth of impingement attack on 70/30 brass when added to sterile seawater, compared with that in sterile seawater alone. The acceleration of attack was similar to that seen when inorganic hydrogen sulfide (H_2S) additions were made to the seawater.

In a separate series of atmospheric exposure tests of brass plates, deep pits occurred when molds formed on the metal panels. Rogers postulated that this was due to organic acids, which are produced during the growth of many molds.[11]

The composition of bacterial exopolymeric films (usually polysaccharides) can vary, with some being aggressive toward copper. Jolley et al. showed that some exopolysaccharides could accelerate the corrosion of copper and anchor the colony to the surface.[12-14] Geesey and colleagues showed that the acidic polysaccharides were the most aggressive.[15-16] However, the effect of acidic polysaccharides was variable, with some being much more aggressive toward copper than others.

In Chapter 5, Type III pitting of copper was discussed, as was the possible involvement of bacteria in the mechanism. Fischer et al. described tests carried out on the biofilms removed from failed copper pipes from a German hospital.[17] These carried water at approximately 50°C and showed the typical pepper-pot appearance of perforations. The biofilm was present between the metal and the oxide-hydroxide layer. This implies the presence of a species very tolerant of copper that can form a biofilm prior to the formation of the oxide layer. Pitting occurred where the biofilm was disrupted. This was believed to be due to a copper concentration cell, due to the large amount of copper in the exopolymeric layer. In addition, peroxide and hydroperoxides were detected, and these may have provided a more rapid cathodic reaction than the reduction of DO, thus accelerating the corrosion process.

The phenomenon of blue water was discussed in Chapter 4, and it occurs in soft, fresh waters when the water is left stagnant in pipes for an extended period (days or longer) so that a protective oxide film does not form. Critchley et al. compared the dissolution of copper from pipes exposed to sterile water with the dissolution of pipes exposed to sterile water to which an addition of bacteria from a problem water had been made.[18] Much higher rates of copper dissolution occurred on tubes exposed to the water with bacteria. This suggests that bacteria can contribute to the problem of blue water and that it need not be solely chemical.

Much of the work in seawater has been concerned with the effect of sulfur compounds, particularly hydrogen sulfide, produced by SRB. Davis et al. carried out tests on copper, 90/10 copper-nickel, and a single-phase aluminum bronze (C60800) in four environments: aerated seawater, aerated sterile seawater, deaerated seawater, and deaerated sterile seawater.[19] Samples in aerated environments corroded more than samples in deaerated ones, and biological activity accelerated the formation of the oxide film in aerated water and the sulfide film in deaerated water. Biological activity was clearly accelerating chemical and electrochemical processes. When samples were transferred from deaerated to aerated conditions, the corrosion rate increased, and the increase was greater in water with bacteria present. This demonstrates the importance of bacteria in seawater to the corrosion of some common copper alloys.

The nature of the protective film on 90/10 copper-nickel, where a thin, chloride-rich layer beneath the copper oxide causes cathodic polarization, was discussed in Chapter 3. This film occurs whether the samples are exposed to natural seawater or 3% sodium chloride solution. However, the cathodic polarization is lost if the seawater is changed from aerated to deaerated, while it is not lost with 3% sodium chloride solution.[20] This suggests that anaerobic bacteria are contributing to the loss of cathodic polarization, probably due to the production of sulfur compounds.

Chamberlain and Garner investigated SRB and found that some species were particularly copper-tolerant and could colonize 90/10 copper-nickel surfaces, as opposed to being secondary colonizers of the biofilm.[9] After 14 days' exposure, substantial quantities of sulfur were detected in the film on the metal surface. When samples were moved from deaerated to aerated seawater, the same acceleration in corrosion was seen as occurs with inorganic sulfide additions. This showed that the sulfide does not have to be produced elsewhere when 90/10 copper-nickel suffers attack. Chamberlain and Garner postulated that the variable performance of 90/10 copper-nickel under polluted conditions in the literature could be due, at least in part, to variable colonization by SRB, depending on the local availability of copper-tolerant species.[9]

Little et al. exposed samples of welded 90/10 copper-nickel piping in natural seawater with and without SRB additions.[21] The welded areas contained crevices and irregularities and were favored sites for SRB colonization. This explains why service failures of piping are often downstream of the welds and close to them.

In an unpublished review, Tuthill described some service failures due to MIC.[22] In two of these, 90/10 copper-nickel was used to replace Alloy 400 that failed by MIC, thought to be SRB. However, in another plant, which used a natural, fresh water for cooling, 90/10 copper-nickel failed, which was thought to be due to MIC, whereas 70/30 copper-nickel was successful. This suggests that 90/10 copper-nickel can suffer MIC due to SRB when particularly aggressive bacteria are present. However, Tuthill knew of no cases of MIC failure of 70/30 copper-nickel.[22]

Nicklin described MIC corrosion problems associated with 70/30 copper-nickel heat exchanger tubes in Royal Navy submarines.[23] The failures were localized pitting, believed to be due principally to SRB introduced during U.K. docking. All submarine coolers had the same type of bacteria present, but only the ones in service in warmer waters suffered corrosion problems.

Rangel et al. reported failures of 70/30 copper-nickel heat exchanger tubes that handled an estuarine cooling water and were also dosed with ferrous sulfate ($FeSO_4 \cdot 7H_2O$).[24] Although there were black spots on the tubes, these were not sulfides. They identified both iron- and sulfur-oxidizing bacteria on the tube surface, and it was thought that one of these was the cause of the problem. No detailed study of the effect of these bacteria on copper alloys has been published.

Alloy 400 has been observed to suffer from severe attack in the presence of sulfides in seawater. Gouda et al. studied Alloy 400 exposed to SRB cultures and also stagnant seawater from the Arabian Gulf at 19, 30, and 50°C.[25] They found that SRB readily colonized the metal surface and that corrosion took the form of localized pitting. Attack was most severe when exposure alternated between deaerated and aerated conditions, as was found for other copper alloys. They also found that attack was most severe at 30°C, with less at 19 and 50°C. This shows that particular bacteria have an optimum temperature range for high activity.

Similar results were reported by Little et al. for Alloy 400 in the Gulf of Mexico, and the LaQue Laboratory, in North Carolina, found pits up to 1.07 mm deep after 3-year exposures in stagnant, natural seawater.[26-27]

Lee et al. carried out tests on Alloy 400 with both inorganic sulfides and SRB.[28] The corrosion with inorganic sulfides was uniform, whereas corrosion with SRB took the form of pitting. They concluded that this was because the sulfide concentration varied across the metal surface, depending on local SRB activity, and this generated local concentration cells. A similar reaction with inorganic sulfides could only occur if the flow was sufficiently turbulent to mechanically disrupt the sulfide film.

In his review of service failures due to MIC, Tuthill recorded three service failures of Alloy 400 heat exchanger tubing due to SRB.[22] Two of these were solved by replacing the tubes with 90/10 copper-nickel, and the other case was solved by replacement with 70/30 copper-nickel.

The nature of biofilms on metals is very variable, and Dexter and LaFontaine reported a case in which biological activity increased galvanic corrosion.[29] They tested 6% Mo austenitic stainless steel (SS) coupled to an equal area of copper in natural seawater. In half of the couples, the SS had been preexposed to seawater for 4 weeks, whereas the others had freshly cleaned panels. The galvanic current was approximately 2 μA for the freshly cleaned samples, whereas the ones with preexposed SS had a galvanic current of 500–1,000 μA. The copper panels coupled to preexposed SS also suffered a higher weight loss than those coupled to freshly exposed SS, but only 2–3 times greater. The reason for the acceleration was the biofilm that forms on SS in aerated seawater and depolarizes the cathodic reaction, the reduction of DO. Without this film, SS are not very efficient cathodes.[30] Although the galvanic current increased greatly with biofilm-coated SS, the weight loss increased by much less. This was because even if the SS is not a very efficient cathode, the copper will still be corroding in natural seawater, and the initial rate will be quite high, until a protective film has formed.

13.3 Bacterial Toxicity

In previous sections, the toxicity of copper toward microbes has been described in terms of reducing fouling on metal surfaces. However, copper is very toxic toward some bacteria that can be very harmful to human beings.

TABLE 13.2 Population of *E. coli* on various pipe surfaces with time[29]

	E. coli count/initial *E. coli* (%)	
Material	5 h	24 h
PVC	68	75
Polybutylene	93	92
Cross-linked PVC	89	99
Copper	1	0

Wells described tests using different materials and tap water containing *Escherichia coli*.[31] The water was dechlorinated and circulated through tubes of three different plastics and copper. The *E. coli* population was determined at intervals, and the results are shown in Table 13.2. These clearly show a rapid and dramatic reduction in the *E. coli* population in water passing through copper tube, whereas that going through the polymer tubing showed very little reduction.

Following the demonstration of toxicity to *E. coli*, Wells did further tests on *Legionella pneumophila* in tap water.[32] The tests were carried out in a similar manner to those on *E. coli*, but the high pH of the water (pH 9–10) caused a rapid reduction of *L. pneumophila* even in glass. The results in Table 13.3 were obtained using water with a biological buffer to keep the pH closer to neutral. It can clearly be seen that there was no difference between the glass control system and PVC tubing, whereas both metal piping systems showed dramatic reductions in *Legionella* populations after only 24 h Extended tests, for up to 6 months, showed no significant biofilm on the copper tube, whereas the polybutylene tube contained fungi.

West et al. also compared the toxicity of copper pipes to *Legionella* and other bacteria compared with several polymeric materials and a glass control.[33] They found that copper was particularly toxic to *Legionella*,

TABLE 13.3 Growth of *Legionella* in Trizma-buffered water in contact with different tubing materials[30]

	Legionella count			
Tubing	0 days	1 day	4 days	8 days
Glass (control)	17,500	66,000	290,000	4,900,000
Copper	17,500	<10	<10	<10
Galvanized steel	17,500	<10	<10	<10
PVC	17,500	64,000	280,000	4,300,000

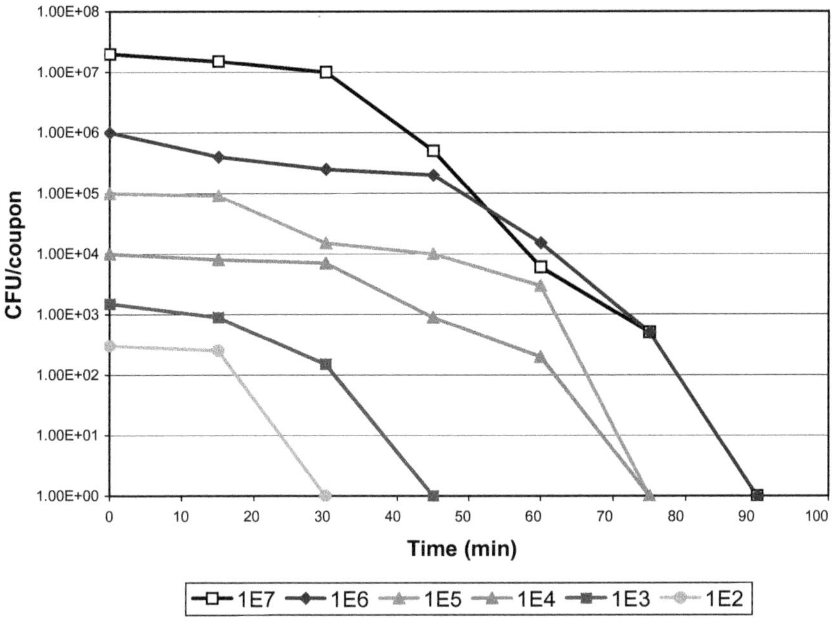

FIGURE 13.6 Effect of innoculum size of EMRSA-16 on the time for total kill when exposed to copper.[32]

reducing or halting its growth, whereas more copper-resistant, but non-harmful, bacteria continued to grow. Polybutylene was readily colonized by *Legionella*, and the balance of the bacterial population was unchanged in polybutylene tubing.

Hospital-acquired infections are a current concern, particularly the so-called superbugs such as methicillin-resistant *Staphylococcus aureus* (MRSA). About 80% of infectious diseases are spread by touch, and a hand contaminated with influenza A virus, for example, will contaminate the next seven surfaces that are touched.

A study was carried out to compare the survival rates of MRSA as a function of innoculum size on copper plates, as shown in Figure 13.6.[34] The results show that as many as 10 million MRSA are inactivated in 90 min. Further work has compared the viability of MRSA on copper, some copper alloys, and Type 304 SS (UNS S30400).[35] The results in Figure 13.7 show that the MRSA count decreases to very low levels in just 90 min on copper and in 270 min on 80/20 brass, whereas there was a much smaller reduction on nickel silver and very little on the SS. The viability of MRSA is clearly reduced as the copper content of the alloy increases. An 18-month trial began in April 2007 in a U.K. hospital to compare copper surfaces

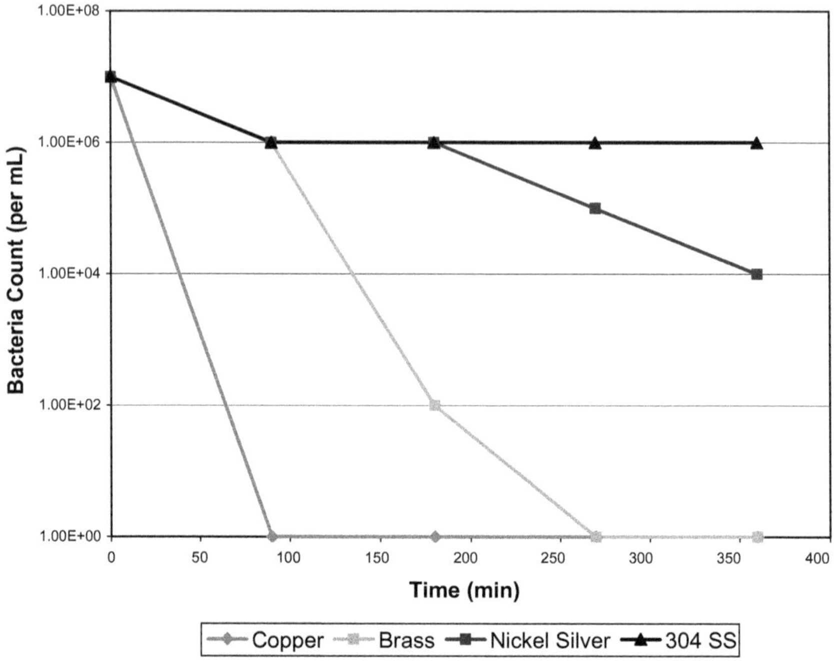

FIGURE 13.7 MRSA viability on some metals at 20°C.[33]

against traditional materials in controlling hospital-acquired infections such as MRSA.

In 2008, 275 copper alloys were registered with the U.S. Environmental Protection Agency as antimicrobial materials that kill these bacteria. The list includes copper, brasses, and bronzes, and these alloys offer a range of properties that make them suitable for antimicrobial surfaces.

13.4 Methods of Prevention

Where the toxicity of copper and copper alloy surfaces to fouling is not sufficient, additional methods of fouling control must be considered. One of the most well-known methods of preventing fouling, either macro or micro, is to dose either intermittently or continuously with chlorine or hypochlorite (OCl^-).[36-38] Chlorine is readily generated electrolytically in seawater or brines and is the most common way of controlling fouling. Dosing may only be needed in the fouling season, which is site-specific, and the dose may be varied as fouling becomes more or less severe. The effect of chlorination on materials and the chemical reactions involved were reviewed by Goodman.[39]

One effect of chlorine is to reduce the resistance of copper alloys to erosion corrosion if the chlorine dose is too high.[40] This is discussed in detail in Chapter 7. If ferrous sulfate dosing is being carried out to help with film formation on heat exchanger tubes, it is important to turn off the chlorine dosing when ferrous sulfate is being added as it prevents the formation of a protective hydrated iron oxide (FeOOH) deposit.[41]

Where a water contains small amounts of hydrogen sulfide, such as from SRB activity, chlorine additions can be used to prevent accelerated corrosion.[42] By using a redox probe in the cooling-water line, chlorine can be injected whenever the redox potential drops, but excessive chlorine dosing can be avoided by stopping dosing before the redox potential gets too high. This method was used to prevent accelerated attack of aluminum brass condenser tubes at a Belgian power station and to avoid a costly retube, as occurred at a sister station.[42]

However, chlorine does not prevent accelerated attack of all alloys. Gouda et al. found that chlorine additions reduced sulfide pitting due to SRB of Alloy 400 but did not prevent it.[25]

Chlorine dosing is less damaging if it is carried out intermittently, rather than continuously.[40,43] The most suitable dosing concentration and frequency depend on the local conditions and must be determined for each plant.

In fresh waters, with much lower chloride concentrations than seawater, copper alloys are more tolerant of chlorine. Tuthill et al. presented data showing that copper and a number of commonly used copper alloys were not affected by up to 2 mg/L chlorine added continuously, provided water velocities were not excessive.[44]

When potable water systems are installed in large public buildings, it is common to disinfect the tubes with a strong solution of hypochlorite, at anywhere from 20 to 100 mg/L free chlorine. Francis and Saxton investigated the effect of strong disinfection regimes on the subsequent performance of copper tubes.[45] Tests up to 30 days under both stagnant and flowing conditions (up to 3 m/s) showed no effect of a pretreatment of 100 mg/L chlorine for 2 h or 20 mg/L chlorine for 16 h. The tests included both a hard well water and a soft, surface-derived water at room temperature and at 60°C. Under continuous chlorine dosing at 1 mg/L, there was no effect on copper corrosion at room temperature under any flow condition. At 60°C, there was a marked increase in general corrosion rates in both waters, an increase in the susceptibility to Type II pitting under slow flows in the soft water, and an increased susceptibility to erosion corrosion at high flow velocities. The effect of chlorine on Type II pitting is discussed in more detail in Chapter 5.

Bond et al. reported the results from tests lasting up to 12 months using the same disinfection regimes, flow rates of 0.1 and 2 m/s, and ambient and 60°C temperatures.[46] After 12 months, the corrosion rates of all the tubes were in the range 1–3 μm/y—a very low rate. The corrosion rates after 4 months were somewhat higher, but all were less than 0.01 mm/y—still a low corrosion rate. Type II pitting was not seen in the hot, soft water with continuous chlorine dosing because the soft water used was different to that used in the 30-day tests and did not support Type II pitting.

There are other ways of controlling fouling in addition to chlorine. Campbell and Searle compared the fouling of various metals and non-metals under slow flow conditions in seawater.[47] One tank was untreated, as a control; another was dosed to give a chlorine residual of approximately 0.5 mg/L; and a third had electrically stimulated copper anodes to provide approximately 60 ppb copper ions in solution. The fourth tank had electrically stimulated anodes of copper and aluminum as this was claimed by some to be superior to copper alone, with levels of approximately 60 ppb copper and approximately 30 ppb aluminum. Fouling and corrosion of a range of copper alloys was greatest in the control tank and least in the tank dosed with chlorine. There was no significant difference between the results from the tanks dosed with copper and with copper and aluminum. The degree of fouling and corrosion was intermediate between that of the control tank and the tank dosed with chlorine.

Where heat exchanger tubes become fouled with deposits, it is possible to restore the heat transfer efficiency by passing sponge balls through the heat exchanger. These are injected into the inlet water box and are collected from the discharge water box. Excessive use of sponge balls can cause accelerated corrosion of copper alloy condenser tubes, and Sato and colleagues described the optimum cleaning regime to maintain heat transfer without accelerating corrosion of copper alloy tubes.[48-50] They found that fewer than six balls per tube every 1–2 weeks was enough to give satisfactory heat exchanger performance.

Carrera and Gabetta reviewed the biocide options available to the offshore oil and gas industry.[51] They considered a wide range of equipment and the best way to control bacteria in each one, as biocides are not always the best solution. They considered both oxidizing biocides, such as chlorine, and a range of organic biocides, listing both the advantages and disadvantages of each. Many, but not all, of MIC problems are due to SRB, and one advantage of some biocides is their toxicity to a wide range of bacteria.

When considering sulfide attack by SRB of 90/10 copper-nickel in seawater, Little et al. observed no beneficial effects on corrosion of iron or ferrous sulfate additions.[21] However, additions of sodium sulfite (Na_2SO_3)

reduced the attack, although they did not prevent it. The ineffectiveness of ferrous sulfate additions in preventing sulfide attack of copper alloys has been noted by many researchers. Sato et al. found that ferrous sulfate did work against sulfides if a protective film was first formed in clean seawater.[52]

In Chapter 5, the attack of copper tubes by Type III pitting was shown to have a microbiological contribution. Fischer et al. described the measures adopted at a German hospital to prevent failures due to Type III pitting.[53] In the hot water pipes, increasing the water temperature above 55°C and eliminating dead water regions prevented attack. In the cold water pipes, it was necessary to use ultraviolet radiation to reduce the active bacteria in the water. In the long term, a plant was built to increase the bicarbonate alkalinity of the supply water, as described in Chapter 5.

References

1. K.D. Efird, *MP* 15, 4 (1976): p. 16.
2. K.D. Efird, D.B. Anderson, *MP* 14, 1 (1975): p. 3.
3. S. Campbell, R. Fletcher, C. Powell, "Long Term Exposure Trials Evaluating the Biofouling and Corrosion Resistance of Copper-Nickel Alloy Sheathing Materials," 12th International Congress on Marine Corrosion and Fouling (Southampton, U.K.: 2004).
4. C.A. Powell, "The Biofouling Resistance of Copper-Nickel—Properties and Experience," CDA Report, July 1997.
5. C. Powell, H. Michels, "Review of Splash Zone Corrosion and Biofouling of C70600 Sheathed Steel During 20 Years Exposure," EUROCORR 2006 (Maastricht, Holland: European Federation of Corrosion, 2006).
6. E.K. Albaugh, "Copper-Nickel Piping Reduces Costs, Biofouling/Corrosion," *World Oil*, November 1984.
7. J.W. Costerton, G.G. Geesey, P.A. Jones, *MP* 27, 4 (1988): p. 49.
8. T.R. Bott, *Effluent Water Treat. J.* 19 (1979): p. 453.
9. A.H.L. Chamberlain, B.J. Garner, *Int. Biodeterioration* 24 (1988): p. 213.
10. J.E. Castle, D.C. Epler, "The Development of Inorganic and Biological Fouling Layers on Copper Based Alloys," Progress in the Prevention of Fouling in Industrial Plants (Nottingham, U.K.: Institute of Corrosion, 1981).
11. T.H. Rogers, *J. Inst. Met.* 75 (1948): p. 19.
12. J.G. Jolley, G.G. Geesey, M.R. Hankins, R.B. Wright, P.L. Wichlacz, *Surf. Interface Sci.* 11 (1988): p. 371.
13. J.G. Jolley, G.G. Geesey, M.R. Hankins, R.B. Wright, P.L. Wichlacz, *Appl. Spectrosc.* 43 (1989): p. 1,062.

14. J.G. Jolley, G.G. Geesey, M.R. Hankins, R.B. Wright, P.L. Wichlacz, *Appl. Surf. Sci.* 37 (1989): p. 469.

15. G.G. Geesey, P.J. Bremer, *MTS J.* 24 (1988): p. 36.

16. G.G. Geesey, L. Lang, J.G. Jolley, M.R. Hankins, T. Iwaoka, P.R. Griffiths, *Water Sci. Technol.* 20 (1988): p. 161.

17. W. Fischer, I. Hanssel, H.H. Paradies, *Microbial Corrosion*, vol. 1 (New York, NY: Elsevier, 1988): p. 300.

18. M. Critchley, R. Taylor, R. O'Halloran, *MP* 44, 6 (2005): p. 56.

19. S.C. Davis, R.M. Mitchell, F.J. Ansuini, International Copper Research Association Project Final Report, 316, November, 1981.

20. J.E. Castle, M.S. Parvizi, A.H.L. Chamberlain, *Microbial Corrosion* (Book 303: Metallurgical Society, 1983): p. 36.

21. B. Little, P. Wagner, J. Jacobus, *MP* 27, 8 (1988): p. 57.

22. A. Tuthill, "Base Metal Resistance of Alloys to Microbially Influenced Corrosion," unpublished report prepared for NiDI, June 1992.

23. G.J.E. Nicklin, "Living with the Threat of Microbially Influenced Corrosion in Submarine Systems: The Royal Navy's Perspective MOD." Ninth International Naval Engineering Conference and Exhibition (Hamburg, Germany: British Crown Copyright 2008).

24. C.M. Rangel, M.L. Quinta, M.E. Simas, Microbial Corrosion–1, Conference held in Sintra, Portugal (New York: Elsevier, 1988): p. 346.

25. V.K. Gouda, I.M. Banut, W.T. Riad, S. Mansour, *Corrosion* 49 (1993): p. 63.

26. B.J. Little, P.A. Wagner, R.I. May, M.B. McNiell, *Marine Tech. Soc. J.* 24, 3 (1991): p. 10.

27. J.R. Davis, ed., *Metals Handbook*, 2nd ed. (Materials Park, OH: ASM International, 1998).

28. J.S. Lee, K.L. Lowe, R.I. Ray, B.J. Little, CORROSION 2003, paper no. 214 (Houston, TX: NACE International, 2003).

29. S.C. Dexter, J.P. LaFontaine, *Corrosion* 54 (1998): p. 851.

30. R. Francis, *Galvanic Corrosion—A Practical Guide for Engineers* (Houston, TX: NACE International, 2001).

31. F.E. Wells, International Copper Research Association Project Report, 348, March 1984.

32. F.E. Wells, International Copper Research Association Project Report, 348, October 1989.

33. A.A. West, J. Rogers, J.V. Lee, C.W. Keevil, J.S. Colbourne, P.J. Dennis, K. Smith, International Copper Research Association Project Report, 401A, July 1989.

34. J.O. Noyce, H. Michels, C.W. Keevil, *J. Hospital Infect.* 62 (2006): p. 289.

35. H. Michels, S.A. Wilks, J.O. Noyce, C.W. Keevil, "Copper Alloys for Human Infectious Disease Control," Materials Science and Technology Conference (Pittsburgh, PA: 2005).

36. D.C. Mangum, B.P. Shepherd, J.C. Williams, 3rd International Congress on Marine Corrosion and Fouling, NBS Special Publication, October 1972.

37. T.J. Lamb, 3rd International Congress on Marine Corrosion and Fouling, NBS Special Publication, October 1972.

38. F. Inui, *Bull. MESJ* 1 (1973): p. 21.

39. P.D. Goodman, *Brit. Corros. J.* 22 (1987): p. 56.

40. R. Francis, *MP* 21, 8 (1982): p. 44.

41. E.B. Shone, G.C. Grimm, *Trans. Inst. Marine Eng.* 98 (1985): paper 11.

42. R. Cigna, K. De Ranter, O. Fumei, L. Giuliani, *Brit. Corros. J.* 23 (1988): p. 190.

43. D.B. Anderson, B.R. Richards, *J. Eng. Power* July (1966).

44. A.H. Tuthill, R.E. Avery, S. Lamb, G. Kobrin, CORROSION/98, paper no. 708 (Houston, TX: NACE International, 1998).

45. R. Francis, H. Saxton, BNF Metals Technology Centre Research Report, R658/6, January 1991.

46. S. Bond, R. Francis, H. Saxton, BNF Metals Technology Centre Research Report, R658/12, August 1992.

47. I. Campbell, N.K. Searle, "Comparative Effects of Fouling and Antifouling Measures on Corrosion," Fouling and Corrosion of Metals in Seawater (Dunstaffnage, Scotland: SMBA, April, 1982).

48. S. Sato, K. Nagata, "Factors Affecting Corrosion and Fouling of Condenser Tubes of Copper Alloys and Titanium," Corrosion in the Power Industry (Montreal, Canada: NACE International, 1981).

49. S. Sato, K. Nagata, *Sumitomo Light Met. Tech. Rep.* 19 (1978): p. 83.

50. S. Sato, K. Nagata, S. Yamauchi, CORROSION/81, paper no. 195 (Houston, TX: NACE International, 1981).

51. P. Carrera, G. Gabetta, European Federation of Corrosion Publication, EFC 19, IOM 1996, p. 44.

52. S. Sato, T. Nosetani, Y. Yamaguchi, K. Onda, *Sumitomo Light Met. Tech. Rep.* 16 (1975): p. 23.

53. W.R. Fischer, D. Wagner, H. Siedlarek, *MP* 34, 10 (1995): p. 50.

CHAPTER 14

Galvanic Corrosion

In the past, galvanic corrosion has caused numerous failures of equipment in service, some of them very costly. This chapter describes the causes of galvanic corrosion, the main factors affecting its severity, and some preventative measures that can be adopted. For further information, the reader is referred to the work of Francis.[1]

When a metal is immersed in a conducting liquid, it takes up an electrode potential (also known as the *corrosion potential*). This is determined by the equilibrium between the anodic and cathodic reactions occurring on the surface and is usually measured with reference to a standard electrode, such as the saturated calomel electrode (SCE).

Galvanic corrosion occurs when two metals with different potentials are in electrical contact while immersed in an electrically conducting, corrosive liquid. Because the metals have different natural potentials in the liquid, a current will flow from the anode (more electronegative) to the cathode (more electropositive) to equalize the potentials. This will increase the corrosion on the anode (Figure 14.1). This additional corrosion is *galvanic corrosion*, which is also known as *bimetallic corrosion, dissimilar metal corrosion*, or *contact corrosion*. These terms are used interchangeably throughout this chapter.

In general, the reactions that occur under bimetallic corrosion are similar to those that would occur on a single, uncoupled metal, but the rate of attack is increased, sometimes dramatically. With some metal combinations, the change in the electrode potential in the couple compared with the uncoupled potential can induce corrosion that would not have occurred in the uncoupled state, for example, pitting. In some environments, the change in potential of the cathode in the couple, compared to the uncoupled state, can also introduce problems, for example, hydrogen embrittlement (HE).

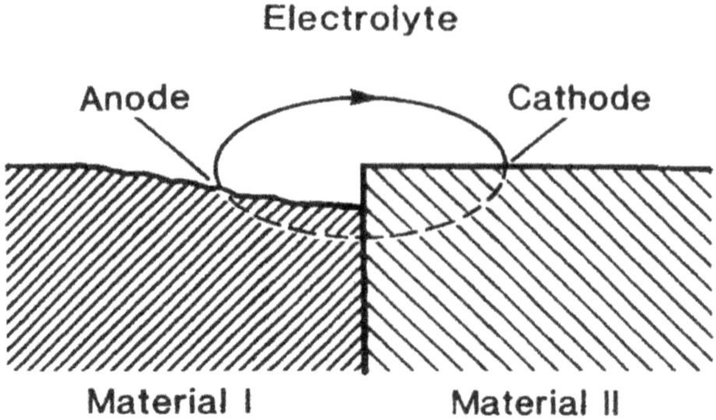

FIGURE 14.1 Schematic of a galvanic corrosion cell.

The effect of coupling two metals together increases the corrosion rate of the anode and reduces, or even suppresses, corrosion of the cathode. Hence, coupling a component to a sacrificial anode can prevent corrosion; this is the principle of *cathodic protection* (CP).

When a metal is corroding, two processes are occurring. One is the dissolution of metal at the anode, for example, iron:

$$Fe \rightarrow Fe^{2+} + 2e \tag{14.1}$$

This must be balanced by a cathodic reaction. Most practical cases of galvanic corrosion occur in solutions containing dissolved oxygen (DO), and the primary cathodic reaction is the reduction of DO:

$$O_2 + 2H_2O + 4e \rightarrow 4OH^- \tag{14.2}$$

Under a few circumstances, the cathodic reaction can be the reduction of hydrogen ions to hydrogen gas:

$$2H^+ + 2e \rightarrow H_2 \tag{14.3}$$

Under uncoupled corrosion conditions, the anodic and cathodic reactions occur at small, local areas on the metal surface. In a bimetallic couple, the cathodic reaction is mostly, or totally, on the electropositive member of the couple, and the anodic reaction is mostly, or totally, on the electronegative member of the couple.

14.1 Conditions Necessary for Galvanic Corrosion

The basic requirements necessary to cause bimetallic corrosion are as follows:

1. An electrolyte bridging the two metals, which may not be aggressive to the individual metals when they are not coupled, and which may be in the form of bulk volume of solution, a condensed film, or a damp solid such as soil, salt deposits, or corrosion products
2. Electrical connection between the two metals, usually arising from direct physical contact but also arising where electrical continuity is established between two metals, for example, by an insulation-coated conductor, by structural metal work, or by electrical earthing systems; it is not necessary for the metal junction to be immersed in the electrolyte
3. A sufficient difference in potential between the two metals to provide a significant galvanic current
4. A sustained cathodic reaction on the more noble (electropositive) of the two metals by one of the mechanisms described previously—in most practical situations, the consumption of DO

14.2 Factors Affecting Galvanic Corrosion

Many factors affect galvanic corrosion, including electrode potential, cathodic efficiency, area ratio, and the nature of the electrolyte.

14.2.1 Electrode Potential

The potential, as defined at the beginning of this chapter, is determined by many factors, but the only environment for which this is really well documented is seawater. Unfortunately, the potential, even in a single liquid, is not a fixed value; rather, it can vary with changes in temperature, liquid flow rate, the level of aeration, and so on. However, the relative ranking of alloys remains largely unaffected by such changes, even though the potential may change by several hundred millivolts for some alloys. Figure 14.2 shows the electrochemical series in seawater at 10°C (50°F), a typical temperature in the North Sea. The difference between the most electropositive and the most electronegative metals is nearly 2 V, and a couple between these materials generates high currents and, hence, high corrosion rates on the anode.

Three metals are not shown in Figure 14.2: platinum, gold, and silver. These are used, for example, in electrical contacts, where bimetallic corrosion can sometimes be a problem. Platinum and gold are similar in potential to titanium, that is, they are very electropositive, whereas silver is similar in potential to nickel.

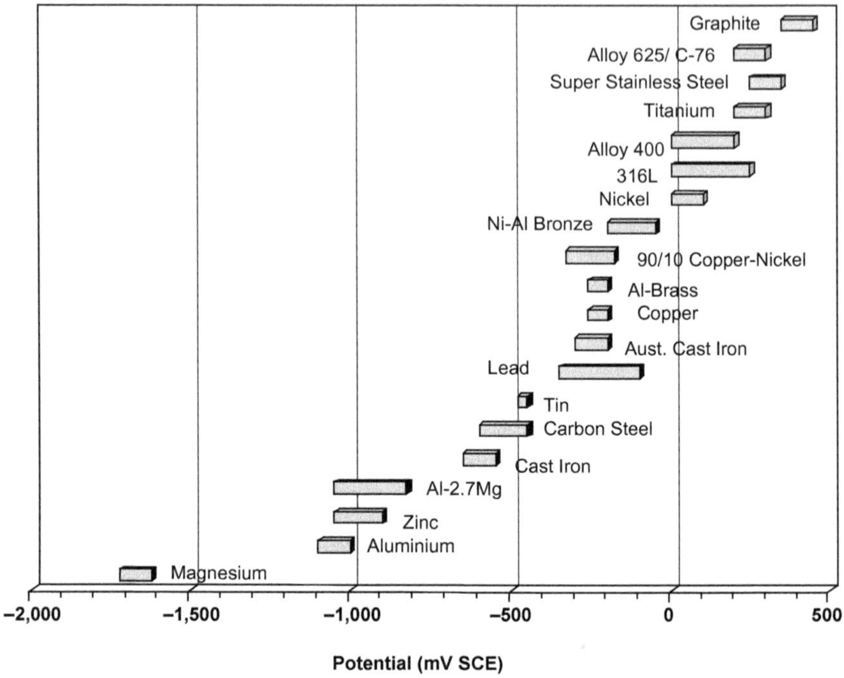

FIGURE 14.2 Galvanic series in seawater at 10°C. (Data courtesy of SINTEF and the author.)

However, the magnitude of the potential difference alone will not necessarily indicate the amount of bimetallic corrosion. For instance, metals with a potential difference of only 50 mV have shown severe bimetallic corrosion problems, while metals with a potential difference of 800 mV have been successfully coupled together. This is because the potential difference gives no information about the kinetics of bimetallic corrosion, which depend on the current flowing between the two metals in the couple.

Figure 14.2 shows the electrochemical series for seawater at 10°C, but the relative position of the metals is very similar for many near-neutral, aerated solutions. Hence, this series has applications outside its initial, anticipated area. It can be seen that copper and its alloys are midway in the galvanic series and so can be anodes when connected to more noble (electropositive) metals and cathodes when connected to less noble (electronegative) metals.

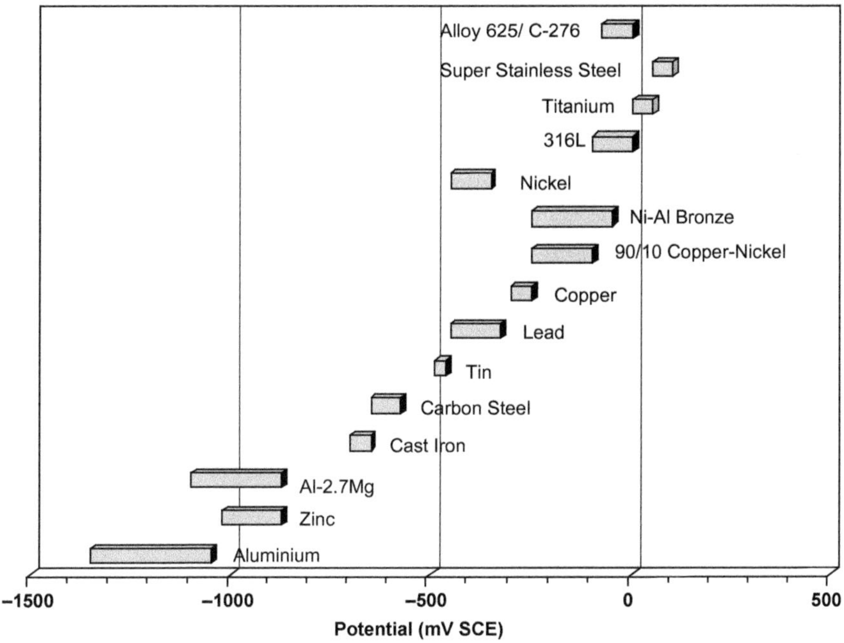

FIGURE 14.3 Galvanic series in seawater at 40°C. (Data courtesy of SINTEF and the author.)

14.2.2 Potential Variability

Changes in the corrosion potential of individual metals can result in polarity changes for some couples. A change in potential can be caused by a change in the kinetics of the anodic and/or cathodic reaction or a change in the nature of the reaction, for example, a change from the reduction of DO to the reduction of hydrogen ions as the cathodic reaction. Potential changes can be caused by complexing ions; changes in pH, temperature, or velocity;[2] and also by intense aeration when it enhances film formation on the hitherto anodic metal.

Changing the temperature of the seawater can have a significant effect on the potential of some alloys. Figure 14.3 shows the galvanic series at 40°C. Some metals, such as aluminum, nickel-chrome-molybdenum alloys, high-alloy stainless steels (SS), and titanium, show significant changes in potential with temperature, whereas metals like zinc and 90/10 copper-nickel (UNS C70600) show only small changes. The reasons for this are mostly connected with the nature of the surface film. For the most electropositive alloys, the potential decrease with temperature increase is due to the absence of a biofilm at 40°C, which forms in natural seawater at ambient temperature.

Temperature has little effect on the potential of copper alloys in seawater and similar conducting liquids. Small additions of oxidizers, such as hypochlorite (ClO_3^-), produce small increases in the potential of most copper alloys. Similar to temperature, high water velocities cause only small changes in potential of copper alloys.[2] If the water is deaerated, or very low in oxygen, then the potential of most copper alloys will move electronegatively some 200 to 300 mV.

Under some conditions, it is possible to get galvanic corrosion with a single metal. This occurs when part of a system is replaced with new metal, while some of the old metal remains. If there is a potential difference between the old, filmed metal and the new, bare metal, significant galvanic corrosion may occur before the new metal forms a protective film. One solution is to clean the old metal to remove the film so that the entire system starts at the same potential. Corbett reported such a problem for copper pipes in a fresh water system with a high carbon dioxide (CO_2) content.[3]

14.2.3 Electrode Efficiency

When a metal forms part of a galvanic couple, the rate of the anodic or cathodic reaction is not the same for all metals. This is because the kinetics of the reaction vary from metal to metal. The rate of the reaction at the metal surface determines its efficiency as an anode or cathode. In galvanic corrosion, a change of efficiency in the cathodic reaction usually has a more significant effect than a change in efficiency in the anodic reaction.

The efficiency can also be explained in electrochemical terms. Current flow between the two metals in contact in a couple is accompanied by a shift in the potential of the anodic member to a more electropositive value and in the potential of the cathodic member to a more electronegative value. These changes, called *polarization*, result in the two metals approaching the same potential, any potential difference between them being equal to the product of the current and the resistance through the electrolyte. The extent of polarization depends on both the metal and the environment. In neutral electrolytes, the cathode is almost always polarized much more than the anode, which accounts for the fact that a small area of sacrificial anode will effectively provide protection to a relatively large cathodic area. The extent of polarization will determine how effective any particular metal may be as a cathode to drive the corrosion of the anode.

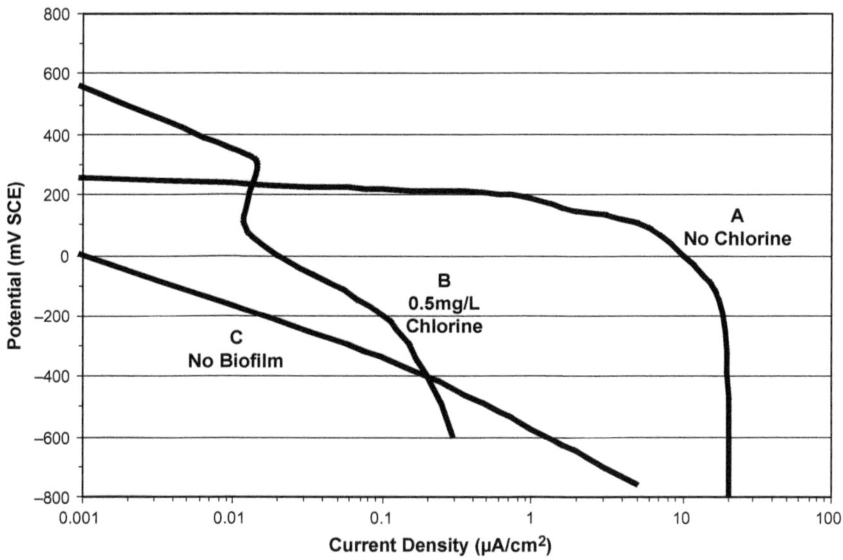

FIGURE 14.4 Cathodic polarization curves for high-alloy SS in three types of seawater.[4]

Some metals, such as titanium, are not very efficient at reducing DO compared with, say, copper alloys, so it is possible for a metal, such as carbon steel (CS), to corrode more coupled to a copper alloy than to titanium, despite titanium being much more electropositive than copper. All copper alloys are efficient cathodes, including the nickel-copper alloys, Alloys 400 and K-500.

The cathodic efficiency can also change under different conditions. An example is high-alloyed SS in seawater. Figure 14.4 shows the cathodic polarization curves for a 6% Mo SS under three conditions.[4] Curve A shows the behavior in natural seawater after the biofilm has formed. The rest potential is about +300 mV SCE, but under cathodic polarization, very large currents are produced for relatively small changes in potential. Hence, the SS with the biofilm is an efficient cathode, and corrosion rates of the anode in the couple can be very high. Both titanium and the nickel-chromium-molybdenum alloys behave in a similar manner in this environment.

Curve B shows the behavior in chlorinated seawater (>0.3 mg/L chlorine). The cathodic reaction is now mainly the reduction of hypochlorite ions to chloride, instead of the reduction of DO. SS in chlorinated seawater is not such an efficient cathode as it is with the biofilm, and

FIGURE 14.5 Galvanic corrosion of a NAB tube plate with titanium tubes.

hence, lower currents are produced at any given potential, compared with natural (i.e., unchlorinated) seawater, despite the more electropositive rest potential.

Curve C shows the behavior in seawater when no biofilm forms. The cathodic reaction is the reduction of DO, but without the biofilm, the reaction on the SS is very slow, and hence, even lower currents are produced at any given potential than occur in chlorinated seawater.

An example that demonstrates this is the use of 6% Mo SS or titanium tubes in heat exchangers with nickel aluminum bronze (NAB) tube plates. Sometimes galvanic corrosion of the tube plates has been severe, as shown in Figure 14.5. However, there have been other cases in which there has been little acceleration of corrosion of the NAB. The author believes that this is due to chlorination of the seawater. At plants where little or no chlorination has been carried out, the cathodic reaction would not be polarized, and severe corrosion could occur. At stations with low-level continuous chlorination, the cathodic efficiency would be much reduced, as shown in Figure 14.4, and corrosion would be less likely.

14.2.4 Electrolyte

Electrolyte factors that have a major influence on bimetallic corrosion are composition, pH, and, in particular, electrical conductivity, which affects both the intensity and the distribution of corrosion.

The severity of corrosion often increases with increasing electrical conductivity of the electrolyte because, in practice, high conductivity is often caused by the presence of aggressive ions, such as chloride, or by acid or alkali.

The electrical conductivity of electrolytes can vary widely; typical values are as follows:

Fresh distilled water	0.5–2 μS/cm
Stored distilled water	2–4 μS/cm
Supply water	50–1,500 μS/cm
Seawater	40,000 μS/cm
Saturated sodium chloride (NaCl) solution	250,000 μS/cm
Sulfuric acid (H_2SO_4)	up to 800,000 μS/cm
Molten salts	20,000–10^7 μS/cm

Except for very critical items of equipment, such as electronic components, and where it is essential to suppress metal dissolution, bimetallic corrosion is seldom a problem when the conductivity is very low such as in pure water.

In a solution of low conductivity, or when only a thin film of a highly conducting electrolyte covers the metals, corrosion is confined to an area near the junction between the two metals. Although the total amount of corrosion may be low, it is likely to be highly localized and, therefore, may be intense. In a highly conducting solution, corrosion will still be concentrated at the metal junction, but it will be more widespread than in a solution of low conductivity. Intense corrosion is likely to arise even with nominally pure water under condensing conditions if the atmosphere is polluted with acid gases, such as sulfur dioxide (SO_2), or with salt, for example, at industrial or sea coast locations.

In common with local cell corrosion, bimetallic corrosion is sensitive to the presence of constituents in the electrolyte that affect stability of the corroded metal ions. In some cases, insoluble corrosion products from the anodic metal can deposit on, and induce crevice corrosion of, the cathodic member of a couple. For instance, corrosion products from an electronegative alloy, such as steel or aluminum, could deposit on a copper cathode and induce crevice corrosion due to the establishment of a copper ion concentration cell between the metal under the deposit and the freely exposed metal (see Chapter 6). It is also possible to modify the composition of the environment by adding inhibitors to control bimetallic corrosion. This practice has proved effective in the treatment of waters in engines and recirculating industrial cooling plants. Bicarbonate, cyanide, and tartrate ions form soluble complexes with copper and zinc, and thus, their presence increases the rate of corrosion. Other metals and alloys can also have their corrosion rates accelerated by the presence of suitable organic or inorganic complexants.

As mentioned previously, in some instances, the cathodic reaction can become hydrogen evolution. In seawater, this becomes significant at potentials of −850 mV SCE and more electronegative. This means that alloys that are susceptible to HE may crack when connected to anodes that produce these low potentials. Materials like Alloys 400 and K-500 can embrittle when coupled to some aluminum alloys, zinc, or magnesium. As an example, Efird reported the failure of Alloy K-500 bolts by HE when used subsea and connected to zinc anodes, which were installed to protect adjacent CS.[5] Other copper alloys are resistant to HE (see Chapter 12).

14.2.5 Area Ratio

The ratio of the exposed areas of the anode and cathode materials in a couple is very important in the consideration of the likelihood of galvanic corrosion. The larger the cathode compared with the anode, the more oxygen reduction, or other cathodic reaction, can occur, and, hence, the greater the galvanic current. This results in more corrosion on the anode.

Adverse area ratios are likely to occur with fasteners and at joints. Weld, braze, or rivet metal should be of the same potential as, or, better still, slightly cathodic to, the base or parent metal. For example, the use of mild steel bolts or rivets to fasten nickel, copper, or SS alloy plates should be avoided. Figure 14.6 shows the effect of area ratio on a joint with various liquid conductivities and potentials.

The effect of area ratio is not confined to general corrosion, but may also affect other types of corrosion. In Chapter 7, the effect on the depth of erosion corrosion of 90/10 copper-nickel at high velocity in seawater by coupling it to other copper alloys was described. A failure from a North Sea platform also demonstrates this effect. The failure occurred at the junction of a 6% Mo SS pipe and one made of 90/10 copper-nickel. In chlorinated seawater, this would not normally result in

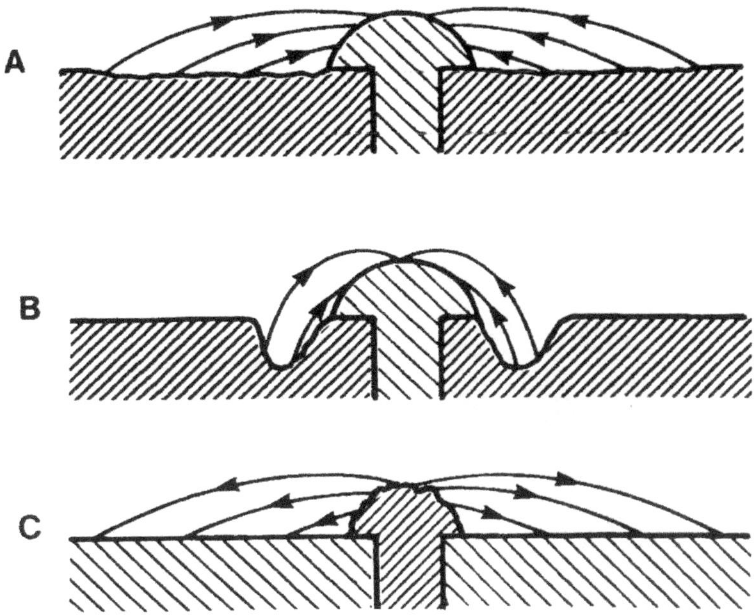

FIGURE 14.6 Galvanic corrosion in a riveted joint for three electrolytic cases: (A) large anode area, small cathode area, and high-conductivity electrolyte; (B) large anode area, small cathode area, and low-conductivity electrolyte; and (C) small anode area and large cathode area.

any significant acceleration of the corrosion of the 90/10 alloy.[1] However, there was turbulence at the junction due to a protruding gasket, and this, combined with the coupling to SS, induced erosion corrosion. This was over a small area, and hence, the cathode-anode area ratio was very high and resulted in rapid penetration of the copper-nickel.[6]

If the electrical conductance of the electrolyte bridging the bimetallic contact is low, either because the bulk conductivity is low or because the electrolyte is present only as a thin film, the effective areas taking part in bimetallic cell reactions are small, and the total amount of corrosion is relatively small, although it may be severe immediately adjacent to the metal junction.

Under immersed conditions in many supply waters, which generally have a relatively low electrical conductivity, adverse effects are uncommon if the contacting metals are of similar area. Thus, galvanized steel pipes can be used with brass or gunmetal connectors, but serious corrosion to the pipe end is likely to result if the galvanized steel is connected directly to a large area of copper such as a tank or cylinder. Similarly, SS and copper tubes

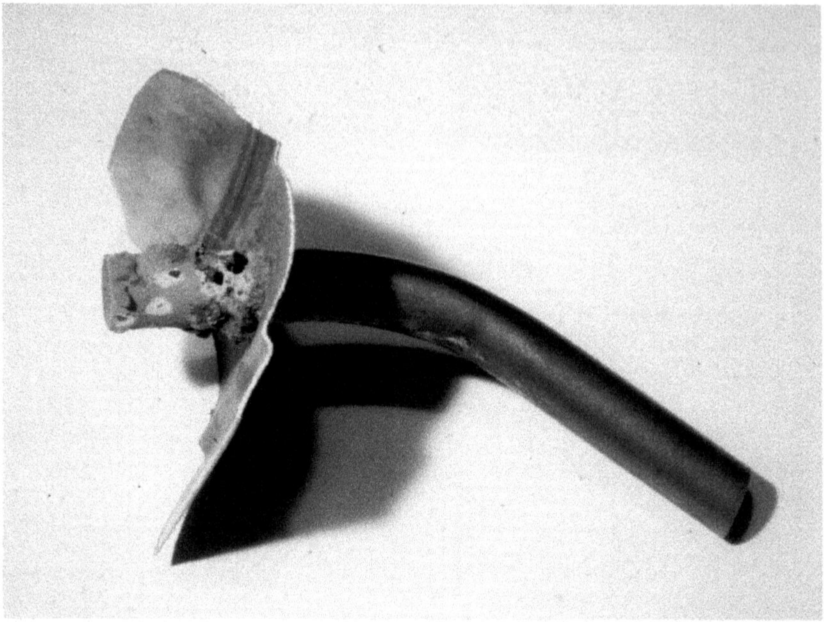

FIGURE 14.7 Severe corrosion of a copper pipe entering an SS hot water cylinder.

can usually be joined without problems resulting, but accelerated corrosion of the copper tube is likely to occur if it is attached to an SS tank, as shown in Figure 14.7.

Under immersed conditions in a highly conducting electrolyte, such as seawater, effective areas will be greater, and severe corrosion may be encountered on small anodic areas of many metals.

Extremely small anodic areas exist at discontinuities, such as cracks or pinholes, in cathodic coatings such as copper plating on steel. In seawater, where electrolyte conductivity is high, it is necessary to apply protection or, in the case of metal coatings, to specify a thickness of adequate integrity. Similar considerations apply to pores or defects in a paint coating if the metal is in contact with a more electropositive metal, that is, the cathode-anode area ratio is dramatically increased compared with unpainted metal. This should be avoided if at all possible, as shown in Figure 14.8, which shows the corrosion on a steel shaft connected to a copper alloy propeller. The lower line shows that the depth of attack on the bare shaft decreases rapidly to the uncoupled corrosion rate only 1 m from the propeller. When the shaft is coated, but there are defects in the coating, the depth of attack

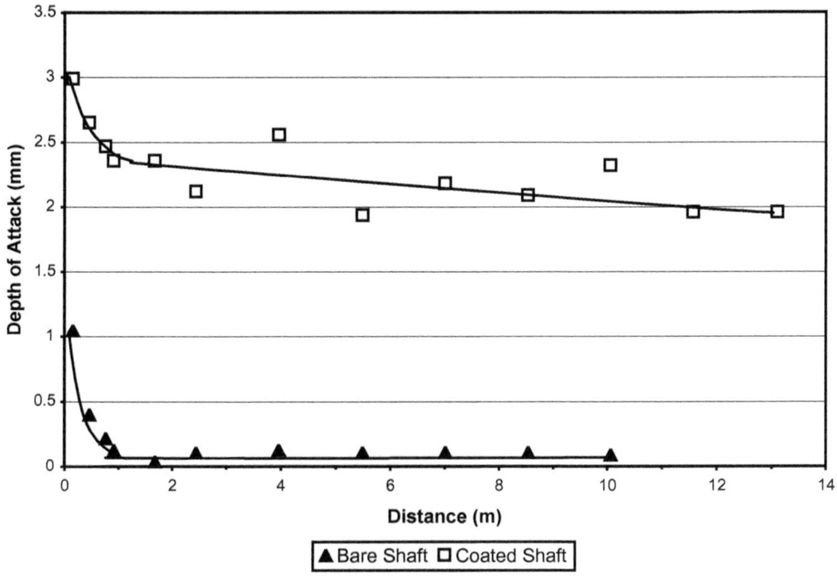

FIGURE 14.8 Depth of attack of a steel shaft coupled to a copper alloy propeller after 43 days in seawater.

is much greater and remains at a high value over 10 m from the propeller, as shown in the upper curve.

By correct selection of area ratio, dissimilar metals have frequently been used successfully together. This is essentially done by making sure that the less noble anode is large and the more noble cathode is small. Note that it is not the absolute area of the components that is important, but the wetted area.

14.2.6 Aeration and Flow Rate

The majority of practical situations involving galvanic corrosion arise in aqueous solutions under conditions in which the cathodic reaction is the reduction of DO. As with single-metal corrosion, bimetallic corrosion is, therefore, partly dependent on the rate at which oxygen can diffuse to the cathodic surface from the bulk electrolyte.

The galvanic corrosion rate of many copper and nickel-based alloys, and of SS in seawater, depends on the flow rate of the water as well as the area ratio. Copper and copper-nickel alloys tend to become more noble, that is, more electropositive, and corrode less as the flow rate increases. In well-aerated, flowing solutions, nickel-based alloys and SS are also likely to become passive and corrode less than under stagnant conditions.

In neutral electrolytes, complete deaeration will, in many instances, suppress single-metal and bimetallic corrosion. However, under such anaerobic conditions, cathodic depolarization and corrosion can occur if sulfate-reducing bacteria are present, that is, an alternative reaction to the reduction of DO is available. In closed, recirculating systems, deaeration is extremely practical, and very low DO concentrations can be maintained with chemical oxygen scavengers. Theoretically, the corrosion of iron and steel in a closed, recirculating system should consume the DO fairly quickly. In practice, there are often places where small quantities of air can be drawn into the system and so corrosion inhibitors and oxygen scavengers become necessary.

A car's cooling system often contains metals such as aluminum, steel, cast iron, copper, brass, and so on, and bimetallic corrosion does not occur because commercial antifreeze solutions contain corrosion inhibitors and/or oxygen scavengers. In once-through cooling systems, it is much more difficult to control oxygen levels, and 100% deaeration is rarely achieved. In land-based installations, such as multistage flash (MSF) desalination plants, deaeration to <20 ppb is routinely achieved. Under these conditions, the corrosion rate of CS is very low. However, on offshore oil platforms, where space and weight are at a premium, deaeration of injection water rarely gets below 50 ppb DO, and levels of 100 or 200 ppb are not uncommon. Under these conditions, corrosion rates are significant for some metals, such as CS, and bimetallic corrosion is also a possibility. Figure 14.9 shows severe corrosion of a CS water box that was coupled to copper alloy heat exchanger tubes in an MSF desalination plant. The deaeration was not well controlled and the corrosion of the water box was so severe that it had to be scrapped. In this case, the copper alloy acted as the cathode.

14.2.7 Metallurgical Condition and Composition

In some cases, differences in corrosion potential can exist between metals or alloys of nominally the same composition. Subjection to cold working often tends to make an alloy more anodic. In some alloys, heat treatment can produce galvanic differences, although this is a rarity with single-phase copper alloys.

In multiphase alloys, it is common for one phase to be anodic to another. Normally, these differences are small and do not give rise to problems,

FIGURE 14.9 Corrosion of a CS water box coupled to copper alloy heat exchanger tubes and a tube plate.

but in some instances, severe bimetallic corrosion can occur. A good example is NAB (C95800), in which the main alpha phase is anodic to the kappa III phase in seawater. The kappa III phase is present in long, thin stringers (see Figure 2.1), and the relatively large area of the alpha (anodic) phase means that there is no significant galvanic corrosion. However, when the pH drops from 8 to 3, the potentials are reversed, and there is now a large cathode-anode area ratio. Severe corrosion then occurs along the kappa III phase, and corrosion rates of >1 mm/y have been observed in crevices and shielded areas.[7]

The Royal Navy has experienced numerous failures of NAB, many associated with welds. However, commercial vessels have not experienced this problem. The reason is that naval vessels usually spend a long time being fitted out, with stagnant and intermittent flow in the seawater system. Commercial vessels are put into operation as soon as possible and rarely spend more than the minimum time fitting out. Rowlands showed that deep pitting along the kappa III occurred easily in NAB under intermittent or very slow flow conditions; however, with a normal flow rate of 3 m/s, a protective film quickly formed, and there was no pitting.[7]

Local changes in composition can also arise at joints made by welding, both in the weld bead and in the heat-affected zone of the parent metal.

Problems are usually avoided by selecting appropriate filler metals and welding techniques (see Chapter 18). Because the weld bead has a very small area compared with the parent metal, it is important that it has the same potential as the parent metal or, preferably, is slightly electropositive. Even a reduction of 50 mV in the potential of the weld metal compared to the parent may cause severe galvanic corrosion in some fluids.

Bimetallic corrosion can occur between alloys of similar type, but somewhat different composition. Thus, 90/10 copper-nickel can be anodic to 70/30 copper-nickel under some circumstances.

14.2.8 Bimetallic Corrosion without Physical Contact

Localized corrosion of a metal may give rise to soluble corrosion products, which deposit by a displacement reaction onto a less noble metal exposed in the same solution and form local, intense bimetallic cells. This is particularly evident with the more electropositive metals, such as copper, and problems have been encountered in water heating systems as a result of copper dissolving from pipe work and depositing onto steel radiators, aluminum alloy pump impellers, and galvanized steel tanks. Insulation of joints between dissimilar metal components is not necessarily a safeguard if cuprosolvent water flows or diffuses from copper to components in susceptible metals. An example is the filling of aluminum saucepans from the hot water tap in an area where the water is slightly cuprosolvent (i.e., a low pH and/or a high free carbon dioxide content). If the copper forms deposits on the aluminum, it can result in localized corrosion that may eventually lead to perforation.

Copper can cause corrosion of steel and galvanized steel, particularly in warm or hot water. Hence, it is usual to specify that copper and its alloys may be used downstream of galvanized steel, but not upstream, in supply waters. An interesting example of this occurred in a block of flats in Denmark with a common heating system. The main feed pipes for the hot water were in galvanized steel, with copper risers to the individual flats. This is as recommended, that is, copper after galvanized steel. However, as the hot water system was not in use all the time, it was possible for copper corrosion products to descend under gravity into the larger galvanized pipe during periods of nonuse. The galvanized steel pitted near the takeoff points due to the copper corrosion products. The solution was to move the takeoff points to the underside of the galvanized steel feeder pipe so that the copper corrosion products would no longer enter it.

In plants and equipment that involve recirculated electrolyte, such as vehicle cooling systems, central heating systems, and industrial heat exchangers, this type of problem can sometimes be avoided by removing oxygen from the system or by adding inhibitors, as described previously.

14.2.9 Stifling Effects

In general, the rate of corrosion of coupled and noncoupled metals decreases with time of exposure. This is partly due to the diminishing rate of diffusion of oxygen (or whatever the cathodic reactant) through the electrolyte and through films of corrosion product to cathodic regions, and partly due to the protection afforded to anodic regions by the corrosion product. To some extent, therefore, the bimetallic corrosion rate is affected by the permeability of the corrosion product. In seawater and, to a lesser extent, in hard supply waters, the alkaline conditions produced at the cathode result in the formation of calcareous deposits of calcium and magnesium carbonates, and magnesium hydroxide, which may decrease the rate of bimetallic corrosion. With steel, for example, corrosion is decreased by one order of magnitude after less than a year's exposure. Where corrosion products are water-permeable or hygroscopic, however, their presence can sometimes increase the rate of bimetallic corrosion, either because they are more aggressive than the bulk environment or because they maintain wet conditions at the metal surface.

14.3 Alloy Groups

When looking at the galvanic compatibility of materials in seawater, it is found that they can be divided into four major groups, according to their corrosion behavior, numbered from 1 to 4. Table 14.1 shows the alloy groups and the common alloys found in each class. Group 1 includes all the passive alloys, which, by and large, do not usually suffer corrosion in seawater at around ambient temperature. Some definitions are required to identify alloy classes. The nickel-chrome-molybdenum alloy class covers all the nickel alloys with more than 7 wt% molybdenum, such as Alloy 625 and Alloy C-276. Titanium covers not only grade 2 (commercially pure), but also the common alloys, such as Ti-6Al-4V and Ti-0.15% Pd. Superaustenitic SS cover all those alloys with a molybdenum content equal to 6 wt% or greater and having a PREN >40 (for which PREN = %Cr + 3.3x%Mo + 16x%N). The superduplex SS cover all duplex alloys with 25% or more chromium and a PREN >40.

In group 2 are the passive alloys that are not totally immune to localized corrosion (for example, crevice corrosion) in seawater. Such alloys

TABLE 14.1 Alloy groupings for seawater at ambient temperature

Category	Type	Alloys
1	Noble; passive	Nickel-chrome-molybdenum alloys (Mo > 7%) 6% Mo austenitic SS Superduplex SS Titanium and its alloys
2	Passive; not truly corrosion-resistant	Alloy 400/K-500 904L 22% Cr duplex Alloy 825 Alloy 20 316L
3	Moderate corrosion resistance	Copper alloys, Austenitic cast iron
4	Poor corrosion resistance	CS, Cast iron
		Aluminum alloys

are mostly SS and iron-nickel-chromium-molybdenum alloys with PREN values below 40 (for example, Type 316L SS [UNS S31603] and alloy 825). Another type of alloy that falls into this group are the nickel-copper alloys, the best known of which are Alloys 400 and K-500. These can suffer crevice corrosion in seawater.[8-9] The common group 2 alloys are shown in Table 14.1.

Group 3 covers alloys of moderate corrosion resistance that do not show true passivity. The most common ones are copper and its alloys and the austenitic cast irons. These cast irons are more noble than conventional cast irons because of their high nickel content.

Group 4 covers alloys with poor corrosion resistance and has been subdivided into an iron and steel group and an aluminum alloy group. The reason for this is that there can be a substantial potential difference between aluminum alloys and iron and steel in seawater, and so these need to be treated separately, as shown in Table 14.1.

The principal reason for grouping the alloys as shown in Table 14.1 is that there is ample evidence that bimetallic couples between any of the alloys in a particular group will not generally give rise to any significant bimetallic corrosion. This does not necessarily mean that no corrosion will occur, but no significant additional corrosion will occur. For example, naval brass will dezincify in seawater, but coupling it to 90/10 copper-nickel will not significantly increase the rate of penetration. There are a few exceptions to this rule, but they are not common, and even then, with a suitable area ratio, bimetallic corrosion need not be a problem. Hence, when there is any doubt about materials compatibility, the selection of an alloy from the same group will prevent problems. There are obviously cases in which alloys from different groups can be connected together safely, and some of these are discussed in the following sections, but selecting alloys from the same group is a good rule of thumb when in doubt.

14.4 Galvanic Corrosion in Seawater

A wide range of copper alloys are used in seawater, but the main groups are the copper-nickels, brasses, tin bronzes, aluminum bronzes, and gunmetals. The latter category is only available as castings. All these have been used extensively, and the data for some of the alloys are reported later. However, there is not much difference between any of the alloys unless selective phase attack is occurring, for example, alpha-beta brasses and two-phase aluminum bronzes.

Walleen and Andersson examined couples of 6% Mo austenitic SS and 90/10 copper-nickel exposed in natural seawater for 1 year.[8] Although the corrosion rate increases with area of SS, the corrosion only becomes significant at area ratios greater than 1:1. However, in chlorinated seawater systems, the biofilm does not form, and the SS is a less efficient cathode. This means that area ratios of 10:1 or more can be tolerated without a significant increase in the corrosion of the copper-nickel.[6] This was demonstrated on a Norwegian offshore platform, where a 90/10 copper-nickel column pipe from an NAB pump was joined to a 6% Mo superaustenitic SS ring main. The seawater had a chlorine residual of ~0.7 mg/L, and after several years' operation, there was no significant acceleration of corrosion of the copper-nickel at the junction.[10]

Two-phase brasses, such as muntz metal and rolled naval brass, are often used for heat exchanger tube plates with copper alloy tubes. The dominant form of corrosion is dezincification, and this is countered by making the tube plates thick (35 mm or more). Coupling to other copper alloys does not significantly change the rate of dezincification. However, coupling to 6% Mo superaustenitic SS or titanium showed increases in the depth of dezincification of 10 to 30 times.[11-13] Clearly an alternative tube plate material or CP is required in such cases.

A copper alloy commonly used in seawater because of its good resistance to corrosion is aluminum bronze. A large number of these alloys are used around the world, but the most common are discussed in Chapter 2.

Examples are Cu-7% Al, a single-phase alloy; Alloy D, a U.S. alloy used mostly for condenser tube plates; and NAB, the strongest alloy of this type and widely used in both cast and wrought forms.

Gehring and colleagues carried out studies on 6% Mo SS and titanium coupled to aluminum bronze, Alloy D, in seawater, with an area ratio of SS to bronze of 400:1, as a function of salinity, temperature, and pollution level.[11-13] The results showed that the corrosion rate of aluminum bronze (Alloy D) is much less than that of muntz metal, particularly at low water temperatures. However, at temperatures of about 20°C or more, the rate of attack was unacceptable. As with muntz metal, the solution is to make the tube plate very thick or to protect it cathodically, either by impressed current or by the attachment of soft iron or mild steel anodes.

When aluminum bronze is coupled to titanium or SS, it shows severe attack in natural seawater and erratic behavior in power stations when used as a tube plate material in steam condensers.[14] This may be due to the presence or absence of a chlorine residual from the antifouling treatment. As previously discussed, chlorine prevents the formation of the biofilm, which results in the cathode being far less efficient.

The author investigated this aspect when NAB valves were being considered for a superduplex SS piping system.[15] Panels of superduplex and cast NAB were bolted together and exposed in natural seawater or chlorinated seawater (0.7 mg/L chlorine) at about 23°C. Two area ratios were investigated, that is, SS to NAB of 1:1 and 10:1. The main problem was localized attack. In natural seawater, this took the form of both crevice corrosion and pitting on open surfaces. The pitting was along the kappa III phase and hence was deep, up to ~0.5 mm after 60 days. In chlorinated seawater, there was no open surface pitting, and crevice corrosion only occurred at an area ratio of 10:1.

These results suggest that NAB valves would corrode rapidly in an unchlorinated SS seawater cooling system, and even in a chlorinated system, some crevice corrosion might occur. The service experience is that severe corrosion of NAB valves in superduplex piping systems occurred on both the Woodside Goodwyn A and BP Andrew platforms. The seawater on Goodwyn A was chlorinated, but the seawater temperature varied from 20 to 40°C, that is, higher than in the author's tests. The seawater on BP Andrew was unchlorinated and about 10°C, and severe corrosion of NAB was seen in service, as predicted by the tests.

14.5 Galvanic Corrosion in Fresh Water

There are two main differences between fresh water and seawater, apart from the much lower total dissolved solids content: conductivity and hardness. The conductivity of a typical fresh water is about two orders of magnitude less than that of seawater. This means that bimetallic corrosion tends to be confined to the point of contact of the two metals. It also means that galvanic corrosion is often less severe than in seawater, unless the area ratios are very different or the potential difference is very large. The example of copper and SS was discussed previously, where copper fittings with SS pipe have been very successful, but a copper pipe entering an SS tank was badly corroded.

Seawater is very hard because of the high magnesium and sulfate contents. Fresh waters are usually subdivided into soft, moderately hard, and hard, for which the measure of hardness is the total hardness, expressed as calcium carbonate ($CaCO_3$). A soft water is usually from 0 to 50 mg/L total hardness, a moderately hard water is from 50 to 150 mg/L, and a hard water has a total hardness >150 mg/L. In fresh waters, much of the hardness is often present as calcium bicarbonate ($Ca[HCO_3]_2$), and many hard and moderately hard waters tend to deposit calcium carbonate scale with time. The scale inhibits the access of water to the metal surface, slowing down corrosion. This is the reason that CS, galvanized steel, and cast iron can last for many years in hard waters.

Soft waters do not usually deposit scale, but they also generally have lower dissolved solids contents, including chlorides, and so corrosion rates in soft waters are not necessarily much higher than in hard waters. One factor that is important in soft waters is pH. Even if the pH is quite high (for example, pH 8.5), there is little buffering action because the bicarbonate content is low and the creation of hydrogen ions, for example, at a corrosion site, can result in a large local decrease of pH.

Linder and Mattsson conducted extensive galvanic corrosion tests both at room temperature (25°C [77°F]) and at 75°C (167°F).[16] Most of the tests were in a water of intermediate hardness (100 mg/L as $CaCO_3$), but additional tests were also carried out in a soft water (hardness = 2.5 mg/L) and a hard water (hardness = 245 mg/L). During these tests, they investigated both coupled and uncoupled potentials as well as galvanic currents for a range of materials commonly used in fresh water systems.

From these data, they produced galvanic series for both hot and cold moderately hard water, which are shown in Figures 14.10 and 14.11. The

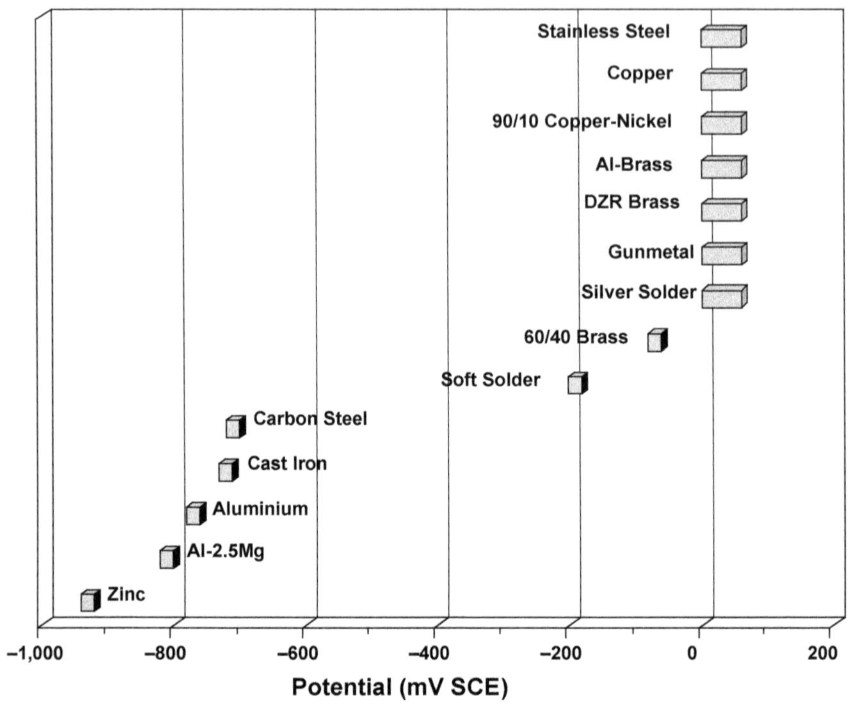

FIGURE 14.10 Electrochemical series for some common metals in a moderately hard fresh water at 25°C.[16]

heading "stainless steel" covers the four types they tested, that is, 13% Cr and 18% Cr ferritic and Types 304 and 316 austenitic SS (UNS S30400 and S31600). There were no significant differences between these alloys, but there were no tight crevices on the samples and the maximum chloride content was only 75 mg/L, with most of the data generated at 11 mg/L chloride. Higher chloride levels (>100 mg/L) occur in some fresh waters, and these might cause crevice corrosion with these alloys, particularly at higher temperatures. The silver solder tested was a copper-silver-phosphorus type, while the soft solder was a tin-silver alloy recommended for potable waters (see Chapter 18). The names of the other alloys are self-explanatory.

Both series show similar potentials for the SS and most of the copper alloys, with 60/40 brass and the soft solder being slightly more electronegative. The order of the other alloys is identical to that in seawater at 25°C, although the potential values are different. However, at 75°C, the relative positions of zinc and aluminum were reversed.

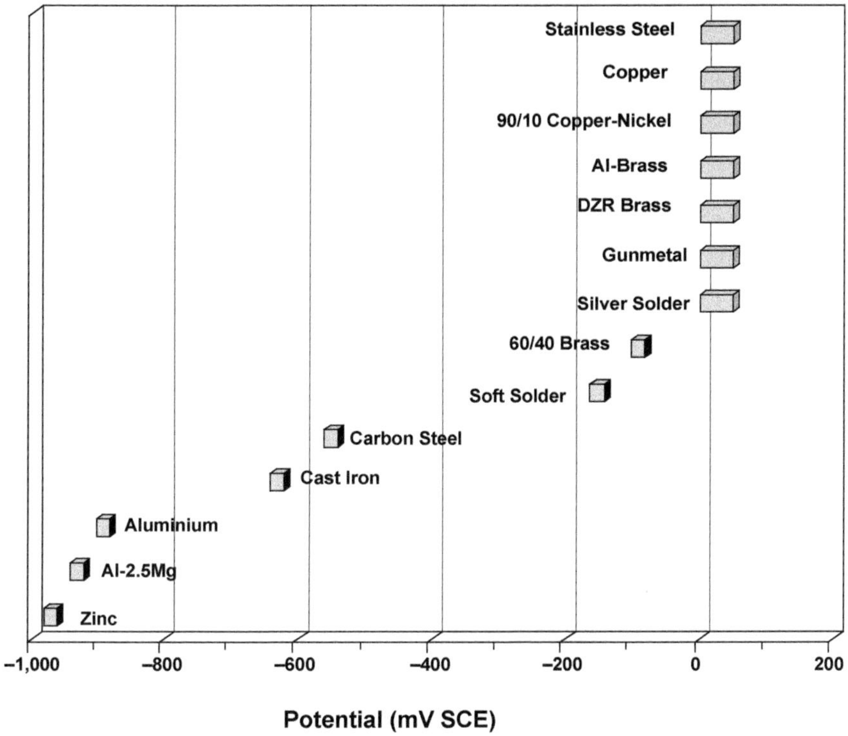

FIGURE 14.11 Electrochemical series for some common metals in a moderately hard fresh water at 75°C.[16]

The corrosion currents that were measured did not always reflect the position of the alloys in the two series. From the corrosion data, the metals have been arranged into groups in a similar manner to that done for seawater in Section 14.4. The different groups are shown in Table 14.2 for the moderately hard water at 25°C.

Although Figure 14.10 shows similar potentials for copper alloys and the SS, it is well known that copper can suffer severe galvanic corrosion when coupled to large areas of SS, as described previously. Hence, the copper alloys have been put into a separate group, while acknowledging that under the right circumstances, they can be coupled to SS. Linder and Mattsson observed some galvanic currents for the soft solder coupled to copper, although service experience shows few corrosion problems with solders at copper joints. Hence, soft solder has been put into a subsection of group 2, meaning that problems may arise in some circumstances.

TABLE 14.2 Grouping of compatible materials in moderately hard fresh water at 25°C

Group	Type	Alloys
1	Passive	Austenitic SS (304, 316) Ferritic SS (13% Cr, 18% Cr)
2	Good corrosion resistance	Copper and its alloys Silver solder
		Soft solder (Sn-Ag)
3	Moderate corrosion resistance	Cast iron CS
4	Low corrosion resistance	Aluminum Zinc

Cast iron and CS have been grouped together because they show similar potentials. When coupled together at 25°C, a little corrosion occurred on the cast iron.

Zinc and aluminum have been grouped together because they behaved similarly when coupled to other metals. The potential of aluminum was more electropositive than the potential of zinc, but when the alloys were coupled together, it was the aluminum that became the anode, although the current was small. The reason for this is believed to be that in the couple, the presence of Zn^{2+} ions modifies the potential of the aluminum, moving it electronegatively.

We can now see what effect increasing the temperature to 75°C has on Table 14.2. There are two effects, one of which is that soft solder becomes more anodic to copper than at 25°C and cast iron becomes cathodic to CS. The higher than expected galvanic current when CS and cast iron were connected together was probably due to the graphite in the cast iron increasing its cathodic efficiency. The other effect was that zinc became more electropositive than aluminum, with the result that there was more corrosion of the aluminum when it was coupled to zinc.

In soft water at 25°C, there were no significant differences in corrosion behavior compared with the moderately hard water. At 75°C, the grouping of alloys did not change, but the currents between incompatible metals were higher than for the moderately hard water, and therefore corrosion rates were higher. This may have been because the pH was not so well buffered and decreased more in the soft water during the exposure tests. Alternatively, it may have been that some scale was being deposited on the cathodes in the moderately hard water, which reduced the corrosion rate. Soft waters generally show no real scale-forming tendency.

In hard water at 25°C, the corrosion between incompatible metals (i.e., from different groups) was much less than in the moderately hard water. Aluminum was less anodic to copper and SS and was cathodic to zinc. Zinc was less anodic to CS, which was itself less anodic to SS. Only aluminum coupled to CS gave large corrosion currents. The reduction in corrosion was probably due to the formation of calcium carbonate scale on the cathode and the buffering of the pH by the bicarbonate ions.

When the temperature of the hard water was increased to 75°C, aluminum became anodic to zinc and corroded substantially when coupled to any other alloy. Couples between the other alloys were similar in behavior to that at 25°C, but generally, there was less corrosion, presumably because of the increased scale-forming tendency of the hard water at high temperatures.

14.6 Galvanic Corrosion in the Atmosphere

The most important factor in atmospheric corrosion is water. Corrosion in direct contact with air at ambient temperature is generally not significant from an engineering point of view. Therefore, water must condense on the exposed metal surfaces or precipitate as rain. This will be followed by evaporation as the components dry out, and thus, a most important parameter for atmospheric corrosion is the time of wetness. Condensation is controlled by the relative humidity, and it is generally accepted that a relative humidity of 80% or more and a temperature greater than 0°C are required for significant condensation.[17] Rainfall is dependent on the local weather and can vary greatly, even in a small country like the United Kingdom.

Corrosion in the atmosphere is generally not so great as corrosion fully immersed in water. However, galvanic corrosion rates can be as high as those under immersed conditions.

As the water evaporates, anything dissolved in it will concentrate. Hence, when these species are aggressive to a metal, the corrosion rate will increase as the water evaporates. The water films formed in the atmosphere are generally thin, and rapid diffusion of oxygen can occur. Hence, there is a ready supply of oxygen that can be reduced as the cathodic reaction.

Ammonia (NH_3) is common in rural areas and is readily soluble, causing increases in pH. This is beneficial for most alloys, although stress corrosion cracking of some copper alloys is possible. Sulfur and nitrogen oxides are common in industrial and urban areas as effluent gases, for example, fossil-fuelled power stations. All these gases dissolve readily, forming strong acids, which are the cause of acid rain.

In addition to these gases, two other factors affect atmospheric corrosion: temperature and chlorides. Temperature is self-explanatory: the warmer the climate, the greater the rate of chemical reactions, including corrosion. Chloride is found adjacent to the sea, and the closer a site is to the sea, the higher the chloride concentration will be in condensed water.

Using all these factors, sites are generally divided into four types by atmosphere: rural, urban, industrial, and marine. Rural sites have low chloride and sulfur dioxide and may have ammonia. The urban atmosphere has some sulfur dioxide, whereas industrial areas have high sulfur dioxide and NO_x content. Finally, the marine atmosphere is characterized by high chloride content.

The cathodic reaction in atmospheric corrosion is usually the reduction of DO, and as described previously, the reaction rates can be high, especially with thin water films. If the dissolved solids and gases content of the water is low, the conductivity will be low, reducing the rate of corrosion, but as the films evaporate, the conductivity will increase.

Galvanic corrosion in the atmosphere tends to be confined to the dissimilar metal joint where the condensation has occurred. Hence, area ratios are usually around 1:1. From the preceding discussion, it can be deduced that atmospheric galvanic corrosion is similar to galvanic corrosion when immersed in a low-conductivity electrolyte. The corrosion rate obviously varies as the species dissolved in the water vary, and as the water films evaporate and condense again. This is not readily simulated in the laboratory, but there is a body of data from long-term exposure trials of galvanic couples that is very useful.

Kucera and Mattsson published bimetallic corrosion data for a variety of atmospheres. In addition, they made an attempt to categorize the corrosion in terms of engineering risk, and the data are summarized below.[18] Table 14.3 shows the atmospheric corrosion rates for 11 commonly used alloys. Weathering steel is a low-alloy CS with ~0.5% copper. This produces an adherent rust layer that can lower atmospheric corrosion rates compared to CS. In addition, aluminum was tested in the anodized condition, as this is commonly used to reduce aluminum corrosion. Three types of atmosphere

TABLE 14.3 Corrosion of some common metals in various atmospheric environments[18]

	Corrosion rate (μm/y)		
Alloy	Rural	Urban	Marine
Magnesium	6.5	11	14
Aluminum	0.1	0.5	0.6
Anodized aluminum	0	0.2	0.2
Zinc	0.5	2.5	1.3
Carbon steel	22	53	36
Weathering steel	22	48	32
Tin	1.4	2.7	10
Lead	0.3	0	0
Copper	1.2	1.4	3.2
Nickel	0.4	2.7	1.0
Stainless steel (Type 304)	0.03	0.04	0.04

are reported: rural, urban, and marine. Some metals suffered little corrosion in any of the environments, for example, lead and SS. However, while the corrosion rate of some metals is low in rural atmospheres, it is much greater in urban and/or marine atmospheres. Zinc is clearly more susceptible to acid sulfur gases than chlorides, that is, it is more susceptible in urban, rather than marine, atmospheres, whereas tin and copper corroded most in marine atmospheres.

Kucera and Mattsson tested many, but not all, of these metals in combination, and the results for marine atmospheres are summarized in Table 14.4.[18] The data from urban and rural areas are similar to those in Table 14.4 and may be found in the work of Francis.[1] The aim has been to show couples with no significant change in corrosion rate, a slight increase in corrosion rate, and a large increase in corrosion rate. This has to be compared with the uncoupled corrosion rate in Table 14.3, as an increase in corrosion rate may still not represent much loss of metal and may be acceptable for a particular application. There are no absolutes here, and some metal combinations may be acceptable in one case and not in another, depending on the criticality of the application. Chromium is included in Table 14.4 because of its widespread use as electroplating in atmospheric applications.

With many metals used uncoated in the atmosphere (steel, zinc, and aluminum), copper is the cathode. Being an efficient cathode, the corrosion rate of the anode is accelerated, but not often dramatically, as the effective area ratio is usually 1:1.

It might be thought that coupling copper, or one of its alloys, to SS may result in accelerated corrosion of the copper. This is not so because

TABLE 14.4 Risk of galvanic corrosion in a marine atmosphere for some common metals[18]

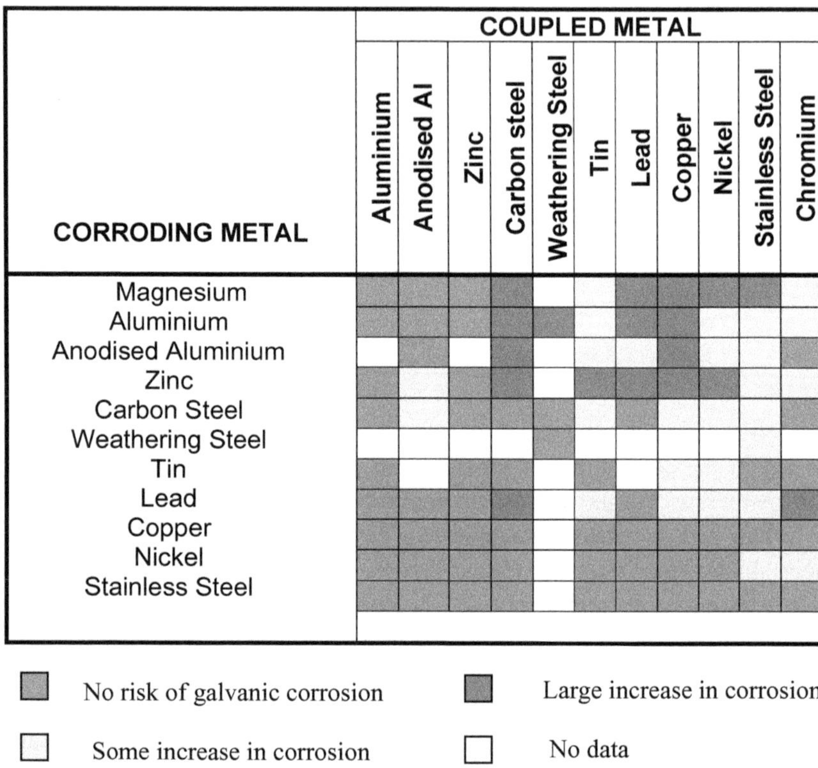

the effective area ratio is usually 1:1, and SS is not a very efficient cathode. This was why SS was used to replace corroded steel internal supports of the copper skin on the Statue of Liberty.[19-20]

14.7 Methods of Prevention

The following sections review the main methods of preventing and combating galvanic corrosion. To do this properly, it is necessary to know as much as possible about the environmental conditions. With more information, more confidence can be placed in the decisions made to mitigate galvanic corrosion.

14.7.1 Materials

Where the design is being completed before any materials are purchased, there is the maximum opportunity to select galvanically compatible metals for the specific application. The following guidelines offer some helpful suggestions.

Use compatible metals wherever possible. Try to select alloys from the same group, as shown in Tables 14.1 and 14.2. Avoid metals that are far apart in the electrochemical series, unless it is well documented that they can be coupled successfully, for example, 90/10 copper-nickel and high-alloy SS in chlorinated seawater.

When welding, pay particular attention to the condition of the weld after fabrication. It is desirable that it be cathodic to the parent metal. When carrying out autogenous welding, choose welding parameters that will achieve this. When using filler metal, select a filler that produces weld metal that has a similar potential or is electropositive to the parent metal. With such a high cathode-anode area ratio as occurs in a weld, a weld metal potential only 50 mV negative to that of the parent metal can result in severe corrosion in some fluids.

Where compatible materials cannot be used, for whatever reason, one of the following methods of prevention of galvanic corrosion should be considered.

14.7.2 Insulation and Separation

If the anode and cathode can be prevented from coming into electrical contact, then no galvanic corrosion will occur. In some situations, this can be done quite conveniently, while in others, it is impractical.

Several methods for achieving this when bolting two flat surfaces together were discussed by Parkins and Chandler, depending on whether it is the bolts or one of the plates that is the anode.[1,21]

Another method of preventing galvanic corrosion between two adjacent metals could involve the use of a thick coat of nonconducting, insoluble mastic at a joint. The mastic may additionally contain chemicals to inhibit corrosion of the metals as they are leached out.

One method for preventing galvanic corrosion of piping at dissimilar metal joints is the isolating flange. This uses a nonconducting insert between the two metal flanges and nonconducting sleeves and washers for the bolts. This means that the two metals are no longer in electrical contact. However, there are two major disadvantages to this technique. First, the sleeves and washers have to be undamaged and correctly assembled for the isolation flange to work properly. On many industrial sites, sufficiently skilled labor is not available, and this is difficult to ensure. Second, there is frequently a requirement by electrical engineers to earth all major items of metal equipment for safety reasons. This has the effect of shorting across the isolation flange and, unfortunately, is a common industrial practice.[22] Pipe hangers or supports can also act as inadvertent electrical shorting between adjacent pipe runs. Where isolating flanges are used it is important that the resistance between the two components be measured prior to start-up, to ensure that isolation has been achieved.

An alternative method of separating the anode and cathode when one is a pipe and the other is a pipe, vessel, valve, etc., is to use an insulated spool piece. This is a length of pipe between the two items that is either internally coated or is made from a nonmetallic, for example, fiber-reinforced plastic (FRP/glass-reinforced plastic). Note that the coating must extend up onto the flange faces as well as along the pipe bore. The metal pipe being coated must be made of the same metal as that forming the cathode so that defects in the coating do not give rise to rapid localized corrosion. If a coating is being used, it should have a life consistent with the planned maintenance intervals for the plant. A coating that needs replacing at more frequent intervals is often not cost-effective, and an alternative solution to the galvanic corrosion problem may be required.

The length of the spool piece depends on the electrochemical properties of the two metals and the particular fluid under consideration. An example for seawater serves to demonstrate this principle. When joining a copper alloy, such as 90/10 copper-nickel, to a group 1 alloy, it is common to use a spool piece approximately six pipe diameters long, provided that there is no undue turbulence to cause localized erosion corrosion of the copper alloy. BP Tankers had an oil tanker with 90/10 copper-nickel piping and a titanium-plate heat exchanger. Galvanic corrosion occurred on the copper-nickel adjacent to the heat exchanger, which was solved by fitting a coated titanium spool piece five diameters long. The length of the spool was limited by the available space in the ship, but it prevented further corrosion of the copper-nickel pipe.

Bardal et al. showed that with high-alloy SS and 90/10 copper-nickel in chlorinated seawater, a spool piece is unnecessary.[6] However, if the chlorination were likely to be turned off at all, a spool piece would be a wise precaution to prevent corrosion at the dissimilar metal joint. In natural seawater, the corrosion rate of 90/10 copper-nickel was ~1.2 mm/y at the junction with an SS pipe.

14.7.3 Paints and Coatings

When trying to prevent a bimetallic corrosion problem, it is sometimes possible to use paints or coatings on the affected area. Preferably, both the anode and cathode should be coated, but it is often sufficient just to coat the cathode. A small defect in the coating will still mean that only a very tiny cathode area is exposed to the seawater. On no account should only the anode be coated. The cathode-anode area ratio will be very high at defects in the coating, and the rate of corrosion at the anode will usually increase dramatically.

It is important to select coatings that have a life similar to that of the components they are protecting, especially if access is limited or difficult. Otherwise, regular maintenance of the coating to prevent galvanic corrosion will be necessary as the coating breaks down.

An alternative method to organic coatings is metallic coatings. The most commonly applied metals are aluminum and zinc, although many other metals can be used as coatings. The most common, and cheapest, method is thermal spraying. However, this produces porous coatings, and for maximum life, polymeric sealers are required after coating. An alternative is to use high-velocity oxy fuel (HVOF) spraying, which produces very dense coatings with no need for sealing, although this process is a little more expensive.

Note that where a metallic coating is porous and no sealer is used, it is essential that the coating be anodic to the base metal; otherwise, accelerated corrosion will occur at the defects in the coating due to the high cathode-anode area ratio. An example of a poor choice of coating is chromium plating on steel, which is unsuitable in seawater.

For the case of atmospheric corrosion only, some additional coating measures can be considered. These all rely on the galvanic corrosion being confined to the joint area with an anode-cathode area ratio of approximately 1:1. One solution is to use a nonconducting mastic at the junction, which will, effectively, increase the resistance between the anode and cathode. A similar effect can be obtained with polymeric tapes, which may contain inhibitors that can be leached out. The tape is usually applied up to about 25 mm to either side of the bimetallic joint. Rubber coating has also been used successfully to prevent atmospheric galvanic corrosion.

14.7.4 Cathodic Protection

This method of protection relies on changing the coupled potential of two dissimilar metals. It is not always necessary to suppress corrosion totally, but only to prevent or reduce the corrosion due to the galvanic couple.

Two methods are used: impressed current and sacrificial anodes. With the former, a current sufficient to change the potential to the desired value is applied. With a sacrificial anode, the final potential is usually close to the open circuit potential of the anode, and so the metal chosen for the anodes is important. Not only must the potential be in the desired range, but the anode must also be of a size and have a life and a cost that are practical.

Some examples will show this. In tube and shell heat exchangers, using seawater for a coolant, muntz metal or rolled naval brass is commonly used as a tube plate material with 90/10 or 70/30 copper-nickel tubes. The brass tube plates dezincify, but the tube plates are usually thick ($\geq$35 mm), and so this may be ignored. In more critical applications, sacrificial anodes have been attached to the tube plates. Zinc and aluminum-indium are common alloys that are readily available. However, the open circuit potential of these alloys is about –1 V SCE. If a large number of anodes are fitted, overprotection results, which can give rise to unexpected problems. A heat exchanger at a Spanish power station was protected in such a manner, and the low potential caused the precipitation of a heavy calcareous scale in the tube inlets. This increased inlet end turbulence, which caused some erosion corrosion of the tubes, which was exacerbated by the presence of sulfides.

The solution is to use soft iron or mild steel (i.e., low carbon) anodes. The potential is now too electropositive for significant calcareous scale to deposit, whereas it is sufficiently low to reduce dezincification of the tube plates. An added benefit is that the supply of iron corrosion products from the anodes aids in the formation of protective films on the copper alloy tubes.

A similar principle can be applied in piping systems by fitting a sacrificial spool piece, usually of CS, five to six pipe diameters long between the two components of the bimetallic couple. In seawater, the spool piece requires replacement every 1 or 2 years, but if access is easy, this can be a cost-effective solution. Remember that iron corrosion products from the spool piece will be carried downstream, and this must be acceptable in the rest of the system. The corrosion products usually form soft deposits of hydrated iron oxides up to $\sim$1 mm thick, which may not always be desirable.

14.7.5 Inhibitors

Inhibitors are rarely used in once-through seawater systems, although they may be used in recirculating closed systems or in seawater injection systems in the oil and gas industry. Note that ferrous sulfate ($FeSO_4 \cdot 7H_2O$), sometimes added to improve the corrosion resistance of copper alloy heat exchanger tubes, is not an inhibitor. The iron hydroxide (FeOOH) deposit that is produced acts as a barrier between the turbulent water and the protective film on the metal surface, as described in Chapter 3.[23]

Where inhibitors are used, it is important to select one that is compatible with all the metals in the system.

In Section 14.7.3, the use of mastics and tapes to prevent atmospheric galvanic corrosion was discussed. In some cases, these can be obtained with inhibitors to provide additional protection against galvanic corrosion. As with fully immersed conditions, it is important that suitable inhibitors be used that are compatible with both metals in the couple being protected.

A slightly different method of inhibition is deaeration, which can usually only be applied to recirculating systems. Provided that hydrogen evolution cannot be the cathodic reaction, the addition of oxygen scavengers to prevent the cathodic reduction of DO is a viable method of preventing corrosion, including galvanic corrosion. It is essential to prevent additional entry of oxygen into the system or regular dosing with a scavenger will be necessary.

14.7.6 Design

If it is addressed early enough in the design stage, galvanic corrosion can be prevented by one of the methods described previously. Once construction has started, changes and retrofits become much more difficult and expensive.

A range of good practices should be adopted to minimize the risk of galvanic corrosion. For structures exposed to the atmosphere, one of these is to prevent water accumulation and the entrapment of debris. Designs that allow water to run off freely pose fewer problems than those that create areas where water can collect (see Chapter 16).

Wherever possible, for fully immersed conditions, a design should be chosen to minimize the area of the cathode and/or maximize the area of the anode. If the cathode-anode area ratio is kept sufficiently small, additional methods of restricting galvanic corrosion may not be necessary.

Finally, one further factor is essential to the implementation of a good design: an understanding by the work force carrying out the construction, fabrication, and so on. If the workers are told the importance of what they are doing and why it must be done a particular way, then they will be more likely to implement the design correctly. Any design to prevent galvanic corrosion must include communication with the work force of the necessary precautions required during assembly.

References

1. R. Francis, *Galvanic Corrosion—A Practical Guide for Engineers* (Houston, TX: NACE International, 2001).
2. T.S. Lee, E.W. Thiele, J.H. Waldorf, *MP* 23, 11 (1984): p. 44.
3. R.A. Corbett, *MP* 36, 12 (1999): p. 63.
4. T. Rogne, U. Steinsmo, "Practical Consequences of the Biofilm in Natural Sea Water and of Chlorination on the Corrosion Behaviour of Stainless Steels," in *Sea Water Corrosion of Stainless Steels—Mechanisms and Experiences,* EFC Publication 19 (London, U.K.: Institute of Materials, Minerals, and Mining, 1996): p. 55.
5. K.D. Efird, *MP* 24, 4 (1985): p. 37.
6. E. Bardal, R. Johnsen, P.O. Gartland, *Corrosion* 40 (1984): p. 12.
7. J. Rowlands, "Studies of the Preferential Phase Corrosion of Cast Nickel Aluminium Bronze in Seawater," Eighth International Congress on Marine Corrosion and Fouling (Mainz, Germany: 1981).
8. B. Walleen, T Andersson, "Galvanic Corrosion of Copper Alloys in Contact with a Highly Alloyed Stainless Steel in Sea Water," Proc. of the 10th Scandinavian Corrosion Congress (Stockholm, Sweden: 1986), p. 149.
9. R. Francis, Weir Materials and Foundries Technical Report, TN1099, July 1994.
10. O. Strandmyr, O. Hagerup, CORROSION/98, paper no. 707 (Houston, TX: NACE International, 1998).
11. G.A. Gehring, J.R. Maurer, CORROSION/81, paper no. 202 (Houston, TX: NACE International, 1981).
12. G.A. Gehring, R.J. Kyle, CORROSION/82, paper no. 60 (Houston, TX: NACE International, 1982).
13. G.A. Gehring, "Inlet End Corrosion Problems in Condensers," Proc. of the Seventh Annual Conference on Materials for Coal Conversion and Utilization (Gaithersburg, MD: National Institute of Standards and Technology, 1982).
14. B. Case, J.O. Davies, W.E. Heaton, J.M. Sketchley, "Assessment of the Corrosion of Tube Plate Materials in Titanium Tubed Condensers," Proc. of the International Colloquium on Choice of Materials for Condenser Tubes and Plates and Tube and Tightness Testing (Avignon, France: SFEN, 1982), p. 171.

15. R. Francis, *Brit. Corros. J.* 34 (1999): p. 139.

16. M. Linder, E. Mattsson, "Galvanic Corrosion of Combinations of Metals in Fresh Water," Proc. of the Seventh Scandinavian Corrosion Conference (Trondheim, Norway: 1975), p. 19.

17. E. Mattsson, *Basic Corrosion Technology for Scientists and Engineers,* 2nd ed. (London, U.K.: Institute of Materials, Minerals, and Mining, 1996): p. 75.

18. V. Kucera, E. Mattsson, "Atmospheric Corrosion of Bimetallic Structures," in *Atmospheric Corrosion,* W.H. Ailor, ed. (New York, NY: John Wiley, 1982): p. 561.

19. R. Baboain, E.B. Kliver, *The Statue of Liberty Restoration* (Houston, TX: NACE International, 1990).

20. R. Baboain, *MP* 35, 11 (1996): p. 5.

21. R. Parkins, M. Chandler, *Corrosion Control in Engineering Design* (London, U.K.: DOI/HMSO, 1978).

22. R. Johnsen, S. Olsen, CORROSION/92, paper no. 397 (Houston, TX: NACE International, 1992).

23. J.E. Castle, M. Parvisi, "Minor Alloying Elements and the Marine Corrosion of Copper Alloys," in Proc. of the International Colloquium on Choice of Materials for Condenser Tubes and Plates and Tube and Tightness Testing (Avignon, France: SFEN, 1982), p. 171.

CHAPTER 15

Underground Corrosion

Corrosion in soils is complex and depends on many variables. A wide range of corrosivity is also possible in soils; failures of buried components have occurred within a year, while archaeological samples have been preserved for hundreds of years. Although a wide range of equipment may be buried, the most common items are pipes (for water, sewage, gas, etc.), cables (electricity and telecom), tanks, pilings, and foundations. These are often required to last for periods of 50 to 100 years. Copper and copper alloy artifacts that are thousands of years old have been recovered from archaeological sites. This shows that given the right soil conditions, copper can resist corrosion for long periods of time.

Soil, essentially, has three components: particles, which may be inorganic (clay, sand, etc.) or organic (peat, muck,[(1)] etc.); moisture, which includes the solids dissolved in the water; and gas. Both nitrogen and oxygen can diffuse into soils from the air, but other gases may be produced by local organic activity, for example, carbon dioxide (CO_2) and hydrogen sulfide (H_2S). The properties of the soil that influence corrosion are as follows:

1. Soil moisture includes the dissolved solids and varies seasonally
2. Oxygen, the quantity of which depends on the ability of the gas to penetrate the soil
3. Redox potential, generally determined by the oxygen content of the soil
4. pH, influenced by the dissolved solids and any chemicals being produced by local organic activity

[(1)] Muck is decaying vegetation, *not* manure.

5. Conductivity, which affects corrosion in a similar way to liquids, that is, the more conductive the soil, the greater the corrosion current
6. Microbial activity, by which a range of chemical by-products may be produced that influence corrosion, for example, H_2S

Soil conditions can vary considerably, even over a small distance, and this can cause large changes in corrosivity. When corrosion in soils is severe, it is similar to aqueous corrosion in seawater or fresh water; when corrosion is mild, it is similar to atmospheric corrosion. In underground corrosion, it is common for the anode and cathode of a corroding metal to be adjacent, as in aqueous corrosion. However, under certain conditions, they may be more widely separated, and in extreme cases, they may be a kilometer or more apart.[1] In such cases, any corrosion products tend to form between the anode and cathode, possibly causing the anode to pit and corrode at a high rate.[1] This is also true of galvanic corrosion underground, although the corrosion is more commonly found at the couple junction.

Although conductivity is used when assessing aqueous solutions, it is more common to use resistivity when considering underground corrosion. This is the inverse of conductivity, that is, 400 μS/cm = 2,500 Ωcm. Typically, soils with a resistivity below 1,000 Ωcm are regarded as aggressive, whereas those with a resistivity >5,000 Ωcm are mildly corrosive.

To assess the corrosivity of a soil, a number of methods have been proposed. One of the best known is that used by the German gas industry for assessing the risk of corrosion of buried pipelines.[2] This involves an index calculated from 10 parameters of the water and soil, and the more negative the index, the more corrosive the soil.

Stray currents from electrical conductors are well known to affect corrosion underground. This is a specialist subject and outside the scope of this book. For more information, the reader is referred to specialist texts on the subject, while Mattsson gives an introduction to the subject.[1]

Note that when equipment is buried and the soil is replaced, the conditions around the buried item can be very different to those in the adjacent, undisturbed soil.[3] The moisture and oxygen content around the buried item can be much higher, leading to greater corrosion than would be expected from general testing of the soil.

15.1 Long-Term Exposure Tests

Between the 1930s and 1950s, several long-term exposure tests were conducted on copper and its alloys buried at a variety of sites in both the

United States and United Kingdom. The U.S. National Bureau of Standards (NBS) data have been summarized by Romanoff, and the highlights, relating to copper, are presented here.[4] The exposures varied from 8 to 14 years in a variety of soils, some of whose properties are listed in Table 15.1, while the corrosion data are summarized in Table 15.2. The NBS study used multiple specimens of both phosphorus-deoxidized and tough pitch coppers. There was no significant difference between the corrosion rates for the two coppers, and so the average rate for all the samples has been quoted.

The resistivity figures in Table 15.1 show a very wide range, as does the drainage. There was more corrosion at poorly drained sites, whereas the corrosion at well-drained sites was slight. The annual rainfall varied from 750 to 1,500 mm at most sites and appeared to have little effect on corrosion within this range. It should be remembered that at sites with heavy rainfall, much of it may fall in a few days and mostly run off the surface, so that the subsoil is comparatively dry for much of the year.

Any effects of temperature are masked by the more marked effects of the other conditions. Very high sulfate (>300 ppm) and/or chloride (>100 ppm) content is associated with poor drainage and usually occurs in soils containing considerable humus. The combination appears to be responsible for most of the worst cases of corrosion (sites 43, 63, 60, and 64). The rate of corrosion, however, bears no direct relation to the sulfate or chloride content. Soils with moderate sulfate and/or chloride content still cause appreciable corrosion when the drainage is bad.

The most acidic soils ($pH \leq 4.0$) all have considerable sulfate content, and generally chloride, too, so the comments in the preceding paragraph apply to these soils as well. Of the alkaline soils ($pH \geq 8.0$), the most corrosive was soil 67, with very poor drainage (category 4 in Table 15.1) and a moderate sulfate content, possibly because this was associated with a particularly aggressive sulfide. The pH figures appear to give only a very rough guide to soil corrosivity, but it must be remembered that the NBS report states that some of the pH values may be incorrect, as they were liable to change on exposure to the air.[4]

Soils with low resistivity (<100 Ωcm) are always very corrosive, but the worst corrosion occurred with soils 67 and 60, which had resistivities in the range 200 to 500 Ωcm and poor drainage. Other sites with resistivities in this range had less corrosion but also had low rainfall and drainage, varying from poor (category 3) to good (category 1). It should be noted that the resistivities are quoted for saturated soils at 16°C and that these soils would have higher resistivities for most of the year. A better correlation

TABLE 15.1 Summary of soil conditions during NBS exposure tests[4]

Soil							Comp. of water (ppm)			
No.	Type	Mean temp. (°C)	Ann. rainfall (mm)	Drainage[(A)]	pH	Resistivity (Ωcm)	Na	HCO_3	Cl	SO_4
67	Cinders	8	762	4	8.0	455	8	6	0.8	29
43	Tidal marsh	11	1,092	4	3.1	60	451	0	433	370
63	Tidal marsh	19	1,143	4	2.9	84	336	0	127	366
60	Peat	9	940	4	2.6	218	29	0	0	567
33	Peat	8	762	4	6.8	800	15	–	23	21
58	Muck[(B)]	21	1,448	4	4.0	712	20	0	5	25
29	Muck	21	1,448	4	4.2	1,270	22	0	17	23
45	Alkali soil	8	381	3	7.4	263	82	2	2	120
64	Clay	14	406	4	8.3	62	281	9	288	3
56	Clay	21	1,245	3	7.1	406	31	8	16	30
61	Clay	21	1,448	3	5.9	943	7	7	1	9
27	Clay	19	1,422	3	6.6	570	5	20	1	15
28	Clay adobe	16	254	3	6.8	408	15	1	10	9
5	Clay adobe	13	584	3	7.0	1,346	9	7	0.3	2
3	Clay loam	16	1,219	1	5.2	30,000	–	–	–	–
8	Clay loam	9	533	3	7.6	350	14	7	0.1	44
25	Clay loam	8	762	2	7.2	2,980	2	10	0.3	1
36	Sandy loam	18	1,346	1	4.5	11,200	–	–	–	–
10	Sandy loam	10	1,041	2	6.6	7,460	–	–	–	–
12	Sandy loam	17	381	2	7.1	3,190	4	4	0	1
16	Sandy loam	19	1,549	2	4.4	8,290	–	–	–	–
37	Fine sand	21	1,194	3	3.8	11,200	–	–	–	–
31	Fine sand	21	1,194	1	4.7	20,500	–	–	–	–
66	Gravelly loam	21	203	1	8.7	232	66	7	28	30
6	Gravelly loam	11	864	1	5.9	45,100	–	–	–	–
4	Loam	12	1,016	2	5.6	6,670	–	–	–	–
35	Loam	17	381	1	7.3	2,060	7	11	0.6	4
23	Silt loam	18	152	3	9.4	278	84	19	11	56
1	Silt loam	9	864	3	7.0	1,215	7	1	1	8
20	Silt loam	9	864	3	7.5	2,870	2	5	0	2
19	Silt loam	10	813	1	4.6	1,970	4	2	0.3	5
18	Silt loam	11	711	1	7.3	1,410	3	9	0	2

(A) Drainage is indicated as follows: 1 = good; 2 = moderate; 3 = poor; 4 = very poor.

(B) Decaying vegetation.

TABLE 15.2 Summary of corrosion results for copper from NBS exposure tests[4]

Soil		Exposure time (years)	Mean corrosion rate (mm/y)	Maximum pitting rate (mm/y)
No.	Type			
67	Cinders	9.2	0.0400	0.150
43	Tidal marsh	8.0	0.0210	(A)
63	Tidal marsh	9.6	0.0160	0.020
60	Peat	9.2	0.0230	0.107
33	Peat	8.0	0.0040	0.025
58	Muck(B)	9.5	0.0070	0.033
29	Muck	8.0	0.0040	<0.020
45	Alkali soil	8.0	0.0010	0.025
64	Clay	9.2	0.0150	0.043
56	Clay	9.4	0.0028	<0.018
61	Clay	9.5	0.0010	0.020
27	Clay	13.6	0.0004	<0.010
28	Clay adobe	8.0	0.0028	<0.020
5	Clay adobe	13.4	0.0010	0.015
3	Clay loam	8.0	0.0010	<0.020
8	Clay loam	8.0	0.0008	0.036
25	Clay loam	8.0	0.0004	<0.010
36	Sandy loam	13.6	0.0066	<0.010
10	Sandy loam	13.2	0.0030	0.015
12	Sandy loam	8.0	0.0102	0.038
16	Sandy loam	8.0	0.0020	0.066
37	Fine sand	8.0	0.0058	0.041
31	Fine sand	13.7	0.0003	<0.011
66	Gravelly loam	9.2	0.0020	<0.023
6	Gravelly loam	13.3	0.0004	<0.010
4	Loam	8.0	0.0008	0.033
35	Loam	8.0	0.0005	<0.018
23	Silt loam	8.0	0.0046	0.041
1	Silt loam	8.1	0.0020	0.028
20	Silt loam	8.1	0.0013	0.036
19	Silt loam	8.0	0.0013	0.023
18	Silt loam	8.0	0.0003	No pitting

(A) For no. 43, depth of pits is uncertain.
(B) Decaying vegetation.

with corrosivity would be expected if the resistivity values were related to the average moisture conditions during the year.

Of the various soil types examined, cinders was the most aggressive, although its sulfate and chloride contents were not high. The high corrosion rate is probably due to the fact that cinders contain a high proportion of

TABLE 15.3 Summary of soil conditions at BNFMRA test sites[5]

Soil		Moisture loss at 110°C (%)	pH	Water extract composition (ppm)			
Site	Type			Cl	SO_3	CaO	Saline NH_3
F	Wet acid peat	41.2	4.2	4.7	372	262	40
C	Moist acid clay	2.4	4.6	1.7	12	17	1.8
G	Made-up ground(A)	–	–	–	–	–	–
E	Dry acid sand	0.9	5.1	2.0	4.5	12	3
A	Moist normal clay	5.8	7.2	1.2	6.2	24	0.4
D	Chalk	2.6	8.4	5.1(B)	1.6	36	0.3
B	Slightly acid sand	2.3	6.2	0.3	3.1	8	0.3

(A) Miscellaneous builder's rubble.
(B) This is the mean of two widely differing values (1.2 and 9.0). Gilbert explained why two such different values should be found for the same soil.[5]

carbon, which is both noble and an efficient reducer of dissolved oxygen (DO) (the most common cathodic reaction). Hence, the attack can be regarded as galvanic corrosion, with a high corrosion rate because of the high cathode-anode area ratio. Soils composed mainly of decaying organic matter are the next most aggressive, particularly soils 43 and 63 (tidal marsh) and 60 (peat). All of these have high sulfate and/or chloride content and a high moisture content. Soils with less sulfate and/or chloride but similar moisture content (soils 33, 58, and 29) caused much less corrosion. Note that *muck* in the United States means a soil largely composed of decaying vegetation, and not "manure," as it does in the United Kingdom.

The important fact from this study was that the corrosion rate of copper is very low in most soils and is not particularly high even in the most aggressive soils. Pitting is of more concern in the most aggressive soils, where penetration of a copper pipe could occur in ~10 years. In the less aggressive soils, little or no pitting occurred, and perforation would not occur in 50 years.

BNF Metals Research Association (BNFMRA) conducted two studies of underground corrosion that included copper. The first examined samples that had been exposed for up to 11 years at seven U.K. sites.[5] The conditions at the sites are summarized in Table 15.3, whereas the corrosion data are given in Table 15.4. The made-up ground (site G) was composed mainly of builder's rubble and included a significant quantity of cement; hence, the water would not be acidic. This explains the low corrosion rates at this site.

As in the NBS tests, the peat soil (F) with high sulfate and moisture content was very aggressive, although the corrosion rate was only about one-third of that seen in the NBS tests. The BNFMRA tests also included

TABLE 15.4 Corrosion of buried copper in BNFMRA exposures[4]

Soil		Exposure	Mean corrosion	Maximum pitting
Site	Type	(years)	rate (mm/y)	rate (mm/y)
F	Wet acid peat	1.25	0.0009	–
		5.67	0.0030	<0.003
		10.5	0.0066	0.018
C	Moist acid clay	1.42	0.0066	–
		6.0	0.0051	<0.003
		10.5	0.0053	0.038
G	Made-up ground[(A)]	1.0	0.0009	–
		3.08	0.0005	0.064
		9.83	0.0003	<0.002
E	Dry acid sand	–	–	–
		5.92	0.0001	<0.003
		11.08	0.0002	<0.002
A	Moist normal clay	1.25	0	–
		5.75	0.0001	<0.003
		11.08	0.0002	<0.002
D	Chalk	1.58	0	–
		5.83	0	<0.003
		10.92	0.0001	<0.002
B	Slightly acid sand	1.25	0	–
		5.75	0.0001	0.003
		10.83	0.0001	<0.002

[(A)] Miscellaneous builder's rubble.

a chalk soil, a type not included in the NBS study. The higher pH of the water in this soil is believed to be the reason for the low corrosion rate.

The highest sulfate contents were at sites C and F, the two most corrosive soils. The corrosion rates at the two sites after 10 years were much the same, despite the sulfate being 30 times higher at site F compared with site C.

The greatest corrosion occurred at the sites with the lowest pH values, but only in conjunction with a high moisture content. This is demonstrated by soils C and E, which had a similar pH, but soil E had a very low moisture content.

Because the BNFMRA tests removed samples at different time intervals, it is possible to examine the corrosion rate with time.[5] Figure 15.1 shows the average loss of wall thickness for sites C and F, the most corrosive. The average metal loss for all the other sites was much lower, at less than 0.3×10^{-2} mm after 10 years. The results show that corrosion at site C (moist acid clay) was more or less constant with time. At site F (wet

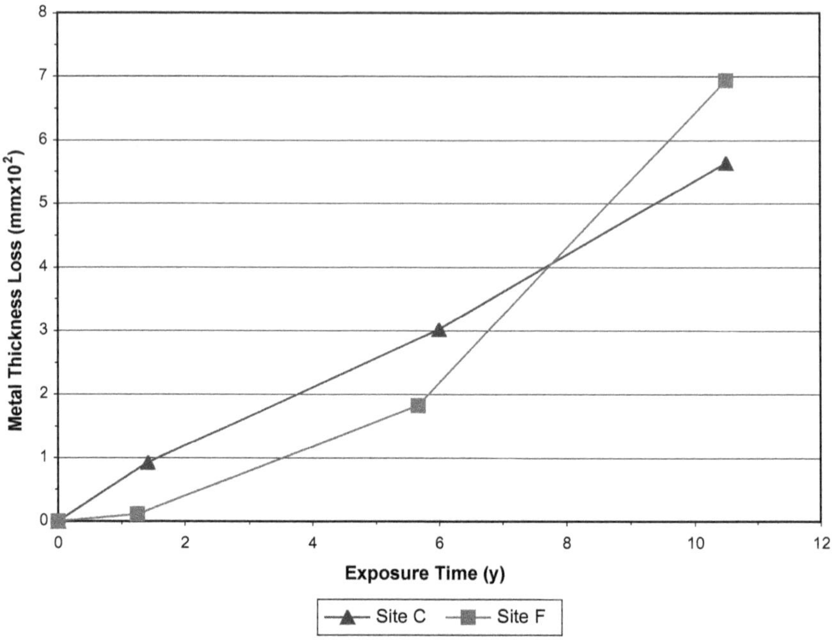

FIGURE 15.1 Effect of exposure time on copper thickness loss.[5]

acid peat), the corrosion rate was low initially, but gradually increased with time. Hence, although corrosion at site F was much lower than at site C after 1 or 2 years, there was little difference in the average metal loss after 10.5 years.

Gilbert and Porter presented the results of 5-year underground tests of copper at various sites in the United Kingdom, one of which was the same as one used in the earlier BNFMRA study.[5-6] Their results also showed low corrosion rates for copper, but the corrosion at Rothhampstead, site C in the earlier study, was much greater than in the earlier study.[5] The average metal loss over 5 years was 5 times that seen in the earlier study, and the pitting was also deeper. These results show just how variable underground corrosion can be, even in the same location. However, despite this variation, the average metal loss of copper was still very low.

Gilbert reviewed the data from all three studies discussed previously and compared the data, as shown in Table 15.5.[7] The results show a good correlation between the mean corrosion rates and general soil corrosivity. The correlation of the maximum pitting rate with soil corrosivity was not so good, but pitting rates were highest in the most aggressive soils.

TABLE 15.5 Comparison of corrosion data for copper from NBS and BNFMRA tests[7]

Soil type	Source	Reference	Exposure (years)	Mean corrosion rate (mm/y × 10^4)	Maximum pitting rate (mm/y)
Mildly corrosive	BNFMRA	Gilbert[5]	10	0.5–2.5	0
	BNFMRA	Gilbert and Porter[6]	5	5.0–25	0.040
	NBS	Romanoff[4]	14	4.0–25	0.043
Moderately corrosive	NBS	Romanoff[4]	14	25–130	0.033
	BNFMRA	Gilbert[5]	10	53–66	0.046
	BNFMRA	Gilbert and Porter[6]	5	66	0.32
Very corrosive	NBS	Romanoff[4]	14	160–355	0.115

Campbell and Sutton tested copper underground for exposure times up to 15 years at a single site.[8] The soil was a clay marl overlying a bed of old peat. The samples were buried in the clay, close to the peat-clay interface. The corrosion rates from the 5-, 10-, and 15-year trenches were all low but showed no consistent trends with time. These results show the variability in corrosion that is possible in the same soil at slightly different locations. The authors ascribed the variations in corrosion rate to the variations in proximity of the samples to the peat-clay interface.[8]

In addition to these research exposure tests, the results of a large number of service experiences with copper pipe underground have also been reported.[9] Copper pipe has been used extensively for underground transportation of both gas and water, particularly in the United States, the United Kingdom, and Australia. The experiences have been very good, with examinations after 10 to 20 years showing little significant corrosion or pitting.[9] These service experiences confirm the results from the long-term exposure tests.

15.2 Copper Alloys

Samples of various copper alloys were included in the NBS exposure tests for periods of 8 to 13 years.[4] The alloys included are shown in Table 15.6. Unfortunately, none of the single-phase brasses included arsenic, and they all suffered dezincification over large areas in most soils, as did the duplex brasses. Thus, the corrosion rates, calculated from weight loss, do not reflect

TABLE 15.6 Copper alloys tested in NBS underground exposure tests for 8–13 years[4]

Type	Name	Composition (wt%)
Brass	Red brass	Cu/15Zn
	Admiralty brass	Cu/30Zn/1.3Sn
	Leaded brass	Cu/31Zn/0.8Pb
	60/40 brass	Cu/40Zn/0.4Pb
	Muntz metal	Cu/40Zn/2.5Pb
Bronze	Tin bronze	Cu/1.8Sn/1Si
	Silicon bronze	Cu/1.5Si
	Silicon bronze	Cu/3.2Si
	Aluminum bronze	Cu/9.4Al/3.5Fe
Nickel silver	Leaded nickel silver	Cu/40Zn/10Ni/2.5Pb
	5/20 nickel silver	Cu/5Zn/20Ni

this and give too favorable an impression of their performance. The most serious dezincification occurred in soils 67 (cinders) and 64 (clay with very poor drainage), as described in Table 15.1. Hence, uninhibited brasses are unsuitable for use underground.

The aluminum bronze that was tested does not represent a modern composition in that it contained no nickel, whereas nickel aluminum bronze (C63200) contains 4% to 5% nickel. The aluminum bronze in the NBS study suffered dealloying, but a single-phase aluminum bronze, such as C61400, would probably have performed much better.[9]

The two nickel silvers also suffered dezincification, as might be expected, but, as noted previously, this is not reflected in the general metal loss rate data. The tin and silicon bronzes performed better and generally gave corrosion rates similar to copper. There were no copper alloys included in the tests that consistently performed better than copper.

15.3 Galvanic Corrosion

Copper is sometimes connected to other metals underground, and it can be seen from the previous chapter that copper will usually be the cathode in such couples. Waters underground are usually similar to fresh waters, and in fresh waters, copper can safely be connected to stainless steel (SS). This was confirmed by Escalante and Gerhold, who found a constant reversal of current flow between coupled copper and SS in an underground exposure test.[10] This shows that the potentials were similar. Hence, copper and SS may safely be coupled together underground in most cases.

In many practical cases, copper will be connected to steel, cast iron, or galvanized steel. Any galvanic corrosion will tend to be close to the metal junction, but in some cases, it may be located elsewhere. This is because the resistance of the soil between two metal items is not as important as might be thought. There is little difference in the circuit resistance through the soil between equal areas 0.3 m apart and two similar areas 1 m apart.[1] This explains why corrosion can take place some distance away from a dissimilar metal junction and why applying a coating to the metals at the junction does not always stop galvanic corrosion underground, although it often reduces it.

There are four conditions under which copper might commonly be connected to steel, cast iron, or galvanized steel underground; in all of these, copper will be the cathode:

1. Copper to bare anode (for example, steel)
2. Copper to partially coated anode
3. Copper to fully coated anode
4. Copper to fully coated anode with coating defects

These possibilities were examined by Hough and Alleyne, who concluded that in cases (1) and (2) the anode would be attacked, but not significantly in excess of the uncoupled corrosion rate because of the relatively large area of the anode.[11] In case (3), the coating would prevent any corrosion from taking place, but case (4) could cause severe corrosion of the steel/cast iron at the defects. Hough and Alleyne exposed a coated steel pipe connected to a bare copper pipe and put defects in the steel coating that were 6 to 25 mm in diameter.[11] After 9 months' exposure, perforation of the steel pipe had occurred at some of the defects. The artificial defects were from 1.2 to 6.7 m from the bimetallic junction, but the severity of the attack was not noticeably affected by the distance from the copper.

It has also been observed that corrosion can occur underground when dissimilar metals are close but not connected. In two of the BNFMRA studies, aluminum was placed parallel to the copper at separations from 75 to 300 mm.[6,8] Only at the closest spacing did accelerated corrosion of the aluminum occur. This was probably due to diffusion of copper corrosion products through the soil to the aluminum. Copper corrosion products are well known to accelerate localized corrosion of aluminum and its alloys.[12]

Concern has been expressed that iron corrosion products might accelerate underground corrosion of copper. Hough exposed copper and steel in an accelerated test simulating underground corrosion.[13-14] Some tests were run in soil alone and some in 75% soil plus 25% iron oxide. The results

TABLE 15.7 Relative corrosion of copper and steel in an accelerated underground corrosion test in the presence of iron oxide[12]

Condition	Metal	Relative weight loss
Soil	Copper	1.2
	Steel	5.8
Soil + rust	Copper	3.9
	Steel	218
	Copper connected to steel	1.0

are shown in Table 15.7, where the weight loss of copper coupled to steel in soil plus oxide is taken as 1. The results showed that copper coupled to steel in the presence of iron oxide corrodes slightly less than when it is uncoupled in soil alone. Even when uncoupled in soil plus iron oxide, the acceleration of corrosion of copper compared with soil alone was only a factor of 3.3. As the corrosion of copper in soil is very low, this magnitude of increase is not significant.

It is also possible that the soil itself may contain a material that can act as one electrode in a galvanic couple. Cinders contain high levels of carbon, which is well known to be noble to most metals and is also a very efficient reducer of DO.[12] In addition, the area of cinders will often be much greater than that of the exposed metal, thus exacerbating attack. Such a case was described in Section 15.1, as part of the NBS study of underground corrosion.[4]

15.4 Methods of Prevention

The corrosion rate of copper underground is generally very low, and service experience shows that long lives are possible for copper pipe used underground.[9] However, the propensity of brasses to dezincify when exposed underground means that they should not be used unless they are inhibited against dezincification. The two alloys most commonly used for fittings underground are LG2 gunmetal (C83600) and dezincification-resistant (DZR) brass (British Standard CZ132). Both of these have performed satisfactorily underground for many years.

In very aggressive soils, coextruded, plastic-coated copper pipe may be necessary. This coating has a high integrity and can give a long, trouble-free life in very aggressive soils.

Where copper is connected to steel or cast iron, then coating of the copper and iron or steel adjacent to the joint will significantly reduce any galvanic corrosion. If the steel/cast iron pipe is fully coated, then coupling to copper could produce rapid corrosion at defects in the coating. It will then be necessary either to fully coat the copper to reduce the cathodic area or to apply cathodic protection to the steel/cast iron.

References

1. E. Mattsson, *Basic Corrosion Technology for Scientists and Engineers,* 2nd ed. (London, U.K.: Institute of Materials, Minerals, and Mining, 1996), p. 75.
2. "Merkblatt für die Beurteilung der Korrosiongefährdung von Eisen und Stahl im Erdboden," Standard DVWG GW9 (Frankfurt, Germany: German Gas and Water Works Engineers Association, 1971).
3. J.O. Harris, D. Eyre, "Soil in the Corrosion Process" in *Corrosion*, 3rd ed., L.L. Shreir, R.A. Jarman, G.T. Burstein, eds. (Oxford, U.K.: Butterworth Heinemann, 1994), p. 2:80.
4. M. Romanoff, "Underground Corrosion," NBS Circulation 579, 1957.
5. P.T. Gilbert, *J. Inst. Met.* 73 (1946): p. 139.
6. P.T. Gilbert, F.C. Porter, *J. Iron Steel Inst. Spec. Rep.* 45 (1952), p. 55.
7. P.T. Gilbert, "Copper and Copper Alloys," in *Corrosion*, 3rd ed., L.L. Shreir, R.A. Jarman, G.T. Burstein, eds. (Oxford, U.K.: Butterworth Heinemann, 1994), p. 4:48.
8. H.S. Campbell, L.J. Sutton, BNF Metals Research Association Research Report, A1385, March 1962.
9. Copper Development Association Publication (U.K.), "Copper Underground: Its Resistance to Soil Corrosion," 40 1967.
10. E. Escalante, W.F. Gerhold, *MP* 14, 10 (1975): p. 16.
11. F.A. Hough, A.B. Alleyne, *Gas* 12 (1936): pp. 18, 90.
12. R. Francis, *Galvanic Corrosion—A Practical Guide for Engineers* (Houston, TX: NACE International, 2001).
13. F.A. Hough, *Am. Gas J.* 144, 1 (1936): p. 31.
14. F.A. Hough, *Am. Gas J.* 144, 2 (1936): p. 17.

CHAPTER 16

Atmospheric Corrosion

Copper and its alloys are often used for architectural applications both outdoors (Figure 16.1) and indoors (Figure 16.2). Copper also has numerous applications in mechanical and electrical equipment exposed to the atmosphere. Engineering equipment frequently has to operate in the open, without protection from the weather, and resistance to corrosion is important. While paints and coatings are often used to prevent corrosion, uncoated metals are also frequently used. As with materials fully immersed in water or some other conducting fluid, corrosion can occur in the atmosphere, and under the most adverse conditions, it can lead to premature failure. Atmospheric corrosion is somewhat different from full immersion corrosion in liquids, such as water, and in the next section, the important factors influencing atmospheric corrosion are discussed.

16.1 Factors Affecting Atmospheric Corrosion

The most important factor is water. Corrosion in direct contact with air at ambient temperature is generally not significant from an engineering point of view. Therefore, water must condense on the exposed metal surfaces or precipitate as rain. This will be followed by evaporation as the components dry out, and thus, a most important parameter for atmospheric corrosion is the time of wetness. Condensation is controlled by the relative humidity (RH), and it is generally accepted that an RH of 80% or more and a temperature >0°C are required for significant condensation.[1] Rainfall is dependent on the local weather and can vary greatly, even in a small country like the United Kingdom.

FIGURE 16.1 Copper-covered spire on St. Magnus Cathedral, Kirkwall, Scotland. (Photograph courtesy of the Copper Development Association.)

FIGURE 16.2 Brass escalator safety balustrades in a U.K. shopping center. (Photograph courtesy of the Copper Development Association.)

Corrosion in the atmosphere is generally not so great as corrosion fully immersed in water. However, galvanic corrosion rates can be as high as rates under immersed conditions.

As the water evaporates, anything dissolved in it will concentrate. Hence, when these species are aggressive to a metal, the corrosion rate will increase as the water evaporates. The common species found in condensation and rainwater are briefly discussed:

1. *Oxygen.* Oxygen is always present in the atmosphere, and when the water films are thin, rapid diffusion of oxygen can occur. Hence, polarization of the cathodic reduction of dissolved oxygen (DO) does not usually occur.

TABLE 16.1 Emission fluxes of atmospheric aerosol particles from natural sources[3]

	Source	Flux (Tg/y)
Direct	Soil dust	1,500
	Sea salt	1,300
	Biological debris	50
	Volcanic dust	33
Indirect	Sulfates from natural precursors	102
	Organic matter from biological volatile organic compounds	55
	Nitrates from NO_x	22

2. *Carbon dioxide* (CO_2). This gas is common in the atmosphere and dissolves readily, causing the water to be mildly acidic. It was not thought to be a major contributor to atmospheric corrosion,[2] but new research suggests that it can play a synergistic role with chloride.
3. *Ammonia* (NH_3). Ammonia is common in rural areas and is readily soluble, causing increases in pH. This is beneficial for most metals, although stress corrosion cracking of some copper alloys is possible.
4. *Sulfur and nitrogen oxides.* These are common in industrial and urban areas as effluent gases, for example, from fossil-fuelled power stations. All these gases dissolve readily, forming strong acids, which are the cause of acid rain.

In addition to these gases, two other factors affect atmospheric corrosion: temperature and particles.[3] Temperature is self-explanatory: the warmer the climate, the greater the rate of chemical reactions, including corrosion.

Particles fall into two groups: fine and coarse.[3] Fine particles range in diameter from ~0.1 to 2 μm and consist chiefly of low-volatility gases condensed onto the surface of atmospheric water droplets. These particles can react with a metal surface but will tend to be removed by rainfall. Coarse particles cover the size range 2 to 100 μm and are generated by a variety of mechanical processes such as wind-blown dust. These particles tend to remain on metal surfaces and are not so readily removed by rain. Table 16.1 shows the emission fluxes of some common atmospheric aerosol particles from natural sources.[3] The table shows that soil dust and sea salt are by far the biggest contributors to the particle count, with natural sulfate forming just 3%.

The closer a site is to the sea, the higher the chloride concentration in condensed water. This has been well demonstrated over the last 50 years by exposures at the LaQue Laboratory in North Carolina, USA, where corrosion rates on the test site 25 m from the sea are higher than those on the test site 250 m from the sea.

Chen investigated the effects of sodium chloride (NaCl) particles on the atmospheric corrosion of copper in the short term.[4] He found that the presence of carbon dioxide had a significant effect on the corrosion rate in the presence of salt. At low salt particle densities, the corrosion rate increased due to the formation of copper chloride. At higher salt particle densities, the corrosion rate also increased initially, but the corrosion product was copper oxide, which was more protective, and the corrosion rate decreased once the oxide layer had formed.

Compared with sea salt, soil particles are relatively inert, but they can create a shielded area if they settle on a metal surface, and crevice corrosion could occur. This can lead to much higher localized metal loss rates compared to open surface corrosion.

Using all these factors, sites are generally divided into four types by atmosphere: rural, urban, industrial, and marine. Rural sites have low chloride and sulfur dioxide and may have ammonia. The urban atmosphere has some sulfur dioxide (SO_2), whereas industrial areas have high sulfur dioxide and NO_x content. Finally, the marine atmosphere is characterized by a high chloride content.

Mattsson has pointed out that with recent environmental concerns and the cleanup of pollution, industrial atmospheres were more corrosive from the mid-1950s to about 1970 in Europe.[1] By 1977, atmospheric corrosion rates in many industrial areas had fallen to pre–Second World War levels. This means that corrosion data from this era may be overly conservative for industrial atmospheres.

The cathodic reaction in atmospheric corrosion is usually the reduction of DO, and as described earlier, the reaction rates can be high, especially with thin water films. If the dissolved solid and gas content of the water is low, the conductivity will be low, reducing the rate of corrosion, but as the films evaporate, the conductivity will increase.

16.2 Atmospheric Aggressivity

At any given site, the factors affecting atmospheric aggressivity can vary from month to month. In addition, there can be huge variations from site to site, even within one atmospheric class, for example, industrial.

TABLE 16.2 Wetness classes from ISO 9226[5]

Class	Wet time (%)	Wet time (h/y)	Example
T1	<0.1	<10	Indoor
T2	0.1–3	10–250	Indoor (unheated)
T3	3–30	250–2,600	Shed storage or cold climate
T4	30–60	2,600–5,200	Outdoors temperate climate
T5	>60	>5,200	Tropical outdoor or surf

In an effort to standardize atmospheric classifications, the ISO CORRAG program tested a range of metals, including copper, at many sites across Europe and North America for periods of up to 8 years. From these data, a classification system was developed, as described in ISO 9226,[5] which uses the variables time of wetness (T), sulfur dioxide content (P), and chloride content (S) and ascribes values to them for different ranges, as shown in Tables 16.2–16.4.

Five corrosion classes exist, with increasing corrosion rates as the class number increases. The corrosion rate for different metals in each class varies, as shown in Table 16.5 for some common metals. Table 16.6 shows how the corrosion class for copper varies depending on the value of the three variables S, T, and P. It is clear from Table 16.6 how important the time of wetness is. With high values of T, the corrosion class is very aggressive for copper, irrespective of the values of S and P. However, even class 5 for copper corresponds to a corrosion rate of $\sim$3 μm/y, which is not a high corrosion rate, although it may be critical in some applications.

One important feature of this classification system is that it allows a wide range of times of wetness and includes not just open exposure outdoors, but also exposures with some level of screening from the elements. Indoor corrosion can be very important, particularly on critical components such as electronic circuit boards. This is discussed in more detail in Section 16.5.

TABLE 16.3 Classes for sulfur dioxide in ISO 9266

SO_2 class	SO_2 deposition rate (mg/m^2/d)	SO_2 concentration (mg/m^3)
P0	<10	<12
P1	10–35	12–40
P2	36–80	41–90
P3	81–200	91–250

TABLE 16.4 Classes for chloride in ISO 9226

Chloride$^-$ class	Cl^- deposition rate ($mg/m^2/d$)
S0	<3
S1	3–60
S2	61–300
S3	>300

While the ISO classification system works well in many cases, there are still some factors that require incorporation into the system.[3] One of these is the frequent difference in corrosion rate between the upper and lower faces of flat standard specimens. Another is the role of pollutants other than chloride and sulfur dioxide.

16.3 Corrosion Rates Outdoors

Kucera and Mattsson conducted atmospheric exposure tests on a range of metals and alloys at various Swedish sites, and Table 16.7 summarizes their data.[6] These results show that the corrosion rate of copper is not very high in any environment, but it is highest in a marine environment. Other metals, such as zinc and nickel, had the highest corrosion rates, after carbon steel, in the urban/industrial environment.

There have been a number of studies of copper exposed for up to 20 years in a variety of atmospheric conditions.[7-9] Table 16.8 summarizes the data from these and other references. The data show reasonable agreement for the different atmospheric types, with the Swedish 16-year data giving the lowest corrosion rates.[8] The variability of the severity of industrial locations is also shown in that some sites show more attack than experienced under marine conditions, whereas other sites were less severe than marine sites.

TABLE 16.5 Corrosion classes in ISO 9226

Corrosion class	Average corrosion rate for first 10 years ($\mu m/y$)				
	Steel	Copper	Zinc	Aluminum	Weathering steel
1	<0.5	<0.01	<0.01	<0.01	<0.1
2	0.5–2	0.01–0.1	0.01–0.5	0.01–0.025	0.1–1
3	2–10	0.1–2	0.5–2	0.025–0.2	1–4
4	10–35	2–3	2–4	0.2–1.0	4–10
5	>35	>3	>4	>1.0	>10

TABLE 16.6 Corrosion classes for different environmental classes for copper and zinc from ISO 9226

	Corrosion class											
	T2			T3			T4			T5		
	S0–S1	S2	S3	S0–S1	S2	S3	S0–S1	S2	S3	S0–S1	S2	S3
P0–P1	1	1–2	2	3	3	3–4	3	4	5	4	5	5
P2	1–2	2	3	3	3–4	4	3–4	4	5	5	5	5
P3	2	3	3–4	3	3–4	4	4–5	5	5	5	5	5

TABLE 16.7 Corrosion of some common metals in various atmospheric environments[6]

	Corrosion rate (μm/y)		
Alloy	Rural	Urban	Marine
Magnesium	6.5	11	14
Aluminum	0.1	0.5	0.6
Anodized aluminum	0	0.2	0.2
Zinc	0.5	2.5	1.3
Carbon steel	22	53	36
Weathering steel	22	48	32
Tin	1.4	2.7	10
Lead	0.3	0	0
Copper	1.2	1.4	3.2
Nickel	0.4	2.7	1.0
Stainless steel (Type 304)	0.03	0.04	0.04

TABLE 16.8 Comparison of copper corrosion rates from different sources

			Corrosion rate (μm/y)			
Country	Time (years)	Reference	Rural	Urban	Industrial	Marine
Sweden	5	5	1.2	1.4	–	3.2
United States	4	6	–	0.9	1.1	2.1
Sweden	16	7	0.3–0.5	0.9–1.3	–	0.5–0.9
United States	2	8	1.2	–	1.9	1.8
	20	8	0.7	–	1.3	1.1
Typical values	–	3	0.5	1–2	<2.5	~1

TABLE 16.9 Comparison of corrosion rates of copper alloys in atmospheric exposure tests

			Corrosion rate (μm/y)			
Country	Time (years)	Reference	Rural	Urban	Industrial	Marine
United States	4	7	–	0.5–1.0	0.9–1.4	0.4–2.7
Sweden	16	8	0.3–0.5	0.9–1.3	–	0.5–0.9
United States	20	9	0.5–1.1	–	1.2–2.3	0.5–1.7

All of the authors of References 6–8 comment that the films formed slowly, with time, on the metal surface and were still forming after years of exposure, although corrosion rates were decreasing with time as the films formed.[7-9] This is shown in Table 16.8 by the data of Costas after 2 and 20 years' exposure.[9]

In addition to pure copper, there are also data on a range of copper alloys. All the studies mentioned previously tested a range of brasses, bronzes, copper-nickels, and nickel silvers.[7-9] The results in Table 16.9 show the range of corrosion rates measured for the various alloys. There were no significant differences between the different copper alloys. It can be seen that these are very similar to the rates for pure copper in Table 16.8; in fact, Holm and Mattsson do not separate them.[8]

All the authors comment that general corrosion is not the main problem with copper alloys and that localized corrosion caused more severe attack.[7-9] Dezincification was seen with many of the uninhibited brasses containing more than 15% zinc, although this was not deep in most cases. Also, the results from the three sets of tests were somewhat variable. For example, Castillo saw deep dezincification only on a manganese bronze (C66900), whereas Holm and Mattsson saw attack on all alpha/beta brasses and Costas saw dezincification only on 70/30 brass. Dezincification of the nickel silvers was slight or nonexistent. Holm and Mattsson reported dezincification depths varying from 50 to 250 μm, which is greater than any metal loss due to general corrosion but may be of little concern from an engineering point of view.[8] Castillo and Popplewell reported pitting and/or crevice corrosion with the phosphor bronzes containing from 4% to 8% tin.[7] The depth of attack varied from 76 to 127 μm, again, much deeper than would occur by general corrosion.

Castillo and Popplewell also exposed U-bend samples of many copper alloys in the cold-worked condition (hard or extra hard).[7] Those that suffered stress corrosion cracking (SCC) failed in a relatively short time. In the industrial environment, failures occurred with cartridge brass (C26000), leaded brass (C35300), and other duplex brasses. The nickel silvers (with the exception of C75200), one aluminum brass sample, and 40% cold-worked admiralty brass also failed by SCC in the industrial environment, although the time to failure for aluminum brass was much greater than for the other alloys. The 10% cold-worked admiralty brass did not suffer SCC at all. In the marine environment, SCC failures occurred only with the nickel silvers and a high beta-phase brass. Copper alloys for atmospheric exposure are more often used in the annealed or half-hard condition. As no alloys in this condition were tested, it is not possible to say how susceptible they might be to SCC over long periods of time. The susceptibility of copper alloys to SCC is discussed in more detail in Chapter 10.

The data presented so far are from northern temperate climates. In tropical climates, conditions can be more aggressive, both because of the higher ambient temperature and the longer time of wetness. This means somewhat higher corrosion rates, but these will still not be very high for copper, although it is important to guard against localized corrosion with copper alloys.

16.4 Galvanic Corrosion

The subject of galvanic corrosion in the atmosphere has been reviewed by Francis, and galvanic corrosion of copper alloys was discussed in Chapter 14.[10] Galvanic corrosion in the atmosphere tends to be confined to the dissimilar metal joint where the condensation has occurred. Hence, area ratios are usually around 1:1. From the preceding characteristics, it can be deduced that atmospheric galvanic corrosion is similar to galvanic corrosion when immersed in a low-conductivity electrolyte. The corrosion rate obviously varies as the species dissolved in the water vary and as the water films evaporate and condense again. With most metal couples, copper is the cathode, and it can be more efficient than well-known cathodic materials such as graphite. Doyle and Wright investigated the galvanic corrosion of aluminum coupled to a range of alloys at several sites in North America.[11] The results show that copper was a more efficient cathode than graphite at three industrial sites and was only a little less efficient than graphite at three marine sites (Figure 16.3). It can also be seen that Alloy 400, with ~35% copper, is much less efficient as a cathode than both pure copper and graphite.

Where copper is connected to more noble metals, such as stainless steels, nickel, and chromium, there is either no increase in the corrosion rate of copper or only a small increase.[6] This is essentially because these noble metals are not very efficient cathodes.

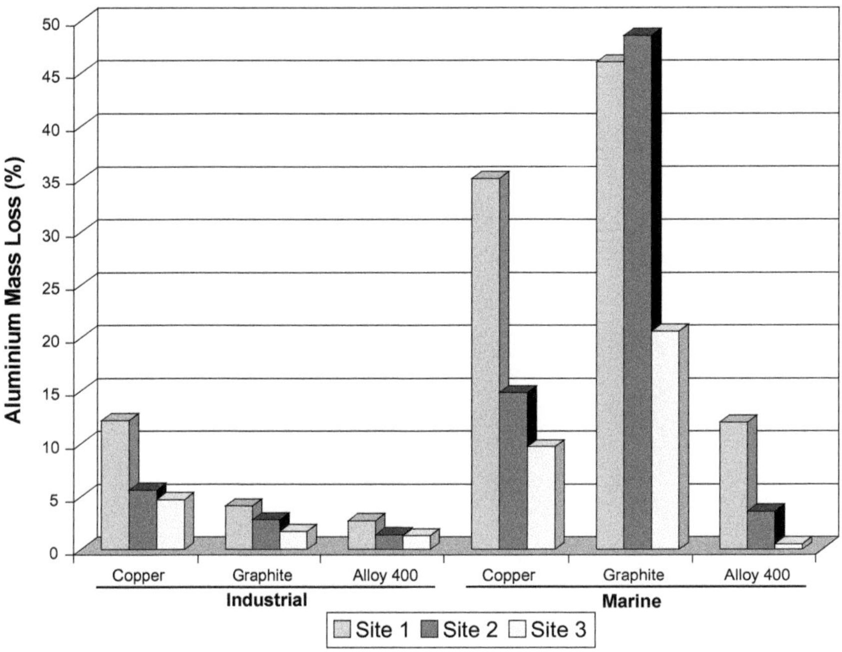

FIGURE 16.3 Atmospheric galvanic corrosion loss of aluminum coupled to various materials.[11]

Even in a tropical atmosphere, copper will not corrode very much when coupled to a noble metal. Beavers et al. coupled a titanium alloy (Ti/6Al/4V) to various metals in the tropical marine atmosphere of Panama, in Central America.[12] The corrosion rate of 70/30 brass was 1 μm/y, no different from the rate for uncoupled brass. This is because titanium and its alloys are not very efficient cathodes in atmospheric galvanic couples.

16.5 Corrosion Indoors

Atmospheric corrosion indoors can also be very important. Copper and its alloys are often used in large buildings, such as offices, hospitals, and hotels, where the appearance of the metal is important, as shown in Figure 16.4. Corrosion can also be important in museums, where further degradation of ancient metal artifacts is undesirable. It should be recognized that the response of an indoor atmosphere to changes in the concentration of external pollutants can be very rapid, of the order of minutes.[3] Hence, it is important to keep the atmosphere under control to minimize corrosion. This involves

FIGURE 16.4 Copper used in a decorative internal curtain wall. (Photograph courtesy of NDM.)

control of both temperature and humidity and, in severe circumstances, the control of aerosols containing aggressive species such as sulfur-containing gases or chlorides.

The Instrument Society of America (ISA) has classified atmospheres indoors by the thickness of the corroded layer after a 1-month exposure. The different levels and their descriptions are shown in Table 16.10. The ISA has also categorized the gaseous constituents of an atmosphere for the various corrosivity levels, as shown in Table 16.11. Note that if the gases in a given environment correspond to different corrosivity levels, then the overall level is determined by the gas concentration at the most severe level. The data in Table 16.11 are valid provided that the RH is less than 50%. Following the ISA classification, the corrosivity class should increase by one level if the average RH increases by 10%.[3]

Leygraf and Graedel reported some corrosion data for copper in Swedish churches and museums.[3] Most of the buildings were at level G1 or G2, but one was in category G3. This shows how severe the environment may be for critical components, even in buildings where one would not expect it.

TABLE 16.10 Corrosivity classification according to the ISA

Corrosivity level	Corroded layer thickness (nm)[(A)]	Description of severity level
G1 mild	<30	Corrosion is not a factor in determining reliability.
G2 moderate	30 to <100	Effects of corrosion are measurable and may be a factor in determining reliability.
G3 harsh	100 to <200	There is a high probability that corrosion will occur. Environmental control or special design and packaging is recommended.
GX severe	>200	Only specially designed and protected equipment is expected to operate. Use of equipment in this class is a matter of negotiation between user and supplier.

[(A)] All thickness values after 1-month exposure.

Atmospheric corrosion is most important for electronic components in which copper is used as the conductor on printed circuit boards. Even a small amount of corrosion can interfere with the correct working of a board, even if it does not stop it working altogether. Corrosion products act as a resistance, which causes a voltage drop. As much of modern electronics operates at low voltages, any voltage drop can affect an instrument's ability to function correctly.

The copper used as a conductor in printed circuit boards is usually coated to prevent corrosion, but contacts and connectors are not. These are often made of copper, or a copper alloy, and are coated with gold or tin. In addition, corrosive elements can penetrate defects in the coatings of integrated circuits and can cause them to malfunction. A detailed description of the corrosion of electronic equipment is contained in the work of Leygraf and Graedel.[3]

TABLE 16.11 Maximum concentrations of gaseous corrodants according to ISA corrosivity classification

	Maximum concentration (ppbv)			
Corrosivity level	H_2S	SO_2	Cl_2	NO_x
G1	<3	<10	<1	<50
G2	<10	<100	<2	<125
G3	<50	<300	<10	<1,250
GX	≥50	≥300	≥10	≥1,250

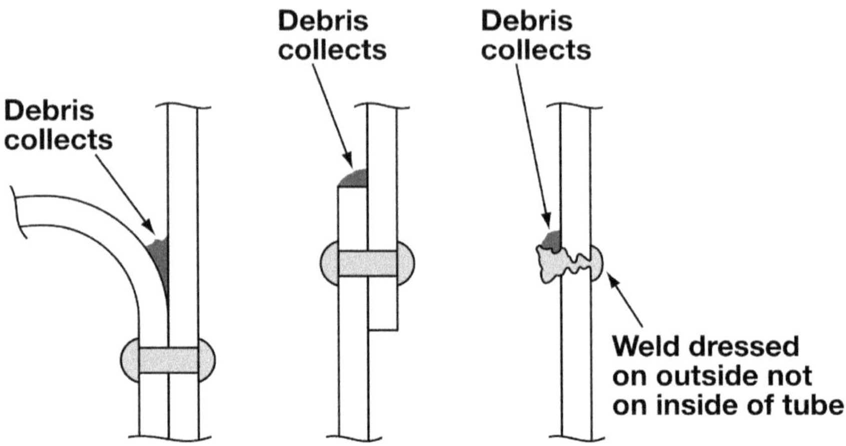

FIGURE 16.5 Design details likely to result in corrosion because of entrapment of debris.[13]

The more severe the operating environment, the more rigorous will be the protective measures that are required to ensure the successful operation of electronic equipment. These are discussed in the following section.

16.6 Methods of Prevention

When selecting a method of corrosion prevention or reduction, it is important to know the severity of the environment and the desired performance life of the component. However, there are some factors that exacerbate atmospheric corrosion.

For components that are to be exposed to the atmosphere, it is important to use a design that removes or minimizes areas where water and debris can accumulate. Figure 16.5 shows some poor designs that can lead to corrosion.[13] Where mixed metals are to be joined and galvanic corrosion could be a problem, use a design that insulates the two metals or apply a suitable coating to the cathode in the couple.[10] The prevention of galvanic corrosion is discussed in more detail in Chapter 14 and also by Francis.[10]

When selecting copper alloys for use in corrosive atmospheres, ensure that an alloy is chosen that will not suffer SCC or be susceptible to pitting in the particular environment. Using alloys in the annealed or stress-relieved condition minimizes internal stresses and the subsequent risk of SCC. SCC can lead to rapid failure, compared to general corrosion, as described in

Section 16.3, while pitting corrosion usually looks unsightly but does not generally cause failure.

Where it is desired to preserve the metallic appearance of copper or a copper alloy outdoors, then a suitable coating system must be chosen. One developed by BNF Metals Research Association for the International Copper Research Association consisted of a clear, cellulose lacquer containing benzotriazole, a well-known inhibitor for copper and its alloys. Known as INCRALAC, this coating has given a long life when properly applied. It is still available using more modern lacquer formulations. Waxes have also been used to preserve the appearance of copper, either alone or in combination with a lacquer.

Where a decorative patina is required, the color of this and the rate of formation depend very much on the composition of the specific atmosphere in which it is exposed. To avoid problems, commercial chemical treatments are available that will create an aesthetically pleasing patina prior to installation.

One area where problems can occur is during transport, particularly on items being transported by sea. Critical components may need to be packed with a desiccant to keep the moisture content low. An alternative is to use a vapor phase inhibitor, ensuring that it is compatible with all the materials used in the component.

It is also important to recognize that under some storage conditions, packing material may release chemicals that are aggressive, resulting in severe corrosion during transport or storage. Such a case was described in Chapter 5, where a packing material used with copper tubes released formic acid (HCOOH) during storage and caused severe formicary corrosion of the tubes.

With indoor applications, a patina will form on copper and its alloys even more slowly than outdoors. Where a shiny, metallic appearance is required, an inhibited, clear lacquer may offer the best method for long-term preservation of the appearance. Components that are handled a lot, such as handrails, require no such treatment as handling alone will keep them clean.

Electronic components may require special protective measures, particularly if they are in a critical application. It is important to check the corrosivity of the atmosphere, both before and after installation, and this should be done at the most corrosive time of the year (usually the most humid months). For a category G2 atmosphere, the selection of materials is important. For categories G3 and GX, additional protective measures, such as a protective housing and/or an organic coating, are required.

TABLE 16.12 Climatic and operating conditions for electronics to Swedish pulp and paper industry standard

	Compound	Maximum level (ppbv)	Average over (h)
Gaseous pollutants	SO_2	30	1
	H_2S	10	1
	NO_x	30	1
	$CE_2 + HF$	10	24
	HF	10	24
	NH_3	500	1
	O_3	5	1
Thickness of corroded copper layer	<30 nm during first month		
Overpressure	20 Pa		
Temperature	±2°C in the range 18°–25°C		
Temperature variations	<0.1°C/min		
Relative humidity	35 ± 5%		
Relative humidity variations	<0.1%/min		
Solar radiation	Protection against direct solar radiation		
Air velocity	<0.2 m/s in rooms with human beings		
Particulate pollutants	<0.01 mg/m^3, average over 5 minutes		
	<0.005 mg/m^3, average over 24 hours		

For electronic components, there are three types of protective measures[3]:

1. Selection of materials and design of components
2. Protective measures during manufacture, transport, and storage
3. Protective measures during installation and service

Design of electronic packages can have an adverse effect on performance by encouraging corrosion. Fans used for cooling can increase the deposition rate of pollutants. Coatings and other organic materials that outgas corrosive substances, such as hydrogen sulfide (H_2S), must be avoided.

The manufacturing process can add corrosion problems to electronics by means such as fingerprints or metallic or organic contaminants. Residues from soldering or cleaning may also lead to corrosion. Hence, care is needed in all cleaning operations.

The atmospheric conditions during transport and storage may be different from those of the service environment. It is important that protective measures are chosen to ensure full protection until installation. This may involve hermetically sealed transport containers, desiccants, or vapor phase inhibitors. Inhibitors must be compatible with all the materials used in the component being shipped. Because corrosion requires water, it is important that all materials are dry during packing.

Critical components in service may require a complete indoor air treatment system, which involves special design of the room. This usually includes hermetically sealed doors or air locks, seals on the ceiling and floor, and seals on all the service entry points. To remove corrosive pollutants, the air may need to be filtered, using activated carbon or aluminum oxide impregnated with potassium permanganate.[3] Remember that the air flow through the filters must be sufficiently slow to allow full removal of the pollutants. Table 16.12 shows a clean air specification for electronics adopted by the Swedish pulp and paper industry. Similar specifications have been adopted in other countries.

References

1. E. Mattsson, *Basic Corrosion Technology for Scientists and Engineers,* 2nd ed. (London, U.K.: Institute of Materials, Minerals, and Mining, 1996), p. 80.
2. D. Fyfe, "The Atmosphere," in *Corrosion,* 3rd ed., L.L. Shreir, R.A. Jarman, G.T. Burstein, eds. (London, U.K.: Butterworth Heinemann, 1994), p. 2:33.
3. C. Leygraf, T.E. Graedel, *Atmospheric Corrosion* (New York, NY: John Wiley, 2000).
4. Z.Y. Chen, "Initial Atmospheric Corrosion of Copper (Influence of Sodium Chloride Particles, Humidity, Carbon Dioxide and Gaseous Pollutants)," Licentiate Thesis (Ph.D.) (Stockholm, Sweden: Royal Institute of Technology, 2005).
5. ISO 9226, "Corrosion of Metals and Alloys—Corrosivity of Atmospheres—Determination of Corrosion Rate of Standard Specimens for the Evaluation of Corrosivity" (Geneva, Switzerland: ISO).
6. V. Kucera, E. Mattsson, "Atmospheric Corrosion of Bimetallic Structures," in *Atmospheric Corrosion,* W.H. Ailor, ed. (New York, NY: John Wiley, 1982).
7. A.P. Castillo, J.M. Popplewell, ASTM International Special Publication, STP 767, 1982, p. 60.
8. R. Holm, E. Mattsson, ASTM International Special Publication, STP 767, 1982, p. 85.
9. L.P. Costas, ASTM International Special Publication, STP 767, 1982, p. 106.
10. R. Francis, *Galvanic Corrosion—A Practical Guide for Engineers* (Houston, TX: NACE International, 2001), p. 81.
11. D.P. Doyle, T.E. Wright, ASTM International Special Publication, STP 978, 1988, p. 161.

12. J.A. Beavers, G.H. Kock, W.E. Berry, *Corrosion of Metals in Marine Environments* (Columbus, OH: Battelle, Metals and Ceramics Information Centre, 1986).

13. R. Parkins, C. Chandler, *Corrosion Control in Engineering Design* (London, U.K.: DOI/HMSO, 1978).

CHAPTER 17

Corrosion in Waters

The previous two chapters dealt with the specific environments of the atmosphere and underground. In this chapter, the use of copper and its alloys in applications associated with waters of one type or another are reviewed. Where specific types of corrosion may occur, the reader is referred to the relevant chapters. The intention of this chapter is to highlight the widespread applications for copper and its alloys in waters and some of the common problems, not all of which are corrosion related.

17.1 Fresh Water

Copper and its alloys are widely used for pipes and fittings to handle fresh and potable waters, for hot waters, and for central heating systems. Although there are occasionally failures due to corrosion, thousands of kilometers of copper and copper alloy piping systems are giving reliable service.

Dezincification of brass fittings, valves, etc., is rarely a problem these days, with the wide availability of dezincification-resistant brass for both forgings and die castings. In some countries, LG2 gunmetal (C83600) is available at lower cost than brass, and so dezincification is virtually unknown there (see Chapter 9).

One problem that still sometimes occurs with brass components in fresh waters is stress corrosion cracking (SCC; see Chapter 10). This is usually from the outside and is often associated with ammonia (NH_3) from phenolic insulation systems. It is important to select insulation systems that cannot release ammonia or formic acid (HCOOH). The latter can lead to formicary corrosion, as described in Chapter 5.

In an attempt to determine the most common causes of corrosion failure at the time of writing, the author had discussions with Phillip Munn, a U.K. consultant who regularly examines copper and copper alloy failures. It appears that there are differences between domestic and commercial properties.

In domestic properties, in the United Kingdom, the most common corrosion problems are blue water and Type I pitting corrosion, sometimes associated with flux residues. The causes of this and the methods to avoid them are discussed in Chapters 4 and 5. In commercial properties with recirculating hot water systems, at least half of the failures are due to erosion corrosion. The causes of this and suitable remedies are discussed in Chapter 7. Such problems can be avoided by attention to water velocity and potential turbulence raisers at the design stage. Microbially influenced corrosion (MIC), or Type III pitting, and deposit attack account for another 30 to 40% of the failures in commercial recirculating hot water systems. The causes of these types of corrosion and their avoidance are discussed in Chapters 5 and 6.

In commercial cold water systems, the two most common causes of failure in the United Kingdom are MIC/Type III pitting (~40%) and Type I pitting (~40%), as discussed in Chapter 5, with erosion corrosion accounting for only 5 to 10% of failures. The remainder of the problems are caused either by excessive chlorination or long periods of stagnation. The ways in which these two factors can affect corrosion in fresh waters are discussed in Chapters 4 and 5.

As mentioned previously (Chapters 4 and 5), there is an increasing tendency to find phosphate deposits associated with problems such as blue water and Type I pitting. Research is needed in this area to determine the role of phosphate in copper corrosion and what levels will not adversely affect corrosion in various types of water.

BNF Metals Technology Centre published a guide on avoiding corrosion problems in domestic central heating systems.[1] Most of the problems occur when air can get into the system. Normally, the dissolved oxygen (DO) from the air is consumed by corrosion of the metal components in the first day of operation, and thereafter, corrosion is negligible. However, leaking joints near the pump suction and overpumping can both lead to the introduction of fresh DO. This does not normally have a significant effect on the corrosion of copper central heating pipes but may cause dezincification of any brass fittings. In severe cases, this may cause blockage of the system with meringue corrosion products (see Chapter 9).

Overpumping occurs if the pressure difference between the points where the vent pipe and cold feed pipe are connected exceeds the head between the highest point of the swan neck on the vent pipe and the water level in the tank supplying the system. Overpumping leads to water being driven out of the vent pipe into the header tank, and fresh, aerated water is drawn into the system from the tank to make good the loss. Hence, the DO content never decreases to low levels. Overpumping is prevented by correct positioning of the vent pipe and cold water feed pipe into the system to keep the pressure difference between them low. This is discussed in more detail in various commercial design guides.

Another possible cause of failure is when central heating pipe work is buried in cement or concrete screed with insufficient allowance for thermal expansion and contraction, as the system heats up and cools down. If the pipe is overly restrained, fatigue failures of the pipe can occur. The incorporation of thermal expansion sections in long pipe runs prevents this problem.

External corrosion can be prevented on small-bore central heating tubing (up to 10 mm outside diameter) by installing it with a coextruded PVC covering. With larger-diameter pipe ($\geq$12 mm outside diameter), it is necessary to leave a duct in concrete when it is being laid if copper central heating tubing is to be installed. This avoids restraining the pipe, such that thermal fatigue might occur, as well as putting an air space around the pipes. The pipes should be insulated using a moisture-resistant coating, avoiding types that may release ammonia or formic acid (see Chapters 5 and 10).

17.2 Seawater

Copper alloys are widely used in seawater, particularly in seawater cooling systems. They are used in both cast and wrought forms for pumps, pipe work, valves, heat exchangers, filters, etc. Copper alloys also find extensive use in firewater systems and for ships' propellers and propulsers. The most commonly used alloys are 90/10 copper-nickel (C70600) for wrought products and nickel aluminum bronze (NAB; C95800) for cast products.

It is not the intention of this section to go over seawater applications in detail; rather, this section intends to point the reader toward sources of further information with regard to design and applications.

A good review of the behavior of 90/10 copper-nickel in seawater was prepared by Parvizi et al., which includes historical data, properties and composition, and corrosion behavior in seawater, including the effects of velocity, sulfides, temperature, etc.[2] The properties of cast NAB, including corrosion resistance in seawater, are compared with other alloys in a publication by Oldfield and Masters.[3] The corrosion resistance of NAB in seawater is also described in Meigh's book on the aluminum bronzes.[4]

Most seawater applications involve surface or near-surface water, where there are changes in flow rate and temperature due to tides and the DO level is close to saturation. Reinhart published corrosion data on a wide range of copper alloys exposed at depths from 760 to 1,800 m (2,500 to 6,000 ft) for periods from 4 months to 3 years.[5] The study examined general corrosion, pitting, and SCC. Many of the results were similar to those seen in surface waters, but some of the alloys corroded differently due to the lower temperature and DO content at great depth.

Firewater systems operate at high velocities when working, up to 10 m/s, and the use of copper alloys might seem problematical. However, firewater systems are only tested for about an hour per week, at most, and spend the rest of the time stagnant. Hence, any erosion corrosion damage sustained during testing will reform protective films when inactive. Although there have been a few problems due to corrosion by sulfides, 90/10 copper-nickel piping and NAB or gunmetal sprinkler heads have given good service in firewater systems all over the world. One advantage of copper alloys over high-alloy stainless steels (SS) is their ability to function without accelerated corrosion at elevated temperatures. In hot climates, firewater piping can be exposed to direct sunlight and surface temperatures may reach 60 or 70°C in the summer. Such high temperatures have caused corrosion of high-alloy SS systems, but the author knows of no failures of copper alloy systems at elevated temperatures.

The author described the selection of materials for seawater cooling systems, including the use of wrought and cast copper alloys.[6] It is also important to stress the use of correct specifications and perform quality-assurance and control procedures to avoid potential problems. Having selected a suitable copper alloy, it is not just sufficient to specify it to be supplied to a standard such as ASTM, EN, or BS. While some standards do cover all the relevant properties, others do not. The problem of carbon films was discussed in Chapter 5; however, at the time of writing, only the old BS 2871, part 3,[7] specifically requires heat exchanger tubes to be free of films in the bore, and even then, a method of testing is not specified. The author's suggestions for detecting carbon films were discussed in Chapter 5.

Another problem can occur with 90/10 copper-nickel. The alloy gives the best corrosion resistance when the iron content is in the range 1.5 to

2.0 wt%, as discussed in Chapter 7. However, many of the common standards for the alloy give the iron limit as 1.0 to 2.0 wt%. To ensure adequate resistance to erosion corrosion, it is necessary to specify that the minimum iron content be 1.5 wt%.

This level of iron exceeds the solubility limit for the alloy at room temperature. However, if the iron is not in solution, the corrosion resistance decreases quite markedly, although not the resistance to erosion corrosion (see Chapter 7). To retain the iron in solution, it is necessary to cool the alloy rapidly from an annealing temperature of $\sim$ 900°C. It has been stated that this problem is well understood, and modern manufacturing methods will always produce good-quality material. However, the author examined failures of 90/10 copper-nickel pipes from the cooling water system of an oil platform in the North Sea. They were suffering a very high rate of general corrosion plus erosion corrosion after the bends, even though the water velocity was only $\sim$3 m/s. The clue to the cause of the problem was at the welds: not only was the 70/30 copper-nickel weld metal unattacked; neither was the parent metal in the heat-affected zone (HAZ). Figure 7.8 (Chapter 7) shows a half-section of pipe, and the cut edge clearly shows the thicker metal in the weld and HAZ. This was because the iron was not in solution in the pipe, but the weld thermal cycles in the HAZ had re-solutionized the iron and restored the corrosion resistance. The weld metal was not affected because iron is more soluble in the 70/30 copper-nickel filler metal. Determination of the magnetic permeability with a simple device, such as the Severn[†] magnetic gauge, showed the parent metal to have a value of $>$1.2, while the weld metal and HAZ were $<$1.05. It is suspected that the attack of the parent metal started as accelerated general corrosion that was particularly rapid because the alloy was being used close to its recommended limit. As the weld and HAZ became increasingly proud of the rest of the surface, they would increase local turbulence such that erosion corrosion immediately downstream became possible. It is recommended that a simple magnetic permeability check be carried out on the parent metal at the same intervals as the other tests (e.g., for carbon films and residual stresses). When asked about this problem, Yorkshire Imperial Metals recommended that material with a magnetic permeability of $>$1.10 be rejected. It should be stressed that this type of failure is not common but can be avoided by additional tests during manufacture.

NAB is a widely used alloy in both cast and wrought forms. NAB has sometimes been observed to suffer dealuminification. It has been found

[†] Trade name.

that keeping the nickel content ≥0.5% above the iron content will prevent dealloying (Chapter 9). This is not covered by most specifications and should be added as a requirement when specifying this alloy, in either cast or wrought form.

At the high temperatures used in hot working or during welding, NAB has a structure that is alpha and beta. During cooling, the beta phase transforms to alpha and kappa. However, if the cooling rate is too fast, the beta phase does not fully transform. Beta phase is highly anodic to the alpha phase, and it is attacked rapidly in seawater (Chapters 3 and 9). The Royal Navy experienced corrosion problems with weld-repaired castings in NAB, and seam-welded pipe in a desalination plant corroded and unzipped along the weld (see Chapter 18).

The author was involved in investigating corrosion at a large multi-stage flash (MSF) desalination plant in the Middle East, where two NAB tube plates out of 128 showed pitting due to the presence of beta phase. It was thought that these two were the first in a batch to be rolled and cooled too quickly off the rolls. As subsequent plates were piled up, the thermal mass increased and cooling rates slowed. The tube plates could not be weld repaired because that would have created more beta phase. The solution was to grind out the corrosion and fill the space with a cold-setting ceramic filler.

It is recommended that all NAB be given a heat treatment after the final hot operation (welding, rolling, etc.). For castings, it is recommended that the Royal Navy heat treatment of 675° ± 25°C for 2 to 6 hours, followed by air cooling, be specified. For wrought products (such as thick plate), a heat treatment of 725° ± 25°C for the same period, followed by air cooling, is recommended. The author has specified this treatment for the tube plates in another MSF plant, and there have been no problems after 10 years in service.

Some nickel alloys often come in a number of different heat-treated forms, each with somewhat different mechanical properties. It is important to make sure that the design has been carried out using the mechanical properties for the required heat treatment of the finished product.

One example is Alloy K-500, which is an age-hardening nickel-copper alloy that is frequently used for marine fasteners. It is usually used in the aged condition to obtain a 0.2% proof stress of 730 MPa or more. However, it can be purchased in both annealed and aged conditions. After manufacturing the component, the incorrect strength may be produced if the material was purchased in the wrong heat treatment form. If the material was already aged and is aged again, the strength will be higher than desired, with a consequent loss of ductility. If the material was purchased annealed and is not aged, the strength will be below the requirement for the project. This demonstrates not only the need for detailed purchasing specifications, but also the need for a QA/QC system that will detect such errors.

A number of authors have detailed their experiences with copper alloys in seawater. However, these tend to discuss failures and frequently neglect to mention the thousands of tons of copper alloys that are giving excellent service.

Birn and Skalski described experiences with copper alloys by the Polish marine industry.[8] Although there have been problems due to dezincification of brasses and erosion corrosion, the principal problem has been accelerated corrosion due to sulfides in polluted waters.

The U.S. Navy also experienced failures of copper-nickel due to polluted water, and these failures, and the extensive research program that followed, are discussed in Chapter 7. This is because, although sulfides cause corrosion that has the appearance of pitting, the rate of propagation is very flow-rate-dependent, with high rates of attack at normal flow velocities (2 to 3 m/s) compared with stagnant conditions.

Johnsen described the experiences with copper alloys in the Norwegian sector of the North Sea by six oil and gas companies.[9] Their experiences were very similar, with the main corrosion problems being erosion corrosion; accelerated corrosion due to sulfides, produced under stagnant conditions; and galvanic corrosion when coupled to higher-alloy materials. These problems led some of the companies to switch to other materials, but others devised strategies to make copper alloys work.[9]

The author has had discussions with oil and gas companies in the U.K. sector of the North Sea, and they have also had some corrosion problems with copper alloys, mostly due to galvanic corrosion, but also due to erosion corrosion. One factor that emerged was that the galvanic problems are not accelerated general corrosion; instead, the copper alloys were suffering from one of several forms of localized corrosion, that was then greatly accelerated by coupling to a higher-alloy material. This is discussed in Chapter 14, with suggestions for alleviation. Despite some corrosion failures, copper alloys are still used in the U.K. sector of the North Sea, where they mostly perform well.

Johnsen pointed out that although the design and operating rules for copper alloys are well known, for example, keeping to velocity limits and flushing out stagnant legs of piping with fresh water, these are frequently ignored, both at the design stage and in operation, at least in the Norwegian sector of the North Sea.[9] The author's discussions with U.K. sector operators suggest that the rules are implemented a little better, as shown by the greater success of copper alloys in this sector.

The author has examined a number of failures of NAB pumps and valves from seawater service over the last 30 years. The main cause of failure has been erosion corrosion due to the presence of sulfides or overchlorination (see Chapter 7). Failures due to corrosion of beta phase (Chapters 9 and 18) are rare, now that the importance of postweld heat treatment is recognized.

Probably the most widely used copper alloy in seawater is 90/10 copper-nickel, and Schleich and Powell give details on how to avoid the typical failures reported previously and also how to use copper alloys to avoid biofouling problems.[10] They stress the importance of getting the conditions at start-up right so that protective films form that give significant resistance to accelerated corrosion under upset conditions, for example, sulfide ingress.

When corrosion problems do occur, or there is significant buildup of scale, it may be necessary to use an acid-cleaning treatment to restore the heat transfer coefficient in copper alloy heat exchanger tubes. The selection of the correct acid and strength is important to avoid doing further damage to the tubes. Le Guyader et al. described the effects of three commercial cleaning solutions on 70/30 copper-nickel and the optimum way to clean them to avoid subsequent problems.[11]

17.3 Swimming Pools

Although copper alloys were once widely used for handrails, grates, and other such features in swimming pools, they have largely been superseded by SS. However, copper alloys, particularly copper-nickels, are still used for heat exchangers, especially in pools utilizing brines or seawater. Brasses are not used because of the risk of SCC from the ammonia (NH_3) and/or chloramines in the pool water. Largely, these units work well, but the author has seen a few failures of 70/30 copper-nickel due to ammonia-induced hot spot attack (see Chapter 9) from the pool water side.[12] The problem can be avoided by keeping the water velocity on the pool side high ($\geq$1.5 m/s). This means that pool water temperature control should be done on the hot product side of the heat exchanger. It may be necessary to modify the heat exchanger to a multipass configuration, rather than single pass, to obtain a sufficiently high flow rate.[12]

17.4 Desalination

With continuing industrial development in many parts of the world, the demand for fresh water, particularly drinking water, is rising sharply. In many areas, there are no further surface or underground sources of water, and desalination, the removal of salt from brackish water or seawater, must be considered. There are three main methods of desalination used on a large, commercial scale: reverse osmosis (RO), MSF distillation, and multiple effect distillation (MED).

Copper alloys have seen little use in RO plants, except that the main seawater pumps are sometimes NAB. However, copper alloys are used extensively in both MSF and MED plants.

17.4.1 Multistage Flash

In the basic MSF process, steam heats the recirculating brine in the heat input section, and the steam condensate produced is returned as feed to the boilers. Figure 17.1 shows a schematic of a typical MSF plant. The hot brine (typically 110°C) then passes through a series of flash chambers in the heat recovery section, each at a lower pressure than the preceding one. In each chamber, some water vapor flashes off and is condensed on heat exchanger tubes. In the last few heat rejection stages, cold, aerated seawater is pumped through the condensers to lower the temperature for further condensation to occur and to remove the heat from the heat input section and complete the operating cycle.

Some of the raw seawater is discharged to waste and some is deaerated and dosed with chemicals to control scaling and foaming, before being added to the recirculating brine as feed makeup. The feed is necessary to replace evaporative losses and to maintain the recirculating brine concentration at acceptable levels. The brine is then heated in the heat recovery condensers, before returning to the heat input section.

It is apparent from the preceding description that an MSF plant is quite complex and presents a range of corrosive environments. However, it can conveniently be divided into four main areas: the heat rejection section, the heat recovery section, the brine heater, and the ejector section, which handles the removal of incondensable gases. The flash chambers are found in the first two areas and generally utilize the same materials of construction. The heat recovery and brine heater sections operate with low oxygen concentrations in the water (<50 ppb), whereas the seawater in the heat rejection side is fully aerated.

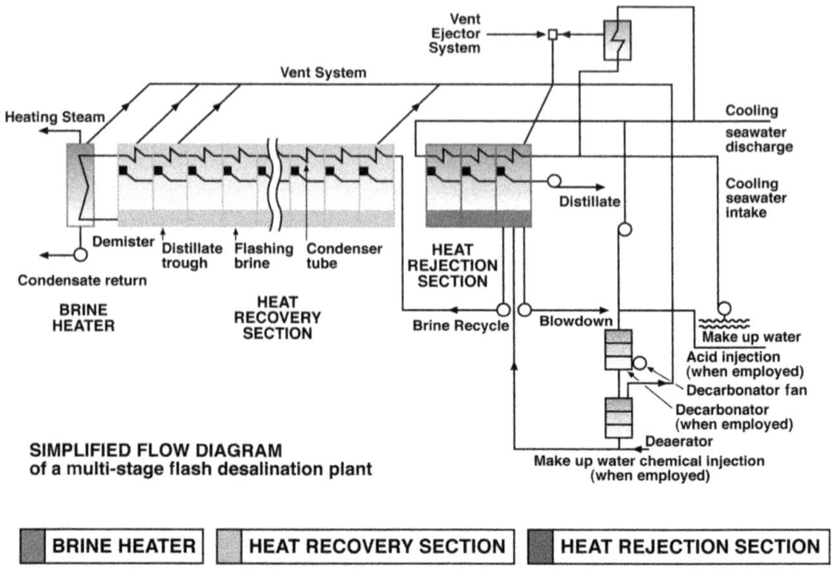

FIGURE 17.1 Schematic of the operation of a typical MSF plant.

Copper alloys are widely used for the heat exchangers in the heat recovery and brine heater sections for tubes, tube plates, and sometimes as linings for water boxes. Aluminum brass and 90/10, 70/30, and 66/30/2/2 copper-nickels are all used for tubing, while both NAB and rolled naval brass are used for tube plates. The 90/10 copper-nickel is also sometimes used as a lining for carbon steel (CS) water boxes.

Although all four tubing alloys have been used in MSF plants, 70/30 copper-nickel has had relatively little use, and aluminum brass tends to be mostly used in the lower-temperature part of the heat recovery section. The 90/10 and 66/30/2/2 copper-nickel-iron-manganese are currently the most widely used copper tubing alloys. The highest temperatures occur in the brine heater (~110°C), and 66/30/2/2 copper-nickel-iron-manganese has been very successful as tubing in this application.

A survey of materials used in desalination plants was carried out under A.D. Little in the 1970s.[13] More recently, in 2003, Oldfield conducted a similar survey, and the results of these surveys are compared in Table 17.1.[14]

In Table 17.1, the materials used for heat exchanger tubes in the brine heater, heat recovery, and heat rejection sections are compared. Note that the 2003 survey divides the heat recovery section into two sections: the high- and low-temperature sections. Table 17.1 averages the data from

TABLE 17.1 A comparison of the use of heat exchanger alloys in MSF plants from the surveys of A.D. Little and Oldfield [13-14]

Alloy	1972[13]			2003[14]		
	Brine heater (%)	Heat recovery (%)	Heat rejection (%)	Brine heater (%)	Heat recovery (%)	Heat rejection (%)
Al-brass	17.9	68	23	0	25.7	5.6
90/10 Cu-Ni	49.0	24.5	34	27.7	52.8	18.8
70/30 Cu-Ni	1.5	1.0	1.5	7.9	11	4.2
66/30/2/2 Cu-Ni-Fe-Mn	31.3	6.0	39.5	51.8	5.2	26.5
Titanium	0.3	0.5	2.0	8.2	3.5	42.5

these two subdivisions. Table 17.2 shows the difference in materials selection for these two subdivisions from the 2003 survey.[14]

The results in Table 17.1 show that the main material for the brine heater tubes has changed from 90/10 copper-nickel to 66/30/2/2 Cu-Ni-Fe-Mn. In the heat recovery section, aluminum brass has remained the most popular alloy in the cooler regions, but 90/10 copper-nickel is the predominantly used alloy in the hotter regions (Table 17.2). In the heat rejection section, there has been a big change from copper alloys to titanium. Although titanium is the most widely used single alloy, the copper alloys together still account for a majority of the tubes. One feature of interest from both surveys is the low use of 70/30 copper-nickel. The 66/30/2/2 Cu-Ni-Fe-Mn alloy appears to be regarded as a significant and useful upgrade from 90/10 copper-nickel in some parts of the plant such as the brine heaters.

The A.D. Little survey also produced data on the failure rates of heat exchanger tubes. Unfortunately, no such data were obtained by Oldfield. Table 17.3 shows the A.D. Little survey results. These numbers need to be compared with the usage data in Table 17.1. For example, the low failure rate of 70/30 copper-nickel in the recovery and rejection sections is not surprising, given its low use. The low failure rate of 66/30/2/2 Cu-Ni-Fe-Mn

TABLE 17.2 Materials used for the heat recovery section heat exchanger tubes[14]

Alloy	Usage (%)	
	Low temperature	High temperature
Al-brass	46.7	4.7
90/10 Cu-Ni	47.1	58.4
70/30 Cu-Ni	0	21.0
66/30/2/2 Cu-Ni-Fe-Mn	0.4	10.0
Titanium	3.5	3.5

TABLE 17.3 Heat exchanger tube failure rates in MSF plants[13]

	Failure rate (%)		
Alloy	Brine heater	Heat recovery	Heat rejection
Al-brass	15.7	1.1	6.8
90/10 Cu-Ni	2.8	0.4	2.3
70/30 Cu-Ni	15.9	0	1.6
66/30/2/2 Cu-Ni-Fe-Mn	1.5	0.02	0.05

is significant, given the alloy's high use. 90/10 copper-nickel is also shown as a reliable alloy, as is aluminum brass in the heat recovery section.

Over the years, a wide range of failures have been described for MSF plants. In the heat rejection section, copper alloy tubes have failed from all the common causes seen in other seawater-cooled plants: erosion corrosion due to lodged obstructions or excessive chlorination and pitting due to the presence of pollutants such as sulfide or ammonia. The first two can be controlled by improved filters and better chlorination control. The effect of pollutants on copper alloys was described in Chapters 3, 7, and 9. The effects of ammonia can be mitigated by ferrous sulfate ($FeSO_4{\cdot}7H_2O$) dosing. When high levels of sulfide are present, ferrous sulfate dosing is not beneficial, and a change of alloy may be required.

17.4.2 Multiple Effect Distillation

MED plants, like MSF plants, require a source of heat, which can be steam or hot water (MED only) or electricity, and so are often run in conjunction with a power station. In most MED plants, a thermocompressor or a mechanical vapor compressor is added to improve efficiency. These are known as thermocompression distillation (TCD) and mechanical vapor compression (MVC) plants, respectively. The higher efficiency of TCD plants means that they are the most common variant.

Figure 17.2 shows a schematic of a typical MED plant, in this case, a TCD plant. A TCD plant consists of a number of evaporator stages (2 to 14), plus a final heat rejection section. Vapor at ~75°C is introduced into the first-stage evaporator tubes, where it is condensed by externally sprayed raw water. In cooling the evaporator tubes, the raw water in stage 1 is heated, and part of it vaporizes. The vapor enters the tubes of the second stage and is condensed by raw water, as in stage 1, forming more condensate. This process repeats itself in subsequent stages. Part of the vapor produced in an intermediate stage is drawn up by the thermocompressor, which increases its pressure and mixes the water vapor with high-pressure steam to feed

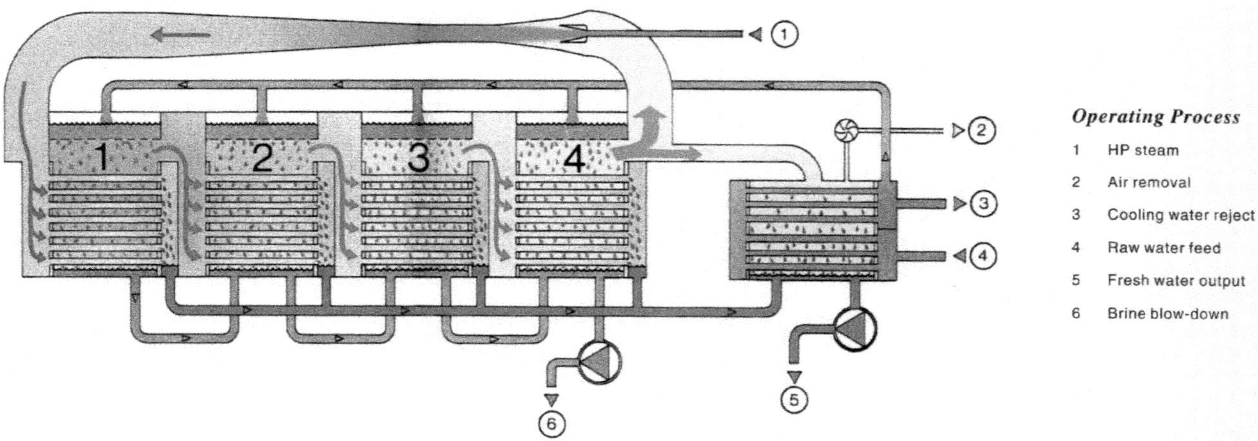

FIGURE 17.2 Schematic of the operation of a typical TCD plant.

the first stage. The remainder of the vapor passes to the following stages and, finally, to a heat exchanger, where it condenses on the outside of tubes cooled internally by raw water. Part of this raw water becomes the feed to the evaporators, and the rest goes to waste. The product water is collected and piped away, while the brine goes to blow-down. An MED plant operates in the same way, but without the thermocompressor; in an MVC plant, the thermocompressor is replaced with a mechanical vapor compressor.

A wide variety of materials are used in an MED plant, but copper alloys are still the principal choice for the tubes in the evaporator section. Titanium and high-alloy SS tend to be the first choice for the final heat recovery heat exchanger, which operates under fully aerated conditions, although 90/10 copper-nickel is sometimes used. In the evaporators, the environment at the top is fully aerated, but oxygen levels decrease dramatically in the lower sections. The highest temperature is currently $\sim$75°C, so aluminum brass and 90/10 copper-nickel perform well and are the most common choice for the evaporator tubes. To increase efficiency, it is necessary to use thin-walled tubes, so a $\sim$0.7-mm wall is common, and it is important to provide adequate tube supports to avoid fatigue problems.

The tube plates in the evaporators may be copper-nickel or naval brass, although Type 316L SS (UNS S31603) has also been used. Because of the lower water temperatures, as compared with MSF plants, water boxes tend to be rubber-lined steel.

Copper alloys have performed well in MED plants, although there is an occasional problem with erosion corrosion in the top two rows of tubes due to excessive velocity from the spray nozzles. This can be cured by redesign of the seawater spray system, although titanium has also been used in the top two rows to prevent this problem.

17.5 Oils and Oily Waters

Most oils are nonconductive and, as such, do not pose a corrosion problem to copper and its alloys. For example, 90/10 copper-nickel is widely used for hydraulic piping on offshore platforms because of its resistance to the external marine atmosphere. In addition, 90/10 copper-nickel has been used for hydraulic brake tubing on cars, in which its resistance to external corrosion means that it will normally last the life of the car. Tungum[†] is a 70/30 brass with 1% aluminum, and this alloy has also been used for

[†] Trade name.

hydraulic tubing because of its resistance to external corrosion. However, Tungum can suffer SCC when ammonia or other aggressive species are present (see Chapter 10).

Only where the oil can break down to release ionic compounds, particularly acids, can internal corrosion occur. Provided that such oils are used within their appropriate temperature range, where decomposition does not occur, corrosion is not a problem.

When water is present with the oil, then two conditions can arise. One is a condition in which the oil and water remain as separate phases, with little mixing. The second case is one in which the oil is miscible with the water or the oil and water are mixed in an emulsion, usually under fast-flowing conditions.

In the first case, in which the oil and water remain separate, corrosion can occur where there is sufficient water to wet the metal surface. This can occur with very small water fractions[15]—exactly how much depends on the nature of the specific oil and the solubility of water in it; however, solubility of water in most oils is low, and as little as 1% water can cause corrosion.[16] Even if the water and oil form an emulsion, the oil and water can separate into their distinct phases if the flow rate decreases, and a series of stratified layers can form.[17] Where a water layer contacts the metal surface, corrosion can occur.

If the water is deaerated or has a very low DO content (<50 ppb), then corrosion will not be significant for alloys such as 90/10 copper-nickel, gunmetals, and aluminum bronzes, provided that there are no other species present that can provide an alternative cathodic reaction to the reduction of DO, for example, carbon dioxide (CO_2). If, however, the water is aerated, then its composition and pH will determine the corrosivity. Any of the forms of corrosion described in Chapters 4–14 may occur, depending on composition, pH, flow rate, stress, and so on. Appropriate precautions to prevent such corrosion should be taken, as suggested in these chapters. For example, if Type I pitting is possible, then the tube bores must be free of carbon films.

Where the oil-water mixture forms an emulsion at high flow rates, the corrosion can be less than in the water alone.[15] One reason is that the conductivity of the emulsion is less than that of the water. Craig reported corrosion of CS in some emulsions with as little as 1% water and no corrosion in others with as much as 50% water.[16] This is because the heavier the oil, the more easily it forms an emulsion.[17] Wetting of the metal by water decreases and wetting by oil increases as the interfacial tension between the oil and water reduces. This is why heavier oil emulsions with water are less corrosive; that is, they can contain more emulsified water before separation occurs.[17]

The lower the conductivity, the higher the oil content, and the less easy it is for fresh reactants to reach the metal surface. However, emulsions normally only form under turbulent conditions so that classic diffusion control is not a rate-determining reaction. Following are the conductivities for some different types of water, without oil, which might indicate the types of corrosion problems that may arise:

Fresh water	50–1,000 μS
Brackish water	1,000–20,000 μS
Brines	>20,000 μS
Seawater	~40,000 μS

Although the effect of oil-water emulsions in reducing the corrosion rate of CS has been studied, there are no published data for copper alloys.[17] While the formation of an emulsion may reduce corrosion compared with the water phase alone, it is advisable to carry out corrosion tests, particularly if the water is aerated. Remember that it may be necessary to carry out tests for more than one type of corrosion, depending on the water composition. In addition, the oil and water will separate if the flow rate decreases sufficiently, and corrosion will then occur wherever the water contacts the metal surface.

References

1. BNF Metals Technology Centre Miscellaneous Publication, MP 586, August 1976.
2. M.S. Parvizi, A. Aladjem, J.E. Castle, *Int. Mater. Rev.* 33 (1988): p. 169.
3. J.W. Oldfield, G.L. Masters, Copper Development Association (U.K.) Publication 115, June 1996.
4. H.J. Meigh, *Cast and Wrought Aluminium Bronzes: Properties, Processes, and Structures* (London, U.K.: Institute of Materials, Minerals, and Mining, 2000).
5. F.M. Reinhart, "Corrosion of Material in Hydrospace Part IV: Copper and Copper Alloys," Naval Civil Engineering Laboratory Report, AD835,104, April 1968.
6. R. Francis, *The Selection of Materials for Seawater Cooling Systems: A Practical Guide for Engineers* (Houston, TX: NACE International, 2006).
7. BS 2871, "Specification for Copper and Copper Alloys Tubes: Tubes for Heat Exchangers" (London, U.K.: BSI).
8. J. Birn, I. Skalski, "Corrosion Behaviour of Non-ferrous Alloys in Seawater in the Polish Marine Industry," in *Corrosion Behaviour and Protection of Copper and*

Aluminium Alloys in Seawater, D. Feron, ed., EFC Publication No. 50 (Cambridge, U.K.: Woodhead and Maney, 2007): p. 3.

9. R. Johnsen, "Experience with the Use of Copper Alloys in Seawater Systems on the Norwegian Continental Shelf," in *Corrosion Behaviour and Protection of Copper and Aluminium Alloys in Seawater*, D. Feron, ed., EFC Publication No. 50 (Cambridge, U.K.: Woodhead and Maney, 2007): p. 62.

10. W. Schleich, C. Powell, "Cu/Ni 90/10: How to Avoid Typical Failures of Seawater Tubing Systems and Marine Biofouling on Structures," in *Corrosion Behaviour and Protection of Copper and Aluminium Alloys in Seawater*, D. Feron, ed., EFC Publication No. 50 (Cambridge, U.K.: Woodhead and Maney, 2007): p. 73.

11. H. Le Guyader, A.-M. Grolleau, E. Lumieux, K. Lucas, T. Wolejszu, "70/30 Copper-Nickel Piping Systems—Use of Descaling Agents and Their Effect on Corrosion Properties," in *Corrosion Behaviour and Protection of Copper and Aluminium Alloys in Seawater*, D. Feron, ed., EFC Publication No. 50 (Cambridge, U.K.: Woodhead and Maney, 2007): p. 95.

12. R. Francis, *Brit. Corros. J.* 20 (1985): p. 167.

13. E.H. Newton, J.D. Barckett, J.M. Ketteringham, "Survey of Materials in Large Desalting Plants Around the World," A.D. Little Report to the U.S. Office of Saline Water, March 1972.

14. J.W. Oldfield, "Novel Material Selection to Improve Corrosion Resistance," Report to the Middle East Desalination Research Centre, Oman, 97-AS-001, 2003.

15. N.G. Galindez Ruiz, B.D. Craig, *MP* 34, 4 (1995): p. 50.

16. B.D. Craig, *Corrosion* 54 (1998): p. 657.

17. C. de Waard, L.M. Smith, B.D. Craig, CORROSION 2003, paper no. 629 (Houston, TX: NACE International, 2003).

CHAPTER 18

Joining and the Corrosion of Joints

In many service applications, there is a need to join metallic components. Where the joint is permanent, it is common with copper and its alloys to use soldering, brazing, or welding. Each of these three processes is considered in this chapter for copper alloys, and the potential corrosion risks of each are described.

18.1 Soldering

For many applications where the stresses are not high, a joint made with a low–melting temperature alloy will be satisfactory. Solders are traditionally lead/tin alloys with typical compositions around the eutectic (63/37 Sn/Pb) to give a low melting temperature (190°C or less).

When joining copper, or its alloys, with solder, it is necessary to use a flux to wet the surface. Traditionally, these have been based on ammonium (NH_4^+) and/or zinc chloride ($ZnCl_2$). The flux residues after soldering are not very soluble and are often acidic, which can result in deep corrosion beneath the flux deposit; alternatively, crevice corrosion may form at the edge of the residue. Fluxes used for electronics applications are typically based on rosin or organic resins, the residues of which tend to be relatively benign. However, if these are supplemented with more aggressive ingredients, the residues may be active enough to cause electromigration and subsequent electrical shorting in humid environments.

In some countries and in some industries, it is necessary to use a soluble flux so that residues can be washed off after soldering. Some of these still use a chloride activator and can be aggressive if not removed. Others use an iodide activator, which is less aggressive than chloride.

A traditional composition for solders has been 60/40 Sn/Pb, with a higher tin content where greater strength or resistance to fatigue is required. Lead is toxic, and so it is no longer permitted in solders for joints in potable water systems and in some areas in the electronics industry. However, tin/lead solders are still preferred for some electronics applications, for example, in aerospace and military applications, where guaranteed long-term reliability is critical to operations.

Where lead-free solders are mandatory, there are several popular options. For high-temperature applications, a tin/copper alloy containing ~3% copper may be used. This has a liquidus above 300°C and tends to be less costly than the lower melting point alternatives that contain silver. Tin/silver and related alloys melt in the range 210 to 240°C and have found use in the electronics industry as tin/lead replacements. When using tin-based solders on copper, it is important to avoid the formation of excessive copper-tin intermetallic compound layers. Many of these are highly cathodic to copper and may cause accelerated corrosion in aqueous solutions and are also relatively brittle. Intermetallics are most likely to grow under prolonged heating such as can occur in rework or high–operating temperature environments.

The high tin content of solders means that in most waters, a stable film of tin oxide (SnO) will form over a wide range of pH values. This oxide is very tenacious and corrosion-resistant. Erosion corrosion is sometimes seen in copper piping systems carrying fresh water, immediately after sharp bends or elbows. If the fitting has been soldered into place, sometimes a protruding bead of solder increases the local turbulence. The erosion corrosion is always on the copper and never on the solder, although the potential of solder is significantly more electronegative than that of copper.[1] Francis and Campbell believed that this was because tin oxide is more stable and mechanically strong than copper oxide in fresh waters. In unpublished work carried out at BNF Metals Technology Centre, they exposed samples of copper coated with solder in the May jet test (see Section B.2 in Appendix B) at 10 m/s, with 3% added air, in a hard borehole water containing 100 mg/L chloride. No measurable erosion corrosion had occurred on the solder after 30 days at 15°C. These results demonstrated the high resistance of solders to erosion corrosion in fresh water. Even in seawater, tin solders have good resistance to both general corrosion and erosion corrosion.

Aluminum bronzes and copper-nickel alloys are difficult to solder because of their protective films. Where they must be soldered, it is necessary either to coat the joint area in copper first, or to use a very aggressive flux that can break down these films.

18.2 Brazing

Brazing involves the joining of copper and its alloys with silver alloys having melting temperatures in the range 600-800°C. Confusingly, the process is sometimes called *silver soldering*, despite the fact that according to conventional nomenclature, the temperature range clearly defines a brazing process. Cadmium is no longer widely used in brazing alloys because of its high toxicity, and so for practical purposes, the minimum melting temperature of brazing alloys is about 640°C. A brazed joint requires a higher temperature than a soldered joint, but it is also much stronger.

A range of copper-silver-phosphorus alloys is commonly used for brazing coppers and brasses, and these alloys contain 0 to 15% silver and 4 to 7% phosphorus, depending on the melting point. The higher the silver content is, the lower the melting temperature will be. These alloys are not suitable for brazing other copper alloys, for which silver brazing alloys free of phosphorus should be used.

As with soldering, it is necessary to use a flux, and the most common contains borates and fluorides. The exception to this is brazing copper with copper-phosphorus alloys; these are self-fluxing and will wet easily if the surface is clean. Any residues when using borax flux are inert, and so the corrosion problems associated with soldering fluxes are absent. However, fluxes containing active ingredients, such as fluorides, must be removed.

Brazing alloys have good corrosion resistance in waters, including seawater. The silver in the alloy generally means that the braze metal is cathodic to the parent metal, and localized galvanic corrosion does not occur.

Heavy oxide films formed during extensive brazing operations usually consist of a thin layer of cupric oxide, with cuprous oxide underneath. Such scales are highly cathodic to copper and can give rise to pitting corrosion at the boundary between the scaled and unscaled parts of the metal surface.[2] The severity of the attack will depend on the water composition, pH, and temperature.

18.3 Welding

Welding involves joints made above the melting point of the parent metal. Such joints involve the most heat but usually have the highest strength. Sometimes the joints are autogenous; sometimes filler is used, often of a matching composition to the parent alloy. For some of the copper alloys, the selection of a filler metal is very important. Copper and its alloys can be welded by all the common arc-welding techniques. In addition, autogenous welding of thick (50 mm), oxygen-free copper cylinders has been achieved by both electron beam welding and friction stir welding. In both cases, weld deposits were produced that were not anodic to the parent metal. This is the most important factor in the corrosion of welds. The weld bead has a much smaller area than the parent metal, and so it is important that its potential be the same as the parent metal or more positive. If this is not achieved, rapid, localized attack of the weld metal can occur.[1]

The copper-nickel alloys are often used in seawater, and welding is preferred for high-strength joints. The iron content of 90/10 copper-nickel exceeds the solubility limit at room temperature (see Chapter 2), and so iron may come out of solution if a matching composition filler is used and the weld cools too slowly. This can reduce the corrosion resistance of the weld metal, particularly the erosion corrosion resistance (see Chapter 7). For this reason, it is common to weld 90/10 copper-nickel with 70/30 copper-nickel filler. The higher nickel content greatly increases the iron solubility limit, and there is no risk of iron precipitation on cooling. Note that slow cooling of thick section welds may cause iron to precipitate out of solution in the heat-affected zone (HAZ). This can be avoided by suitable joint design and the selection of welding parameters when qualifying the welding procedure.

The 70/30 copper-nickel alloys, such as C71500, and the single-phase copper-nickel-chromium alloys are usually welded with a matching composition filler. This produces welds that are not anodic to the parent metal and so will not corrode preferentially. Alloy C71500 has been welded with Alloy 400 filler, but this is generally unnecessary as it is more expensive than matching filler and does not generally confer any significant corrosion benefits.

Little et al. examined the corrosion of welded copper-nickel alloys in seawater polluted by sulfides.[3] They concluded that any corrosion would occur, not on the weld, but downstream of a girth weld, on the HAZ.

In an examination of the corrosion of welded Alloy 400, Lee et al. carried out tests in seawater containing inorganic sulfide or sulfate-reducing bacteria (SRB) cultures.[4] The welds used a filler that was similar to the parent metal, but with an increased silicon and manganese content plus $\sim$2% titanium (UNS N04060). In the inorganic sulfide, there was, more or less, uniform corrosion with no pitting anywhere, even after 20 weeks' exposure. Samples exposed to seawater containing SRBs had suffered pitting of the parent metal after 12 weeks' exposure. After 14 weeks' exposure, there was attack at the weld fusion line, but there was no attack of the weld metal after

20 weeks. This shows that the welds are no more susceptible to localized attack in this environment than the parent metal.

The welding of the aluminum bronzes is somewhat more complex, particularly for the multiphase alloys. The subject is discussed in detail by Meigh and also by Spiller.[5-6] Here a brief overview is presented, readers wishing more information should consult the preceding references.

One problem with welding the aluminum bronzes is the ductility dip. The alloys suffer a loss of ductility around 500°C, and the width of this dip varies with alloy composition but can be extensive for some of the aluminum bronzes.[5] This means that cracking may occur on cooling, particularly when the weld is under severe restraint. Meigh gave recommendations for minimizing the risk of cracking, involving both the choice of welding parameters and the selection of the filler metal composition.[5]

Spiller reported that there is a risk of HAZ cracking when welding aluminum bronzes.[6] However, this has not been seen for many years, and it is believed that modern manufacturing methods produce alloys that are no longer susceptible to HAZ cracking.[7]

When welding the single-phase aluminum bronzes, it is usual to use matching composition filler. The alloy containing 8% aluminum and 3% iron is sometimes preferred because iron increases the strength of the weld metal and reduces the risk of cracking.[5] Iron also prevents the formation of gamma 2 phase, which is anodic to the alpha phase and can corrode preferentially.[8]

When welding nickel aluminum bronze (NAB), it is common to use a matching filler. However, the weld bead cools very quickly, and not all the beta phase present at high temperature may transform to alpha and kappa phases. Beta phase is highly anodic to alpha phase and will corrode rapidly in seawater (Chapter 9). Hence, a postweld heat treatment is desirable to ensure that all the beta phase transforms. The most common heat treatment is that specified by the Royal Navy for all welds to NAB. This involves heating at 675°C for 6 hours, followed by air cooling. Where postweld heat treatment is not possible, then the use of Cu/8Al/3Fe filler is an alternative. Although no postweld heat treatment will be required, the joint will not be as strong as one made with matching filler and may not be as strong as the parent metal. The use of a single-phase filler does not result in any preferential weld corrosion in seawater as the potentials of both aluminum bronzes are similar. Figure 18.1 shows a seam-welded NAB pipe that had not been postweld heat treated. After exposure to seawater, the beta phase corroded rapidly, with the result that the seam weld unzipped.

FIGURE 18.1 Corrosion of the beta phase in an NAB seam-welded pipe that had not been postweld heat treated. (Photograph courtesy of the Nickel Institute.)

NAB has also been successfully joined by friction stir welding.[7] This has the advantage of requiring no filler, but an assessment of the corrosion resistance of such joints has not been made to date.

Provided that they have been heat treated, NAB welds do not generally cause a corrosion problem in seawater. However, the Royal Navy has seen corrosion of NAB, both welded and unwelded, when it was exposed to slowly flowing seawater or alternate stagnant and flowing conditions for extended periods. The work of Maselkowski and Francis, described in Chapter 9, showed that attack was most pronounced in the HAZ of NAB welds, with corrosion taking the form of selective attack of the kappa III phase.[9] As this phase is in long, thin stringers, the attack could be very deep. However, if the NAB is put into flowing seawater such that a protective film forms, then no selective phase attack occurs. This explains why NAB welds give few problems in industrial service, where there is pressure to get equipment into service rapidly.

References

1. R. Francis, *Galvanic Corrosion—A Practical Guide for Engineers* (Houston, TX: NACE International, 2001).

2. H.S. Campbell, BNF Metals Research Association Miscellaneous Publication, MP 574, August 1972.

3. B.J. Little, P.A. Wagner, J. Jacobus, CORROSION/88, paper no. 81 (Houston, TX: NACE International, 1988).

4. J.S. Lee, K.L. Lowe, R.I. Ray, B.J. Little, CORROSION/03, paper no. 214 (Houston, TX: NACE International, 2003).

5. H.J. Meigh, *Cast and Wrought Aluminium Bronzes—Properties, Processes, and Structures* (London, U.K.: Institute of Materials, Minerals, and Mining, 2000).

6. K.R. Spiller, "Practical Aspects of Welding Aluminium Bronze," *Welding and Metal Fabrication*, November 1971, p. 401.

7. M.F. Gittos, The Welding Institute, correspondence to author, 2006.

8. H.S. Campbell, *Aluminium Bronze Alloys Corrosion Resistance Guide*, Publication 80 (Potters Bar, U.K.: Copper Development Association, 1981).

9. C.R. Maselkowski, R. Francis, BNF Metals Technology Centre Research Report, R506 (Supplementary Report), July 1985.

APPENDIX A

Nominal Compositions of Some Commonly Used Copper Alloys

The following tables give the nominal compositions of some copper alloys commonly used in engineering. They are grouped by type, as described in Chapter 2.

TABLE A.1 Brasses

			Nominal composition (wt%)						
Form	Common name	UNS no.	Cu	Zn	Pb	As	Sn	Mn	Other
Wrought	Commercial bronze	C22000	90	Bal[(A)]	–	–	–	–	–
	Red brass	C23000	85	Bal	–	–	–	–	–
	Cartridge brass	C26000	70	Bal	–	–	–	–	–
	Arsenical brass	C26130	70	Bal	–	0.04	–	–	–
	Admiralty brass	C44300	70	Bal	–	0.04	1	–	–
	Aluminium brass	C68700	70	Bal	–	0.04	–	–	2Al
	Muntz metal	C28000	60	Bal	0.5	–	–	–	–
	Naval brass	C46400	61	Bal	–	–	1	–	–
	Hot stamping brass	C38000	58	Bal	2	–	–	–	–
	DZR brass	CZ132[(B)]	62	Bal	2	0.06	–	–	–
	Manganese bronze	CZ114[(B)]	58	Bal	1	–	0.7	1.5	0.8Fe
Cast	DZR brass (die cast)	BS EN 1982[(B)]	62	Bal	2	0.06	–	–	0.5Al
	Manganese bronze	C86550	58	Bal	0.5	–	0.5	2	1Fe
	Beta brass	C86200	60	27	–	–	–	3	3Fe 4Al

[(A)] Bal = balance.
[(B)] British Standard quoted as no UNS number.

TABLE A.2 Copper-nickel and nickel-copper alloys

			Nominal composition (wt%)					
Form	Common name	UNS no.	Cu	Ni	Fe	Mn	Cr	Other
Wrought	90/10 copper-nickel	C70600	Bal[(A)]	10	1.5	1	–	–
	70/30 copper-nickel	C71500	Bal	30	0.7	0.7	–	–
	66/30/2/2 Cu/Ni/Fe/Mn	C71640	Bal	30	2	2	–	–
	Alloy 722	C72200	Bal	16	0.7	0.7	0.5	–
	Alloy 719	C71900	Bal	30	0.3	0.7	2.5	–
	Alloy 729	C72900	Bal	15	–	–	–	8Sn
	Marinel	C72420	Bal	18	0.8	4	–	2Al 0.7Nb
	Alloy 400	N04400	Bal	65	2	2	–	–
	Alloy K-500	N05500	Bal	65	1.5	1	–	3Al 0.4Ti
Cast	IN768	[(B)]	Bal	30	0.7	0.6	1.6	–

[(A)] Bal = balance.
[(B)] No UNS number.

TABLE A.3 Tin bronzes and gunmetals

			Nominal composition (wt%)					
Form	Common name	UNS no.	Cu	Sn	Zn	Pb	P	Other
Wrought	Phosphor bronze	C50900	Bal[(A)]	3	–	–	0.2	–
	Phosphor bronze	C51000	Bal	5	–	–	0.2	–
	Phosphor bronze	C52100	Bal	7	–	–	0.2	–
	AP bronze	[(B)]	Bal	8	–	–	–	1Al 0.1Si
Cast	CT1 phosphor bronze	C90700	Bal	10	–	–	0.1	–
	LB2 leaded bronze	C93700	Bal	10	–	10	0.05	–
	Admiralty gunmetal	C90600	Bal	10	2	1.5	–	–
	LG2 gunmetal	C83600	Bal	5	5	5	–	–
	LG4 gunmetal	C92410	Bal	7	2	3	–	–
	M bronze	C92200	Bal	6	4	1.5	–	–

[(A)] Bal = balance.
[(B)] No UNS number.

TABLE A.4 Silicon bronze

			Nominal composition (wt%)				
Form	Common name	UNS no.	Cu	Si	Mn	Zn	Other
Wrought	Silicon bronze	C65500	Bal[(A)]	3	1	–	–

[(A)] Bal = balance.

TABLE A.5 Aluminum bronzes

Form	Common name	UNS no.	Nominal composition (wt%)				
			Cu	Al	Fe	Ni	Other
Wrought	Copper-aluminum	C60800	Bal[A]	5.5	–	–	0.3As
	Alloy D	C61400	Bal	7	2.5	–	–
	Alloy E	C63300	Bal	10	3	5	–
	NAB	C63200	Bal	9	4	4.5	–
Cast	AB1	C95200	87	9	3	–	–
	NAB	C95800	Bal	9	4	4.5	–
	CMA1	[B]	Bal	8	3	2.5	12Mn

(A) Bal = balance.
(B) No UNS number.

TABLE A.6 Nickel silvers

Form	Common name	UNS no.	Nominal composition (wt%)			
			Cu	Ni	Zn	Pb
Wrought	45/10/43	C77400	45	10	Bal[A]	1.5
	40/14/44	[B]	40	14	Bal	1.5
	63/10/27	C74500	63	10	Bal	–
	63/12/25	C75700	63	12	Bal	–
	63/15/22	C75400	63	15	Bal	–
	63/18/19	C75200	63	18	Bal	–
	55/18/27	C77000	55	18	Bal	–
	63/20/17	[B]	63	20	Bal	–
	58/25/17	[B]	58	25	Bal	–

(A) Bal = balance.
(B) No UNS number.

APPENDIX B

Laboratory Tests to Evaluate the Erosion Corrosion Resistance of Copper Alloys

When examining laboratory erosion corrosion data, it is often difficult to compare results from different test methods, employing different velocities. This appendix briefly summarizes the most commonly used test methods and what can be gleaned from the data.

B.1 Test Methods

The test most representative of service is a flow loop or model condenser, operating at normal flow velocities or maximum design velocities. Although this reproduces field conditions, exposures need to be long (6 months to 2 years) to generate useful data. Sometimes these tests are accelerated by increasing the flow velocity to, say, twice design to produce data more quickly. Erosion corrosion usually then occurs anywhere that local turbulence is increased such as after sharp bends. This must be taken into account when considering the data, for example, not rejecting an alloy because it showed a little attack. At normal flow rates, this alloy could be perfectly satisfactory.

Most erosion corrosion data have been generated by a method involving much higher flow velocities. The oldest of these is the May, or BNF jet test, in which water is fed to a circular water box designed to avoid cavitation.[1] Water is fed via flexible hose from the water box to multiple jet assemblies (12 or 24), which are submerged in a tank of the water being considered. These rigs can be used in either once-through or recirculating mode. The jet assembly consists of a block holding a nozzle, with the test sample fixed a few millimeters from the jet (Figure B.1). The distance from

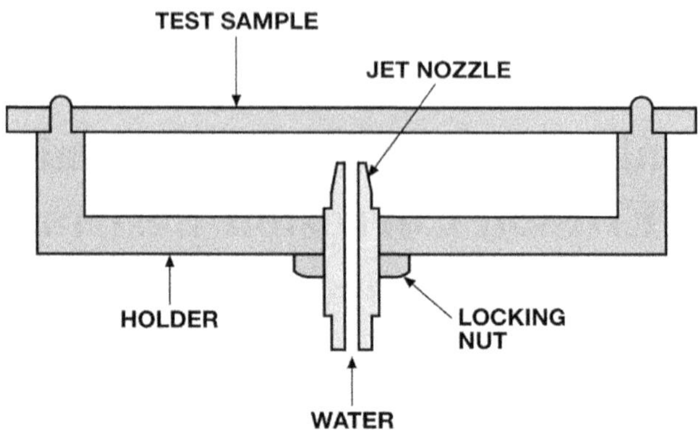

FIGURE B.1 Schematic of the sample holder in the BNF jet test.[1]

the jet to the sample must be set at exactly the right distance to ensure that the sample is in the most turbulent part of the jet.[1] Jet velocities are usually in the range of 5-10 m/s, with tests lasting for 30 days. Note that under recirculating conditions, it is usual to inject 3% air into the water. This is discussed in more detail in the next section.

The LaQue Laboratory, in the United States, used a variant of the BNF jet test, which involved using a cylinder as a high-pressure water box. The jet nozzles were fastened around the base of the cylinder, with the samples held a fixed distance from the jet. The whole assembly was immersed in a tank of seawater. The advantage of this test rig design was that it could operate to even higher velocities than the BNF jet test, without cavitation in the water box.

The jet test is used to simulate turbulent conditions that can occur at condenser tube inlets or at partial blockages within tubes. Sato and Nagata showed that flow velocities around a blockage could be up to five times the nominal flow velocity.[2-3] Thus, for a nominal flow velocity of 2 m/s, the velocity around a blockage could be up to 10 m/s.

The parallel flow rig was designed to study the shear forces that are required to remove protective films and initiate erosion corrosion. Samples are placed in the center of a long pipe, parallel to the axis, with nonconducting spacers to separate each sample. By using long, straight pipes, uniform flow is produced so that the shear force can be calculated.

Campbell designed a test rig that includes not only an impinging jet, but also crevices, slow flow, and heat transfer.[4-5] The condenser tube test rig uses a water box identical to that of the BNF jet test, but this feeds 10 nozzles that have short (200-mm) lengths of 25.4-mm-outside-diameter

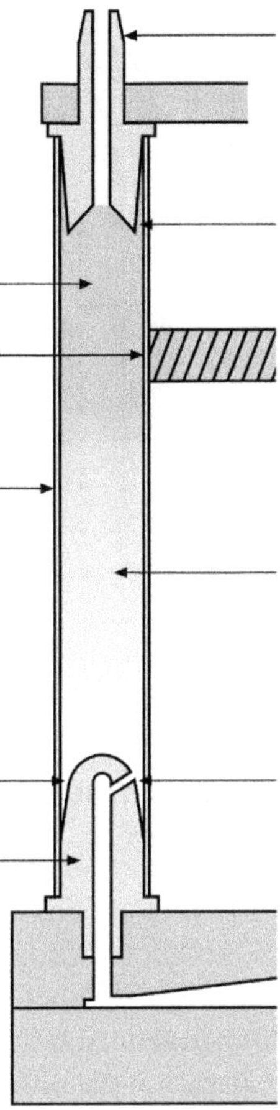

FIGURE B.2 Schematic showing the various corrosion zones in the Campbell condenser tube test rig.[4,5]

condenser tube over them, as shown schematically in Figure B.2. Figure B.3 shows several rigs in parallel at the BNF test site. In this test, the heaters were not fitted. These rigs have mostly been used with jet velocities in the range of 5 to 10 m/s in a once-through mode.[6-7] The inclined angle of the jet, compared to the BNF jet test, does not seem to produce levels of erosion corrosion very different to the levels seen in the jet test under the same operating conditions.

When it is desired to specifically study the effects of a partial blockage on the performance of heat exchanger tubes, Gaffoglio described a plastic insert that can be fixed in a condenser tube to simulate this effect.[8]

FIGURE B.3 Campbell condenser tube rigs at BNF test site.

B.2 Recirculating Vs. Once-Through Testing

Some test rigs use recirculating water and some use once-through water, but it is important to recognize that erosion corrosion testing in synthetic seawater (e.g., ASTM D1141[9]) is *not* the same as testing in natural seawater. This has been clearly demonstrated and is thought to be due to a combination of the organic material and fine suspended solids (typically 50 mg/L) present in natural seawater.[10-11] In Chapter 3, the different films that form on some alloys in natural and synthetic seawater are discussed.

Gilbert and LaQue carried out extensive comparative tests with the BNF jet test under both once-through and recirculating conditions.[10] They found that once-through seawater was more aggressive to some copper alloys compared with recirculating tests. This was thought to be due principally to very fine abrasive particles from plankton—either their bodies or their hard shells. Tests with once-through seawater gave similar results to recirculating tests if the jet velocity in the once-through tests was reduced to about one-third the jet velocity in the recirculating tests. The results from the recirculating tests gave similar depths of attack to those sometimes seen at the inlets of copper alloy condenser tubes. Although the conditions were more erosive under once-through conditions at jet velocities of 5 m/s or more, Gilbert and LaQue felt that these velocities were useful for studying the more erosive conditions that can sometimes develop, for example, at a partial blockage in a condenser tube.

It was also found that the addition of 3% air only had an effect when no erosion corrosion occurred without it. The extra turbulence introduced by the addition of air can then induce some erosion corrosion. As entrained air is quite common in seawater cooling systems, it is useful to add it when conducting jet tests in a recirculating mode. It would only be necessary to add air under once-through conditions if it was being conducted at a very low jet velocity.

References

1. R. May, R.W. DeVere Stacpoole, *J. Inst. Met.* 77 (1950): p. 331.
2. S. Sato, K. Nagata, "Factors Affecting Corrosion and Fouling of Condenser Tubes of Copper Alloys and Titanium," Conference on Corrosion in the Power Industry (Montreal, Canada: NACE International, 1977).
3. S. Sato, K. Nagata, *Sumitomo Light Met. Tech. Rep.* 19 (1978): p. 83.
4. H.S. Campbell, BNF Metals Research Association Miscellaneous Publication, MP577, February 1973.
5. H.S. Campbell, "The Campbell Condenser Tube Apparatus," Third International Congress on Marine Corrosion and Fouling (Gaithersburg, MD: 1972).
6. R. Francis, *MP* 21 (1982): p. 44.
7. R. Francis, *Brit. Corros. J.* 20 (1985): pp. 167, 175.
8. C.J. Gaffoglio, in *Power Engineering*, April 1982, p. 60.
9. ASTM D1141, "Standard Practice for the Preparation of Substitute Ocean Water" (West Conshohocken, PA: ASTM International).
10. P.T. Gilbert, F.L. LaQue, *J. Electrochem. Soc.* 101 (1954): p. 448.
11. P. Gallagher, R.E. Malpas, E.B. Shone, *Brit. Corros. J.* 23 (1988): p. 229.

Index

www.ingramcontent.com/pod-product-compliance
Ingram Content Group UK Ltd.
Pitfield, Milton Keynes, MK11 3LW, UK
UKHW020201290926
15471UKWH00005B/1007

9 781575 902258